Nenad Mitrovic · Goran Mladenovic ·
Aleksandra Mitrovic
Editors

Current Problems in Experimental and Computational Engineering

Proceedings of the International Conference of Experimental and Numerical Investigations and New Technologies, CNNTech 2021

Springer

Editors
Nenad Mitrovic
Faculty of Mechanical Engineering
Department for Process Engineering
and Environmental Protection
University of Belgrade
Belgrade, Serbia

Aleksandra Mitrovic
Faculty of Information Technology
and Engineering
University Union - Nikola Tesla
Belgrade, Serbia

Goran Mladenovic
Faculty of Mechanical Engineering
Department for Production Engineering
University of Belgrade
Belgrade, Serbia

ISSN 2367-3370 ISSN 2367-3389 (electronic)
Lecture Notes in Networks and Systems
ISBN 978-3-030-86008-0 ISBN 978-3-030-86009-7 (eBook)
https://doi.org/10.1007/978-3-030-86009-7

This Springer imprint is published by the registered company Springer Nature Switzerland AG
The registered company address is: Gewerbestrasse 11, 6330 Cham, Switzerland

Scientific Committee

- Miloš Milošević (chairman), University of Belgrade, Faculty of Mechanical Engineering, Serbia
- Nenad Mitrović (co-chairman), University of Belgrade, Faculty of Mechanical Engineering, Serbia
- Aleksandar Sedmak, University of Belgrade, Faculty of Mechanical Engineering, Serbia
- Hloch Sergej, Technical University of Košice, Faculty of Manufacturing Technologies, Slovakia
- Dražan Kozak, University of Osijek, Faculty of Mechanical Engineering in Slavonski Brod, Croatia
- Nenad Gubeljak, University of Maribor, Faculty of Mechanical Engineering, Slovenia
- Monka Peter, Technical University of Kosice, Faculty of Manufacturing Technologies, Slovakia
- Snežana Kirin, University of Belgrade Innovation Center of Faculty of Mechanical Engineering, Serbia
- Ivan Samardžić, University of Osijek, Faculty of Mechanical Engineering in Slavonski Brod , Croatia
- Martina Balać, University of Belgrade, Faculty of Mechanical Engineering, Serbia
- Ludmila Mládková, University of Economics Prague, Czech Republic
- Johanyák Zsolt Csaba, Athéné University, Faculty of Engineering and Computer Science, Hungary
- Igor Svetel, University of Belgrade, Innovation centre of Faculty of Mechanical Engineering, Serbia
- Aleksandra Mitrović, University of Belgrade, Faculty of Technology and Metallurgy, Serbia
- Valentin Birdeanu, National R&D Institute for Welding and Material Testing—ISIM Timişoara, Romania
- Danilo Nikolić, University of Montenegro, Faculty of Mechanical Engineering, Montenegro

Preface

The book is a collection of high-quality peer-reviewed research papers presented at the International Conference of Experimental and Numerical Investigations and New Technologies (CNN Tech 2021) held at Zlatibor, Serbia from 29 June to 02 July 2021. The conference is organized by the Innovation Center of the Faculty of Mechanical Engineering, Faculty of Mechanical Engineering at the University of Belgrade and Center for Business Trainings. Over 90 delegates were attending the CNN Tech 2021—academicians, practitioners and scientists from 12 countries—presenting and authoring more than 90 papers. The Conference Program included eight Invited Lectures, five sessions (oral and poster), one workshop—Regional Innovation Forum 2021 and B2B meetings. Nineteen selected full papers went through the double-blind reviewing process.

The main goal of the conference is to make positive atmosphere for the discussion on a wide variety of industrial, engineering and scientific applications of the engineering techniques. Participation of a number of domestic and international authors, as well as the diversity of topics, have justified our efforts to organize this conference and contribute to exchange of knowledge, research results and experience of industry experts, research institutions and faculties which all share a common interest in the field in experimental and numerical investigations.

The CNN Tech 2021 was focused on the following topics:

- Mechanical Engineering,
- Engineering Materials,
- Chemical and Process Engineering,
- Experimental Techniques,
- Numerical Methods,
- New Technologies,
- Clear sky,
- Sustainable Design and New Technologies,
- Advanced Materials and Technology,
- Artificial intelligence and
- Student session.

We express our gratitude to all people involved in conference planning, preparation and realization, especially to:

- All authors, specially invited speakers, who have contributed to the high scientific and professional level of the conference,
- All members of the Organizing Committee,
- All members of the International Scientific Committee for reviewing the papers and Chairing the Conference Sessions,
- Ministry of Education, Science and Technological Development of Republic of Serbia for supporting of the Conference.

Belgrade, Serbia

Nenad Mitrovic
Goran Mladenovic
Aleksandra Mitrovic

Contents

Contributors

Mohamed Al Kateb Faculty of Mechanical Engineering, University of Belgrade, Belgrade, Serbia

Martina Balac Faculty of Mechanical Engineering, University of Belgrade, Belgrade, Serbia

Marija Baltić Faculty of Mechanical Engineering, University of Belgrade, Belgrade, Serbia

Dragana Barjaktarević Faculty of Technology and Metallurgy, University of Belgrade, Belgrade, Serbia

Aleksandra Božić Department of Belgrade Polytechnic, The Academy of Applied Technical Studies Belgrade, Belgrade, Serbia

Dominik Brkić Department of Belgrade Polytechnic, The Academy of Applied Technical Studies Belgrade, Belgrade, Serbia

Ivan M. Buzurovic Harvard Medical School, Harvard University, Boston, MA, USA

Bojan Ćudić University of Maribor, Maribor, Slovenia

Gradimir Danon Faculty of Forestry, University of Belgrade, Belgrade, Serbia

Djordje Dihovicni The Academy of Applied Technical Studies Belgrade, New Belgrade, Serbia

Zoran Dimic LOLA Institute, Belgrade, Serbia

Andjelija Djordjevic Department of Management and Project Management, Faculty of Organizational Sciences, University of Belgrade, Belgrade, Serbia

Milos Djukic Faculty of Mechanical Engineering, University of Belgrade, Belgrade, Serbia

Veljko Đokić Innovation Centre of the Faculty of Technology and Metallurgy in Belgrade, Belgrade, Serbia

Luka Đorđević Technical Faculty "Mihajlo Pupin", University of Novi Sad, Zrenjanin, Serbia

Nenad Đorđević The Academy of Applied Technical Studies Belgrade, Belgrade, Serbia

Tatjana Đurkić Faculty of Technology and Metallurgy, University of Belgrade, Belgrade, Serbia

Marija Đurković Faculty of Forestry, University of Belgrade, Belgrade, Serbia

Igor Dzincic Department of Wood Science and Technology, Faculty of Forestry, University of Belgrade, Belgrade, Serbia

Mladen Furtula Faculty of Forestry, University of Belgrade, Belgrade, Serbia

Zorana Golubović Faculty of Mechanical Engineering, University of Belgrade, Belgrade, Serbia

Aleksandar Grbovic Faculty of Mechanical Engineering, University of Belgrade, Belgrade, Serbia

Svetlana Grujić Faculty of Technology and Metallurgy, University of Belgrade, Belgrade, Serbia

Eleonora Gvozdić Innovation Center of the Faculty of Technology and Metallurgy, Belgrade, Serbia

Aleksandra Jelić Department of Belgrade Polytechnic, The Academy of Applied Technical Studies Belgrade, Belgrade, Serbia

Rastko Jovanović VINČA Institute of Nuclear Sciences-National Institute of the Republic of Serbia, University of Belgrade, Belgrade, Serbia

Gordana Kastratovic Faculty of Transport and Traffic Engineering, University of Belgrade, Belgrade, Serbia

Snezana Kirin Innovation Center, Faculty of Mechanical Engineering, Belgrade, Serbia

Matjaž Klemenčič University of Maribor, Maribor, Slovenia

Branko Kokotovic Department for Production Engineering, University of Belgrade, Belgrade, Serbia

Danijela Kovačević The Academy of Applied Technical Studies Belgrade, Belgrade, Serbia

Nada Ratković Kovačević The Academy of Applied Technical Studies Belgrade, New Belgrade, Serbia

Dragan Kreculj The Academy of Applied Technical Studies Belgrade, New Belgrade, Serbia

Sara Lahlou Laboratory of Transport Phenomena, Faculty of Mechanical and Process Engineering, Algiers, Algeria

Zoran Lalić The Academy of Applied Technical Studies Belgrade, New Belgrade, Serbia

Steva Lević Faculty of Agriculture, University of Belgrade, Belgrade, Serbia

Ivana Vasović Maksimović Lola Institute, Belgrade, Serbia

Katarina Maksimović City Administration of City of Belgrade, Secretariat for Utilities and Housing Services Water Management, Belgrade, Serbia

Mirko Maksimović PUC Belgrade Waterworks and Sewerage, Belgrade, Serbia

Stevan Maksimović Military Technical Institute, Belgrade, Serbia

Ivana Matić Bujagić Academy of Applied Technical Studies Belgrade, Belgrade Polytechnic College, Belgrade, Serbia

Đorđe Matić Department of Computer-Machine Engineering, The Academy of Applied Technical Studies Belgrade, Belgrade, Serbia

Bojan Međo Faculty of Technology and Metallurgy, University of Belgrade, Belgrade, Serbia

Marko Mihic Department of Management and Project Management, Faculty of Organizational Sciences, University of Belgrade, Belgrade, Serbia

Miloš Milošević Faculty of Mechanical Engineering, Department of Information Technologies, University of Belgrade, Belgrade, Serbia

Milan Milutinovic Department of Traffic, Mechanical and Protection Engineering, Academy of technical vocational studies, Belgrade, Serbia

Nikola Mirkov VINČA Institute of Nuclear Sciences-National Institute of the Republic of Serbia, University of Belgrade, Belgrade, Serbia

Vladimir Misic University of Pittsburgh Medical Center, Pittsburgh, PA, USA

Aleksandra Mitrović Department of Computer-Machine Engineering, The Academy of Applied Technical Studies Belgrade, Belgrade, Serbia

Seif Eddine Ouyahia Laboratory of Transport Phenomena, Faculty of Mechanical and Process Engineering, Algiers, Algeria;
Sonatrach–Direction Centrale Recherche et Développement, Boumerdes, Algeria

Ognjen Peković Faculty of Mechanical Engineering, University of Belgrade, Belgrade, Serbia

Jasmina Perisić Faculty of Entrepreneurial Business and Real Estate Management, UNION "Nikola Tesla" University, Belgrade, Serbia

Aleksandar Petrovic Belgrade, Serbia

Milada Pezo VINČA Institute of Nuclear Sciences-National Institute of the Republic of Serbia, University of Belgrade, Belgrade, Serbia

Slaviša Putić Faculty of Technology and Metallurgy, University in Belgrade, Belgrade, Serbia

Ljiljana Radovanović Technical Faculty "Mihajlo Pupin", University of Novi Sad, Zrenjanin, Serbia

Marko Rakin Faculty of Technology and Metallurgy, University of Belgrade, Belgrade, Serbia

Marko Ristić Institute Mihajlo Pupin, Belgrade, Serbia

Slavisa Salinic Faculty of Mechanical and Civil Engineering, University of Kragujevac, Kraljevo, Serbia

Lajos Sarvas Belgrade, Serbia

Aleksandar Sedmak Faculty of Mechanical Engineering, University of Belgrade, Belgrade, Serbia

Mirjana Stamenić Faculty of Mechanical Engineering, Department of Process and Environmental Engineering, University of Belgrade, Belgrade, Serbia

Marina Stamenović Department of Belgrade Polytechnic, The Academy of Applied Technical Studies Belgrade, Belgrade, Serbia

Jelena Svorcan Faculty of Mechanical Engineering, University of Belgrade, Belgrade, Serbia

Srđan Svrzić Faculty of Forestry, University of Belgrade, Belgrade, Serbia

Srdjan Tadic Innovation Center, Faculty of Mechanical Engineering, Belgrade, Serbia

Nikola Tanasić Department of Transportation, Mechanical and Safety Engineering, The Academy of Applied Technical Studies Belgrade, Belgrade-Zemun, Serbia

Vladimir Tanasić PUC, Belgrade District Heating Company, Belgrade, Serbia

Milan Travica Innovation Center, Faculty of Mechanical Engineering, University of Belgrade, Belgrade, Serbia

Zorana Trivković Faculty of Mechanical Engineering, University of Belgrade, Belgrade, Serbia

Vukašin Ugrinović Innovation Center, Faculty of Technology and Metallurgy, University of Belgrade, Belgrade, Serbia

Goran Vasilic Department of Traffic, Mechanical and Protection Engineering, Academy of technical vocational studies, Belgrade, Serbia;
Department for Production Engineering, University of Belgrade, Belgrade, Serbia

Zoran Vasić Military Technical Institute, Belgrade, Serbia

Ivana Vasović Lola Insititute, Belgrade, Serbia

Sasa Zivanovic Department for Production Engineering, University of Belgrade, Belgrade, Serbia

Nemanja Zorić Faculty of Mechanical Engineering, University of Belgrade, Belgrade, Serbia

A FEM Analysis of Mortise and Tenon Joint Within Chairs

Igor Dzincic

Abstract This paper presents an analysis of a basic model of wooden chair. The aim of the paper is to optimize chair itself as joints using finite element analysis. Contact forces and deformations of two mostly used joints in furniture industry: oval tenon and a mortise, and dowel and a hole was analysed. The theoretical analysis was performed using the finite elements method, whereas experimental verification of the results, was obtained by invasive methods by testing samples on rigidity and durability according to European norms. Verification of the obtained stresses and strains was carried out using strain gauges positioned on chair samples. During the production of test pieces types of fit and types of joints were varied. Presented results showed that connection between side rail and rear leg presents the most critical spot in chair construction. Comparative analysis of the results obtained by finite element method and rigidity tests, showed that setup of parameters during finite element analysis was well done and presents recommendations for further research.

Keywords Chairs · FEM · Joints · Glue line · Rigidity

1 Introduction

The development of a new product in modern production is a very expensive activity. Whether it is an industrial or craft production, the engineer must have information about materials, properties of the product, environmental influences on furniture design as well as numerous technical knowledge that are combined through product development. In the process of constructing furniture, there is still a lack of sufficient exact data relating to the dimensions of individual parts, as well as the dimensions and strength of individual structural joints. From the work of Eckelman [2] to the present day, it can be concluded that a lot has been done in the field of static and dynamic analysis in furniture production. Previous research indicates the possibility

I. Dzincic (✉)
Department of Wood Science and Technology, Faculty of Forestry, University of Belgrade, 11030 Belgrade, Serbia
e-mail: igor.dzincic@sfb.bg.ac.rs

N. Mitrovic et al. (eds.), *Current Problems in Experimental and Computational Engineering*, Lecture Notes in Networks and Systems 323,
https://doi.org/10.1007/978-3-030-86009-7_1

of applying the finite element method (hereinafter FEM) when designing furniture. In the last ten years, research has been extended to the analysis of the strength of joints in furniture. All researchers who have dealt with this topic agree that joints are critical spots in the construction of chairs. The durability and strength of chairs are influenced by a large number of factors that, in correlation with each other, make the overall stress state of the chairs very complex. Analysis of the factors influencing joint strength is covered by papers including the influence of type of fit [10, 29, 32, 33], joint geometry [10, 26], gluing regimes [10, 21, 22, 24], and wood species [1, 21].

This paper presents an attempt to cover in one research, analytical analysis and experimental verification of the entire chair, including the joints. The results of this paper should contribute to the optimization of the shape and volume of details, that are today, mainly constructed and dimensioned based on experience and engineering practice. The results obtained on the basis of FEM analysis were compared with the results of durability tests on real chair samples and critical corner joints using measuring gauges. Tests were performed in order to confirm the choice of the finite element model throughout the test.

2 Material and Methods

Taking into account the complex stress state in chairs, several levels of testing will be conducted within this work. In the first part of the paper, the analysis of chair was performed using FEM in order to minimize the material. Due to the complexity of the problem during model optimization, due to the anisotropy of the base material and stress state, in the first phase it was not possible to include joints so they were analyzed in the third phase. In order to confirm the selected model in the ANSYS environment, in the second phase of work, chair samples were tested according relevant EN using measuring gauges. The measuring gauges were placed on chair elements where critical stresses was registered during the modeling of the chair in the first phase. Based on the results obtained in the first and second phase, in the third phase, the analysis of stresses and strains in the applied joints was conducted. The aim of the third phase of work is to define the stress state in the joints in order to optimize their dimensions in accordance with the defined load values. Based on the results of all previous phases, models of chairs were made in the fourth phase, which were tested for strength and durability.

During the loading of the models, their response to the effect of external loads, which are defined in accordance with European norms, was studied. Loads of virtual and real models in all phases of work correspond to the values defined by European standards EN 1022: 2019 [37] and EN 1728: 2013 [36].

2.1 Initial Model

The dining chair without lower rails and additional corner brackets was chosen as the test model, Fig. 2. Increase of the gluing surface, will undoubtedly led to a increasment of the durability of the structure. This basic model of the dining chair should provide insight into the limit state at chair critical points. The researchers, who undoubtedly made the greatest contribution to the introduction of FEM in the construction of chairs, also based their research on the basic chair model Gustafson [5-9], Ekelman [2], Kasal et al. [11]. Unlike Gustafson and Ekelman, a group of Polish scientists [3, 23, 25] based their research on chair which had additional rails and stretchers. The analysis conducted on more complex models does not give a realistic picture at the critical points of the chair.

As can be seen from Fig. 2, the front and rear legs of the chair have a square cross section with dimensions of 45 × 45 mm. As part of their work, Skakić and Džinčić [19] state that in 63% of dining chairs, the width of the front and rear legs ranges from 32 to 45 mm, so that the size of 45 mm was chosen for the initial dimension of the cross section of the legs. In the same paper, the width of the side rails, as an element of the chair that participates in the formation of critical points, was analyzed. Based on the insight into the technical documentation of the Institute for Furniture Quality Control, it was concluded that in the period from 2010 to 2020, 52% of the examined dining chairs had a sarg width of less than 45 mm, while 48% had a sarg width of more than 45 mm.

Based on the tests performed by Skakić and Džinčić [17], it could be concluded that the width of the tenon has a significantly greater impact on the strength of the joint than its length. By analyzing the torsional moment calculation pattern, the Saint–Venant pattern can only be applied in the form when the width of the plug is raised by a square. As the width of the plug has a significantly greater influence on the strength of the joint, a dimension of 50 mm was chosen for the initial width of the side sarg. The length of the rear leg was harmonized with the chair model used by Eckelman, and according to Gustafson [5, 6, 8] for easier comparison of the obtained results. Seat width, seating height as well as the angle of inclination of the chair back are selected in accordance with SRPS D.E2.100:1990 [35]. In this phase of work, all joints were considered as rigid, i.e the model was made with the assumption that the glued joints were well made.

2.2 Joints

After the completion of the second phase of the work, the analysis of the joints was done. In this phase of the work, the analysis of the joints was performed on the model side rail and back leg. The dimensions of the bearing elements correspond to the dimensions obtained in the first phase of work. Based on previous research and practical experience, it was concluded that the two most commonly used

Table 1. Groups and batches of samples with varied factors

Group	Joint	Batches	Machining accuracy	Type of fit according to DIN 68101:2012-02
A	Type	A_1	TD15	K/p
		A_2		K/m
B	Dowels	B_1		K/p
		B_2		K/m

joints in the construction of chairs are oval mortise and tenon and and dowels and hole. Analyzing the results of research obtained by different researchers (Hunker, according to Rudiger [15], Skakić and Džinčić [17, 18], Džinčić [27]), it can be concluded that the mentioned joints provides the limit values of joint strength, 250 Nm (Rudiger [15], according to Hunker). In the third and fourth phases of the work, samples was produced using those two types of the joints. The chairs of the first group were connected by oval mortise and tenon, while the dowel-hole connection was used as a joint in the second group of chairs. Within each group, two batches of samples were made. Between batches, the type of fit was varied, while the gluing surface, wood type, material moisture, glue quality and gluing regimes were controlled and kept constant, Table 1. Before making the samples, the accuracy of the machines for joint production was tested.

Within each group, boundary cases of types fit were analyzed, while the length of the joints was tolerated as a free measure. The definition of the material is done by setting the Young modulus of elasticity and Poisson's coefficients, in the same way as in the first phase of work. Based on the results obtained by other researchers [21, 22, 24, 32-34], among other things, it was concluded that during modeling the joints using FEM Youngs modulus of elasticity at the limits of proportionality should be used.

Only in the case of the appearance of the upper overlap, for the value of the Joungs modulus of elasticity, a value above the limit of proportionality should be taken. For all joints with a loose fit, the joint is compact along the entire gluing line with a clearly visible difference between the glue and joint. Using the loose fit in joints, which corresponds to the tolerance field K/m (DIN68101: 2012-02), the average thickness of the adhesive in the joint was 0.095 mm.

The discretization of the model was performed using a finite element of type NODE 187. The clamping of the "T" joints was performed on the foreheads of the leg, Fig. 1, in order to avoid bending of the "T" joint and to obtain a clean stress in the joint. The free front of the rail was deprived of four degrees of freedom, respectively the possibility of translational movement along the width and length of the sarg was left. This way of clamping the model reflects the real situation that arises during the normal usage of the chair.

The value of the vertical force has the highest values during the chair durability test, so that the deformations and stresses in the models were tested according to this load. The load was applied as pressure over the entire upper surface of the rail. The

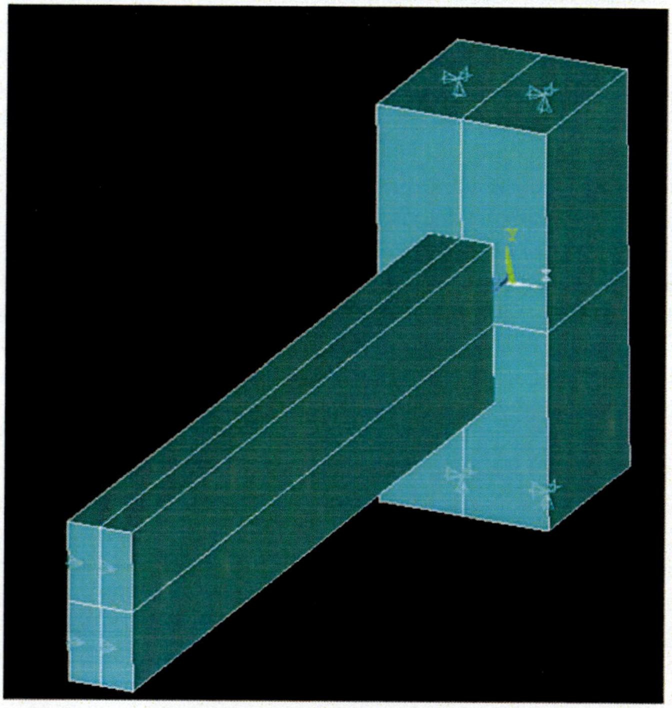

Fig. 1. T joint

value of the load corresponded to one eighth of the value of the total pressure on the seat according to SRPS EN 1728: 2013. In process of determination of the value of the load, it was started from the assumption that the pressure is transferred evenly from the seat to all four rails.

2.3 Materials

Based on research conducted by Skakić and Džinčić [19], among other things, it was concluded that about 72% of chairs are made of beech. Since beech wood in the forests of Serbia covers 50.4% of the volume of all tree species and 43% of the volume increment [30], there was a need to make samples from this tree species.

The density test of beech wood in the absolutely dry state was performed according to SRPS ISO 13061-2:2015 [38]. The analysis of the obtained values showed that the density of beech wood, from which the samples were made, is 0.69 g/cm^3, which corresponds to the average value of density in Serbia. The coefficient of variation is 6.15%, which indicates a relatively small variability of the sample members.

According to Popović [14], and based on data from the literature, the average coefficient of variation, that is taken into account when researching density, is about 10%. Based on this data, it can be concluded that the chairs were made of sawn timber that represents a well-chosen homogeneous pattern. As the obtained value of wood density is in the above range, we can assume that the values of mechanical properties will have a similar distribution. Based on this, the values of the modulus of elasticity and Poisson's coefficients according to Kollman and Cotte [12], pp. 295 ÷ 298) were used in the analysis, namely Ea = 13,739 N/mm^2; Er = 2.236 N/mm^2; Et = 1.138 N/mm^2; Gar = 1.608 N/mm^2; Gat = 1059 N/mm^2; Grt = 0.461 N/mm^2; μtr = 0.36; μrt = 0.71; μat = 0.52; μta = 0.043; μra = 0.073; μar = 0.45. For the purposes of this work, the elements for making the chair model will be cut from radial planks, where the angles that the tangents drawn on the annual rings make with the wider side of the workpiece will be less than 10°. The logs from which the boards were cut had a small drop in diameter, thus avoiding the cutting of fibers.

PVA-c glue from producer Rakoll type Express 25D was used for gluing the joints. According to the resistance to climatic conditions, the used glue is classified in group D2. The glue is applied to the joints on both sides, by hand. When applying the glue, the wetting of the connecting surfaces was controlled. After applying the glue on the joints, the quantity of the applied glue was measured by the gravimetric method. The average amount of applied glue was 185.6 g/m^2. In the modeling phase related to the determination of material characteristics, the adhesive is defined as an isotropic material.

2.4 Model Preparation

In order to optimize the volume, the initial model of the chair is divided into segments. Models (products) with a smaller number of independent variable, does not requires product divisions into segments. In the case of a chair due to the complex stress state, as well as due to the fact that basic material is an anisotropic material, it was necessary to divide the constituent elements. The chair is divided into twenty segments marked with letters from A to T, Fig. 2. The papers of Polish researchers [3, 23, 25], and papers done by Gustafsson [5-9] were used as a starting point for determining the number and length of segments. Within this paper, the length of the segments varies depending on the element to which the segment belongs. When determining the length of the segments, care was taken that during the optimization process, changing the dimension of the segment does not affect the geometry of the joint. Due to the complexity of the analysis, the chair segments carrying the connection elements were not taken into account during the optimization, so that their dimensions remained the same as the initial dimensions. The segments carrying the joints (both segments on legs and rails) had a cross section of 50 × 17 mm, while the segments carrying the openings had cross section dimensions of 45 × 45 mm.

The final element NODE 187 was used to discretize the chair model. The selected element is a spatial 3-D element of higher order. The element is defined by ten nodes

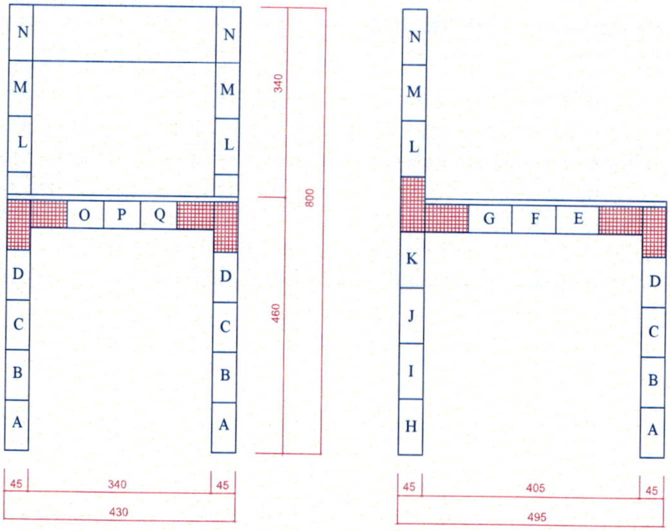

Fig. 2. Initial model with segments

(I, J, K, L, M, N, O, P, Q, R) which have three degrees of freedom each, namely translations along the x, y and z axes. The geometry of the element, the location of the nodes and the coordinate system of the element enable its use in orthotropic and anisotropic materials.

When setting the boundary conditions in stability tests, the case of sliding the legs of the chair on the ground was observed, so that the line of contact between the stop and the foot allowed rotation about one axis, and the leg support itself is allowed to translate in two axes. Due to the large number of elements analyzed in each iteration, as well as due to the anisotropy of the material, the tolerance value generated by the program itself was not acceptable, resulting in the inability to perform optimization. When defining the independent variables, the value of their tolerance was taken to be 0.5 mm, which corresponds to the value of the free measure for the accuracy class TD60 (according to DIN68101: 2012-02) [39].

During the analysis of the segments of the chair legs, the value of the stress on the pressure parallel to the fibers was taken for the value of the dependent variable, while the value of the bending stress perpendicular to the fibers was used during the analysis of the sarg segments. The shear strength of beech perpendicular to the fiber is defined as a dependent variable. As the goal of modeling is volume minimization, the lower limit value at the limit of proportionality, which is about 12 N/mm^2, was taken into consideration (Forest Encyclopedia [31], pp. 432 ÷ 433, tab.26).

3 Results and Discussion

Calculations and results obtained on the basis of static calculation of the chair, diagrams of normal and transverse forces as well as bending moments, are presented according to the applied tests.

3.1 *Optimization of the Initial Chair Model*

The optimization of the initial model was performed only with the aim of obtaining a model on which the joints can be analysed. Based on the defined conditions of the chair construction analysis, and according to the loads given within the European norms, seven chair models were obtained. Due to the complexity of the problem, and in order to shorten the time required for optimization, only those elements that were on the side of the chair that was loaded were analyzed. Due to the limited work space, Fig. 3 shows the optimization results according to the three most critical tests: front seat edge load, side seat edge load, and durability test. The values shown are Von Mises stresses within ANSYS and are expressed in N/m^2.

When optimizing the model through all seven tests, the value of stress on the rails was significantly lower than the allowable value of bending stress at the limit of proportionality. Since the compressive stress parallel to the fibers has almost twice the value of the bending stress [13], the allowable value of the bending stress on the rails was not reached in any test. In the stability tests and the static load test of the front edge of the seat, where the vertical load is dominant over the horizontal, critical points occurred on the legs due to the fact that the value of compressive stress parallel to the fibers is less than the bending stress. When testing the chair for static load of the seat and back, as well as for tests of static load of the legs forward and sideways, horizontal forces acted on the legs in addition to the vertical load, so that the legs that were loaded with pressure were additionally loaded with bending effect horizontal forces.

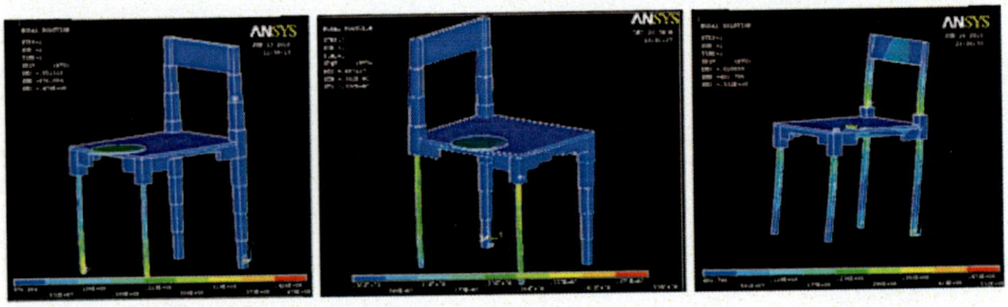

Fig. 3. Optimization of the initial model

3.2 Tests with Measuring Gauges

Comparative analysis of chair element deformations measured with measuring gauges and deformations achieved on the virtual model was performed in order to confirm the applicability of the FEM in chair analysis, as well as to confirm parameter settings during modeling. Due to the lack of recommendations when positioning the measuring tapes on the furniture elements, two measuring bridges were placed at each measuring point. On the side rail, the bridges are placed from the lower and upper side of the rail next to the rear leg, while on the rear leg the measuring bridges are placed on its back side, below and above the position of the side rail, Fig. 4. On the rear leg, readings were performed on the measuring bridge located above the rail, while on the side rail, the deformation was measured on the measuring bridge, which was located on the underside of the side rail.

Comparative analysis of the values obtained from the measuring points and the deformations obtained on the virtual model, shoved that the positions of the measuring gauges are well chosen. In the first second of the cycle (from the 1st to the 150th reading), only the vertical cylinder acted on the chair. The horizontal cylinder acted on the back of the chair for ninety seconds (from the 150th to the 375th reading), but it started its activity after the expiration of the first second, only after the maximum force was reached on the vertical cylinder. The values that will be placed in relation to the values read from the virtual model are the maximum values of deformations that occurred during the period of action of both pistons.

By reviewing the values of deformations on the bottom side of the side rail, the value of $-0.308 \cdot 10^{-3}$ m or -308 μm can be read from, Fig. 6 (left). Values with a negative sign in ANSYS indicate deformation due to stretching. Examining the values due to the elongation of the measuring bridge, Fig. 6, it can be seen that the maximum value of deformation is in the interval between 300 and 340 μm. The shapes of the

Fig. 4. Positions of measuring gauges

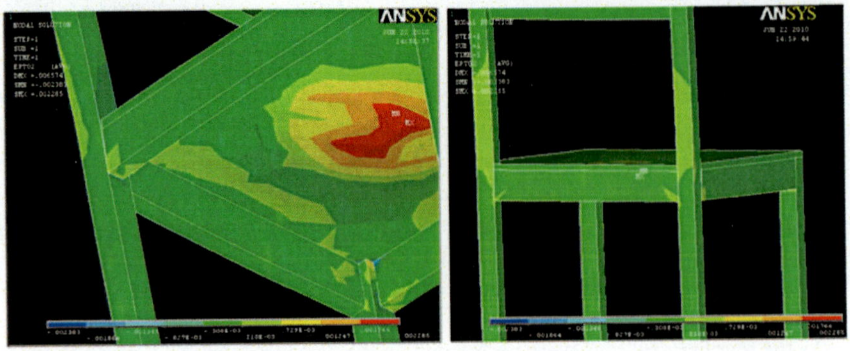

Fig. 5. Size and allocation of dilatations

line graphs of all three measurements are approximately identical, which confirm the stability of the measuring system and the accuracy of the measurements (Fig. 5).

By analyzing the values of the read deformations on the back side of the rear leg of the chair, Fig. 6 (right), it can be seen that the maximum values of deformations of both measurements are about 200 μm. During the measurements on the rear leg, almost identical results were obtained, which led to the matching of line graphs. By reviewing the achieved deformations on the back side of the rear leg of the virtual chair model, the deformation value of +0.21 · 10-3 m and +210 μm can be read from Fig. 5. Deformation values with a positive sign in ANSYS indicate the deformation is caused by pressure.

This analysis confirmed the values of deformations that occurred during the modeling of the chair in the ANSYS environment. Based on the obtained results, it can be seen that the type of finite element, mesh density, mode of load transfer, as well as the values of modulus of elasticity and Poisson's ratios when making a chair model was well chosen.

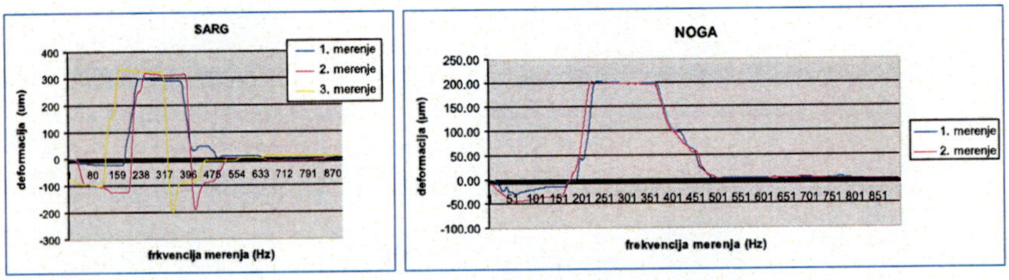

Fig. 6. Results obtained on the basis of deformations of measuring gauges

3.3 Joint Optimization

In accordance with the aim of the work, models of joints were analyzed. Due to the lack of input data through which the material is defined, the analysis of the joints was conducted only on models in which a gap occurred. In joints where the type of fit K/p is applied, there is also a possibility that an upper gap will occur. However, the probability that the plug, which has the value of the lower limit measure, and the opening, which has the value of the upper limit measure, will merge during the gluing of the beams, is 1.96%, so that case was not considered. Figure 7 show the values and distribution of stresses at the joint of the oval mortise and tenon, while Fig. 8 show the results obtained based on the load of the dowels joint.

Based on the performed optimization of the oval tenon and mortise joint, the following can be observed:

After the joint optimization, the calculated width of the side rail is 35 mm. During optimization, the tenon height was reduced from an initial value of 46 mm to 31 mm. Along with the height of the tenon, the value of the width of the rail was also reduced. When defining the width of the rails, a value of 2 mm was adopted for the shoulder height of the tenon, in order to increase the gluing surface as much as possible. According to this design rule, the width of the rail after the optimization process was 35 mm. At the connection with oval tenon and mortise joint, the highest stress pressure was registered on the cheeks of the tenon near the support, as well as on the lower bearing of the tenon and it is about 9 N/mm^2, Fig. 7. Tensile stress occurred on the cheeks of the tenon. The value of tensile stress ranged from 1.6 to 17 N/mm^2. This value is relatively close to the values obtained by Wengang et al. [33]. The observed deviation can be attributed to different dimensions of the connection elements. The lowest values of tensile stress occurred around the axis of the tenon, while its value increases with increasing distance from the center of rotation. The highest value of the tensile stress was registered on the lowermost bearing of the tenon and its value is 39 N/mm^2. This correspond with results presented by Wengang and Huiyuan [34], and Gavronski [4]. In the base of the tenon, Fig. 7 (right), the pressure on the shoulder

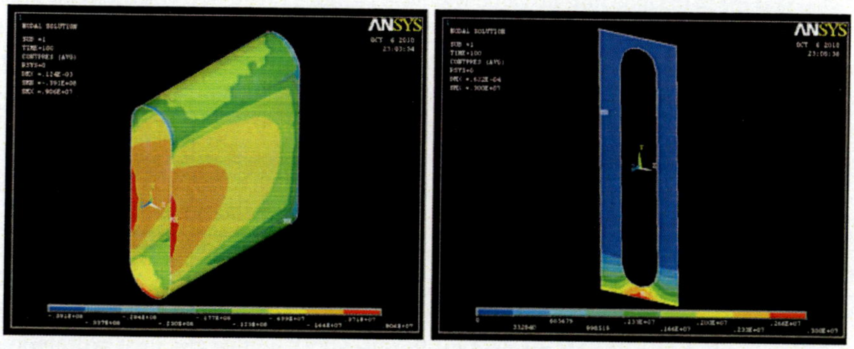

Fig. 7. Distribution of stress and strain on the oval mortise and tenon joint

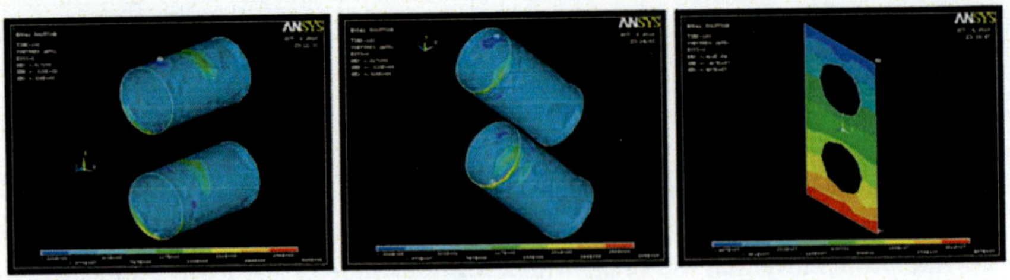

Fig. 8. Distribution of stress and strain on the dowel joint

of the tenon was registered: The value was 3 N/mm^2. Such a small registered value of the pressure on the shoulder of the tenon can be explained by its small size. Namely, during joints dimensioning, the shoulders of the tenon are reduced in order to increase the width of the tenon, and thus the gluing surface. At the base of the tenon, a stress whose value was 11.5 N/mm^2 was registered, Fig. 7. The registered stress represents the shear stress perpendicular to the fiber, whose lower limit value at the level of proportionality is 12 N/mm^2.

Based on the performed optimization of the dowel and hole joint, the following can be observed.

After the joint optimization, the calculated width of the side rail carrying the dowel element is 35 mm. In the base of the dowels, whose diameter was 10 mm, shear stress perpendicular to the fiber was registered, Fig. 8. The value of the registered stress ranged from 10.9 to 12.3 N/mm^2.

During process of optimization, the width of the side rail is reduced from 50 to 35 mm. Along with the width of the rail, the value of the distance between the axes of the dowels was also reduced. According to the recommendation, this distance should be twice the diameter of the dowels. In order to optimize the volume of the rail, this distance is reduced from 20 to 16 mm, where the width of the segment that carries the joint element is 35 mm. Due to the complexity of modeling the contact problem with dowels, only the parts of the dowels that are positioned in the leg were analyzed. The highest value of stress was registered on the lower dowel, at the transition from the rear leg to the side rail. Its value was 32.8 N/mm^2, Fig. 6 (left and middle). This result deviates from results presented by Sjodin et al. [16]. The differences in the results can be explained with number of dowels. In their study they perform investigation on single dowel joint, so that is a reason for difference in presented results. As the recorded value far exceeds the value of the shear stress perpendicular to the fiber, there will undoubtedly be an intersection of the dowel at its base. A similar stress distribution on the plug and groove was obtained by Kasal et al. [11]. The value of the stress on the dowel shell, which represents the surface on which the glue is applied, has a negative sign, which suggests that it is a tensile stress. The value of the tensile stress on the dowel shell ranges from 9.7 to 52 N/mm^2. The registered value of the tension on the dowels is up to five times higher than the value of the tensile strength of the dowels reached by Rudiger [15], so that the joint will undoubtedly loosen along the gluing line.

The value of the stress on the shoulder of the joint with dowels is slightly higher than with the connection with an oval mortise and tenon and its value is 4.8 N/mm^2. This difference can be explained by the slightly higher value of the shoulders in dowel-hole joint. In order to make the connection with the dowels correctly, the minimum value of the shoulder was 4.5 mm, while the value of the shoulder of the plug was only 2 mm.

3.4 Experimental Verification of the Final Model of the Chair

Testing the real model of the dining chair, according to the European standards, was conducted on two groups of samples. The shape of all details was obtained on the basis of approximation of the shape of details obtained in the process of volume optimization. In the case of chairs of all groups, the structure loosened due to the weakening of the joints. As a critical point in the construction, the connection of the side rail of the rear leg appeared. Regardless of the type of joint as well as the type of fit, the weakening of the connection elements occurred due to the loosening of the joint on the gluing line. Table 2 shows the results of the durability test.

Based on the conducted testing of four groups of chairs, it can be noticed that the none of the groups accomplish the durability condition according to EN 12,520. In order to meet the durability conditions, the oval mortise and tenon joint cannot be made in K/p or K/m with the used gluing surface. The small gluing surface was conditioned by the small width of the rail. In order to increase the durability of the construction, it is necessary to increase the width of the rail in the zone of the connection element or to add additional reinforcements in the form of corner brackets.

The highest number of cycles was achieved by the chairs of batch A1, where an oval mortise and tenon was used with the type of fit of K/p. The average number of realized cycles within this party was 3242.4 cycles. Significantly less durability was shown by the chairs of group A2, where the same connection element was used, but with the K/m fit. The average number of cycles within this batch of samples was 1407 cycles, which is 43.4% of the durability of the previous group. Within the group where the dowel-hole joint was used as the, the highest number of cycles was achieved by the chairs of the B1 batch, which were connected in the form of a K/p fit. The average number of realized cycles in group was 1352.4. The chairs of the party within which the K/m fit was applied, achieved significantly lower durability, with the average number of cycles of 856.2. The average realized number of cycles is 36.7% less than the number of realized cycles of lot B1. This results deviates from the results presented by Smardzewski and Gavronski [20]. Differences can be explained by existence of additional lower side rail which definitely increased gluing surface and durability of sample.

If we compare groups A1 and B1, in which the elements are connected with different joints, but with the application of the same type of bearing K/p, it can be seen that the durability of the structure with dowels makes 41.7% of the durability of

Table 2. The results of the durability test

Joint	Batch	Type of fit	Sample no	Achieved no. of cycles	$\bar{x}$ (ciklusa)	σ (ciklusa)
Oval mortise and tenon	A1	K/p	1	3004	3242.4	323.13671
			2	3955		
			3	2632		
			4	3350		
			5	3271		
	A2	K/m	1	1295	1407	286.7629683
			2	11,118		
			3	1334		
			4	1404		
			5	1884		
Dowel and hole	B1	K/p	1	1270	1352.4	257.2504616
			2	1501		
			3	1621		
			4	953		
			5	1417		
	B2	K/m	1	824	856.2	140.3805542
			2	1028		
			3	969		
			4	773		
			5	687		

the structure where the connection element oval mortise and tenon. The difference in durability between batches A2 and B1, where an oval mortise and tenon was used, as a connection element, in the form of K/m fit, and the group where dowels in the form of K/p fit were used, is small and amounts to only 3.9%. A comparative analysis of durability within the groups shows that the K/p fit gives a significantly more durable construction compared to the constructions where the K/m fit was applied.

4 Conclusions

In this study, some factors that have influence on the strength and durability of chairs using FEM were investigated. The performed investigations revealed following conclusions.

Based on the results of comparative analysis of chair deformations measured with measuring gauges, and deformations achieved on the virtual model, it can be concluded that the type of finite element, mesh density, load transfer method, and parameters used in defining the material are well chosen.

Due to small deviations between the results obtained on the use of measuring gauges and the results obtained when modeling the chair for the values of the modulus of elasticity and Poisson's ratios, values presented by Kollman and Cote can be used in FEM analysis.

Applying a commercial software package based on FEM, it is possible to perform analysis and dimensioning of joints in chairs. In case that glued joint is performed correctly, the highest value of the stress will be registered in its base, with both types of connection applied. The registered stress is the shear stress perpendicular to the fiber that tends to cut the connection element at its base.

The minimum width of the rails carrying the tenon is equal to the minimum width of the rail carrying the dowels. The minimum width of the tenon, observed from the point of view of material bearing capacity, is 31 mm. In order to reduce the width of the rail, when a dowel was used as a connection element, the distance between the axes of the dowel was reduced from the value $2 \times d$ to the value $1.6 \times d$.

From the point of view of the analysis of joints, it can be concluded that the side rail-back leg joint appears as a critical point in the construction. A stress value of 46.6 N/mm^2 was recorded on the side rail. The tension on the rail occurs in its lower zone, where the joint resists with its support the moment of action that tends to knock it out of the mortise. By analyzing the stress on the front rail, which is loaded on bending, it can be seen that the highest value of stress is about 6.6 N/mm^2, where the joint side rail-front leg in this model is not a critical point in the design.

In accordance with the aim of the work, after the analysis of the corrected model of the chair, the analysis of the joint elements was performed, using the commercial software package ANSYS. During the development of the model, the type of joint and the type of fit were varied, and through the type of fit and the accuracy class. The modeling was performed on the elements of the oval mortise and tenon and dowel-hole connection, made in the accuracy class TD15 and with the K/m fit. When connecting with an oval plug and groove at a fitting K/m, the value of the tensile stress ranges from 1.6 to 17 N/mm^2. In the case of a dowel and hole joint at a fit K/m value, the tensile stress ranges from 9.7 to 52 N/mm^2.

On the basis of comparative analysis of the results obtained by loading the joints and the results obtained on the basis of experimental verification, for groups of chairs where the joints were made with K/m type of fit, it can be seen that the strengths of the joints corresponding to the real state can be obtained during the experimental test.

References

1. Derikvand, M., Smardzewski, J.: Withdrawal force capacity of mortise and moose tenon T-type furniture joints. Turk. J. Agric. For. **37**, 469–478 (2013)
2. Eckelman, C.A.: A look at the strength design of furniture. For. Prod. J. **16**(3), 21–24 (1966)

3. Gavronski, T.: Multiobjective optimisation of a skeleton furniture construction, Monography. Roczniki akademii rolniczej w Poznaniu, Rozprawy naukowe. Poland (2005)
4. Gavronski, T.: Rigidity-strength models and stress distribution in housed tenon joints subjected to torsion. EJPAU **9**(4), (2006)
5. Gustafsson, S.I.: Furniture design by use of finite element method. Eur. J. Wood Wood Prod. **53**, 257–260 (1995)
6. Gustafsson, S.I.: Stability problems in optimised chairs. Wood Sci. Technol. **30**, 339–345 (1996)
7. Gustafsson, S.I.: Finite element modelling versus reality for birch chairs. Eur. J. Wood Wood Prod. **54**, 355–359 (1996)
8. Gustafsson, S.I.: Optimising ash wood chairs. Wood Sci. Technol. **31**, 291–301 (1997)
9. Gustafsson, S.I.: Solid mechanics for ash wood. Eur. J. Wood Wood Prod. **57**, 373–377 (1999)
10. Hill, M.D., Eckelman, C.A.: Flexibility and bending strength of mortise and tenon joints. Furniture Des. Manufact. **45**(1 and 2), Dun-Donnelly Corp., Chicago, IL (1973)
11. Kasal, A., Birgul, R., Erdil, Y.Z.: Determination of the strength performance of chair frames constructed of solid wood and wood composites. For. Prod. J. **56**(7–8), 55–60 (2006)
12. Kollman, F.F.P., Cote, W.A.: Principles of Wood Science and Technology – Volume I: Solid Wood. Springer Verlag, Berlin, Heidelberg, New York, Tokyo (1984)
13. Marjanov, M., Popović, Z.: Deformacije i naprezanja na pritisak i zatezanje hrastovog, bukovog i borovog drveta u radijalnom anatomskom pravcu. Drvarski glasnik **2–3**, 7–10 (1992)
14. Popović, Z.: Uticaj vlažnosti i temperature na modul elastičnosti i savitljivost bukovog drveta. Master thesis. Šumarski fakultet. Beograd (1990)
15. Rudiger, A., et al.: Grundlagen des Mobel – und Innenausbau, DRW – Verlag, Stuttgart (1995)
16. Sjodin, J., Serrano, E., Enquist, B.: An experimental and numerical study of the effect of friction in single dowel joints. Holz als Roh und Werkstoff, Springer – Verlag, **66**, 363–372 (2008)
17. Skakić, D., Džinčić, I.: Uticaj dimenzija čepa i vida naleganja na čvrstoću spojeva kod stolica. Prerada drveta **2**, 25–30 (2003)
18. Skakić, D., Džinčić, I.: Uticaj vida naleganja na čvrstoću spoja čep-žljeb kod stolica. Prerada drveta **15–16**, 12–15 (2006)
19. Skakić, D., Džinčić, I.: Grupisanje i analiza kvaliteta stolica i njihovih sastavnih delova. Glasnik Šumarskog fakulteta **99**, 147–154 (2009)
20. Smardzewski, J., Gavronski, T.: FEM algorithm for chair optimisation, EJPAU **4**(2), (2001)
21. Smardzewski, J.: Strength of profile-adhesive joint. Wood Sci. Technol. **36**(2), 173–183 (2002)
22. Smardzewski, J.: Effect of wood species and glue type on contact stresses in mortise and tenon joint. J. Mech. Eng. Sci. **222**(12), 2293–2299 (2008)
23. Smardzewski, J., Gavronski, T.: Gradient optimisation of skeleton furniture with different connections. EJPAU **6**(1), (2003)
24. Smardzewski, J., Prekrat, S.: Effect of glue line shape on strength of mortise and tenon joint. Drv. Ind. **61**(4), 223–228 (2010)
25. Smardzewski, J., Papuga, T.: Stress distribution in angle joints of skeleton furniture. EJPAU **7**(1), (2004)
26. Tankut, A.N., Tankut, N.: The effect of joint forms (shape) and dimensions on the strength of mortise and tenon joints. Turk. J. Agric. For. **29**, 493–498 (2005)
27. Džinčić, I.: Uticajni faktori na čvrstoću i trajnost stolica, master thesis. Univerzitet u Beograd, Šumarski fakultet (2006)
28. Džinčić, I., Skakić, D.: Influence of type fit on strength and deformation of oval tenon mortise joint. Wood Res-Slovakia **57**(3), 469–477 (2014)
29. Džinčić, I., Živanić, D.: The influence of fit on the distribution of glue in oval tenon mortise joint. Wood Res-Slovakia **59**(2), 297–302 (2014)
30. Šoškić, B., Skakić, D.: Svojstva i namenska prerada bukovine, monography, Šumarski fakultet, Beograd (1995)
31. Šumarska enciklopedija, Jugoslovenski leksikografski zavod, Zagreb, drugo izdanje; knjiga 1 (1970)
32. Wang, Y.R., Lee, S.H.: Design and analysis on interference fir in the hardwood dowel glued joint by finite element method. Procedia Eng. **79**, 166–172 (2014)

33. Wengang, H., Huiyuan, G., Jilei, Z.: Finite element analysis of tensile load resistance of mortise and tenon joints considering tenon fit effects. Wood Fibre Sci. **50**(2), 1–11 (2018)
34. Wengang, H., Huiyuan, G.: A finite element model of semi-rigid mortise-and-tenon joint considering glue line and friction coefficient. J. Wood Sci. **65**(14), (2019)
35. Serbian standard SRPS D.E2.100:1990 Furniture for sitting-Technical requirements
36. SRPS EN 1728:2013 Furniture – Seating – Test methods for the determination of strength and durability
37. SRPS EN 1022:2019 Furniture – Seating – Determination of stability
38. SRPS ISO 13061-2:2015 Physical and mechanical properties of wood
39. Deutsche Industrie Norm. 2012. DIN 68101: 2012-02: Grundabmaße und Toleranzfelder für Holzbe - und –verarbeitung

How to Improve Energy Consumption and GHG Emissions in the Wood Pellet Production in Serbia

Mladen Furtula, Gradimir Danon, Marija Đurković, and Srđan Svrzić

Abstract As Serbia is currently dependent almost entirely on its own forest resources, consumption of raw wood biomass, for all purposes, should be limited within current boundaries, about 9,000,000 m^3. This situation requires from the wood processing industry to be responsible for the raw material and to reduce energy consumption and environmental pollution at all stages of the life cycle to a minimum. This also applies to the production of wood pellets, which are considered as a (nearly) purely renewable fuel and one of the potential substitutes for fossil fuels for heat production. The attention of this paper was paid to the wood pellet production, because this segment has the greatest impact on overall energy consumption and the influence on the environment in wood pellets life cycle. For the necessary calculations the "GHG Balance of Pellet Production" program with some upgrades, was used. The analysis showed that the investigated Serbian pellet plants, for different reasons, used more raw material and energy for production of 1 ton of pellets than plants elsewhere in Europe and America. In order to change this situation several different improvement methods have been examined and appropriate solutions suggested. The analysis of possible solutions showed that, if the proposed measures were implemented, the possible raw material would be between 7 and 10%. The analysis of possible solutions showed that, if the proposed measures were implemented, the following savings could be achieved: savings in wood raw materials between 7 and 10%, energy savings 12% and reduction in greenhouse gas emissions 23–45%.

Keywords Wood pellet · Energy efficiency · GHG emissions · Natural drying · CHP

M. Furtula (✉) · G. Danon · M. Đurković · S. Svrzić
Faculty of Forestry, University of Belgrade, Belgrade, Serbia
e-mail: mladen.furtula@sfb.bg.ac.rs

N. Mitrovic et al. (eds.), *Current Problems in Experimental and Computational Engineering*, Lecture Notes in Networks and Systems 323,
https://doi.org/10.1007/978-3-030-86009-7_2

Nomenclature

CHP	Cogeneration or combined heat and power is the use of a heat engine or power station to generate electricity and useful heat at the same time
EPS	Electric Power Industry of Serbia (Elektroprivreda Republike Srbije)
GHG	Gas in an atmosphere that absorbs and emits radiation within the thermal infrared range
GWP	Global-warming potential is a relative measure of how much heat a greenhouse gas traps in the atmosphere. A GWP is calculated over a specific time interval, commonly 20, 100 or 500 years
Srbijašume	PE "Srbijašume" has been entrusted with the protection, utilization and management of state forests, forestland and other forest potential and professional-technical work performance in privately owned forests
Turboden	European company in development and production of ORC (Organic Rankine Cycle) turbo-generators

1 Introduction

For reducing human dependence on fossil fuels and mitigation of climate change through the reduction of greenhouse gas emissions into the atmosphere it is necessary to stop or slow down the increase in primary energy consumption, and to turn mostly or completely to the use of renewable energy sources with minimum or no associated greenhouse gas emissions.

Currently, renewable sources accounted for a record 2.8% of global primary energy consumption [1] but this is still very far from the targets, and the question remains when they will be reached. Some researchers are predicting, based on the detailed analysis of the supply potential and the use of biomass, a transition to a fully renewable global energy system by 2050 [2]. Others are not so optimistic and expect that by 2050 fossil fuels will still account for 60% of the primary energy required in the world [3].

In these predictions wood fuels has been recognized as one of the possibilities that might, along with wind and sun, assist in reduction of the use of fossil fuels. Wood fuels have certain advantages over fossil fuels. They represent a renewable source of energy, and their GHG (greenhouse gasses) emissions, taking into account the entire life cycle, according to currently valid opinions, are less than 10% of that of fossil fuels. To this should be added that wood fuels cannot cause acid rains because they contain minimal amounts of Sulphur. On the other hand, biggest drawback of the wood fuels are particulate emissions, which represent a real problem of the wood combustion, but which can be easily controlled/reduced today.

The situation in Serbia regarding the use of renewable energy sources is as follows: according to [4] the share of renewable sources, in gross final energy consumption, amounted to 7.21% in 2020 with a significant proportion of biofuels and waste (7.09%) [4]. In Serbia, various types of wood fuels are used such as firewood, wood chips, wood briquettes, wood pellets and charcoal. The wood is mostly derived directly from the forest, less from the wood processing plants, and symbolically from energy forests. There is also used wood, mostly wooden pallets, as raw material for the production of wood pellets. There are no credible researches on the use of other types of used wood in Serbia.

In order to exploit the potential of the resources mentioned, in a successful, sustainable, efficient and environmentally friendly manner, it is necessary to take into account several conditions that have to be fulfilled:

- Sustainable use of wood resources implies: limiting the exploitation of forest biomass below the annual allowable cut, taking care of biodiversity, protection of areas of particular interest, etc. Annual volume increment in the Serbian forest fund and non-forest areas for the year 2011, based on research conducted as part of the TCP/FAO project, is slightly over 11,000,000 m^3 [5], and did not change much in the following years. The main reasons are the effects of climate change, the slower growth and dieback of forests). This suggests that the consumption of raw wood biomass, for all purposes, should be contained within current limits (about 9,000,000 m^3 as it was in 2012), and the focus of future actions in this area should be directed towards increasing energy efficiency in production, transportation and use, especially when talking about wood fuels.
- Environmental and energy factors throughout the life cycles of wood fuels are defined by the type of production and utilization of wood fuels and the effects of energy consumption and GHG emissions. The most influential factors are the moisture content and calorific value of the raw material, the level of necessary transformation, the transportation of raw materials and final products, as well as other characteristics, especially the energy efficiency of boilers and furnaces specialized for burning wood fuels. Under raw materials, in some cases fresh raw material was considered, and in others raw material with a moisture content of 10% (dry based).
- Economic factors are primarily market dependent, but it is possible to influence them by incentive or punitive measures by the Government. Regulations of the wood fuel market should stimulate better quality of wood fuels, care of the environment, etc. Support of the state is expected and welcomed, but success on the market is still more dependent on the manufacturers of wood fuel and furnaces.

The paper focuses on the wood pellet production in Serbia. Possibilities of on the opportunities for increasing the efficiency of raw materials and energy use, as well as on the possibilities of reduction of fossil fuels usage were.

The research was based on the assumption that the available quantities of wood from the forest are limited and that their volume will not change in the coming years. The possible increase in the production of wood pellets will require redistribution of consumption of firewood among rural households, greater supply of wood residues from wood processing plants, raising energy crops and last but not least better usage of available wood raw material. Only the last option is very likely in the following time period.

Increasing purchases from private forest owners is possible but unpredictable, because in Serbia the wood market has not yet been formed. The largest part of the wood cut does not appear on the market because rural households use it for their own purposes. Higher market surpluses could be provided only if the habits of the rural households change, if they improve the insulation of their houses and the efficiency of the furnaces they use. It should be noted that chipboard factories represent a great deal of competition for wood fuel producers when acquiring raw wood materials.

Characteristics of wood industry in Serbia are fragmentation, poor technical equipment and low efficiency in the use of raw material. Wood residues from plants are used mainly for their needs (dryers, heating etc.), large residues are sold to employees and the rural population in the surrounding area and only a small part comes on the market.

Energy forests represent great potential but there is not enough experience in Serbia with the establishment of intensive plantations of fast growing species for energy needs. A pilot project should be launched in the near future to serve as an experimental polygon for monitoring the "behavior" of selected species in different ecological conditions in Serbia. Only after the completion of this experiment will the potential of these plantations and financial indicators for the given conditions be obtained.

The collection of recycled wood is not widespread in Serbia. In addition recycled wood is mainly used as woodfuel for households.

Efforts to increase energy efficiency in the pellet production chain should focus on those parts of the cycle where the results would be biggest and where they could be achieved without too much investment. Based on our own research and the research of other authors, it can be asserted with confidence that in the whole pellet life cycle, the process of pellet production, in a narrow sense, is the most energy-intensive and therefore has the greatest impact on the environment [6–8] (Table 1).

In pellet production the main energy consumers are the wood chip dryers [6, 8–10]. The amount of energy used for drying clearly depends on the energetic efficiency of the equipment, but much more on the moisture content of the input material. Providing raw materials with lower moisture content would mean greater savings in

Table 1. Participation of pellet production in energy consumption and GHG emissions in pellet life cycle

No	References	Pellet production	
		Participation in life cycle energy consumed, [%]	Participation in life cycle emissions of GHG, [%]
1	[6]	81.86	–
2	[7]	86.70	82.00
3	[8]	88.50*	84.00*

*LCA without transportation to customer

energy consumption but would not lead to significant changes in the greenhouse gas emissions in the pellet production process. For this, it would be necessary to reduce the specific consumption of electricity in production and to switch to a renewable source of electricity supply.

In the conclusions, on the basis of analysis of the results of research on the production of pellets in two factories in Serbia, proposals were made. They contain measures for better utilization of the existing raw wood material, an increase in the energy efficiency of the production of wood pellets and a reduction in the share of fossil fuels.

2 The Basics About the Wood Pellets

Wood pellets are a type of biomass produced from chopped wood and they are suitable both for heat and power applications, with co-firing in coal-fired power stations currently being their main large-scale application. They are cylindrical in shape with 6 mm. to 8 mm. in diameter and 10–30 mm. in length. The wood used for pellets comes either in the form of residues from the wood-processing industry, mostly from sawmills, or straight from the forests.

Wood pellets have a moisture content of less than 10%, have good structural strength, and low dust and ash content. The typical net calorific value of wood pellets is between 4.4 and 4.7 kWh/kg [11]. The studies [11–13] showed differences in calorific value for different species. This is due to the different content of lignin, cellulose, extractive (especially resin) and ash [12] in wood species, and softwoods generally have higher calorific value than hardwoods. Pellets are being produced for industrial and residential use. The characteristics of these two kinds of pellets are different because the logistic facilities and the combustion conditions differ. The EN plus certification scheme defines three pellet quality classes based on standard ISO 17225-2: EN plus A1; EN plus A2 and EN plus B.

The main properties of wood pellets to be controlled are the species and the origin, the durability, the number of tiny pieces in bags, the moisture content, the quantity and composition of ash, the ash deformation temperature, and the content of additives, Sulphur and metals.

2.1 *The Production Process in the Pellet Factory*

The production process begins with the delivery of the raw materials (round wood, wood waste, sawdust, whole trees) to a pellet mill, mostly by trucks. In Serbia, the raw material is usually in the form of long firewood or meter long green or freshly cut firewood [14]. Raw wood material needs to be processed in such a way that the end product has consistent moisture content, heat value and ash content. All pellet manufacturers should produce pellets to the same standard (ISO 17225–2) so

Fig. 1. The scheme of production of pellets from firewood

that pellet appliances burn and heat consistently. Figure 1 shows a pellet production pattern typical for Serbia.

Explanation of the process and power consumption of pellet production shown in Fig. 1:

- Wood splitting—Production starts with splitting logs if necessary so that they can be chopped more easily (part of other power consumption).
- Chipper & Hammer Mill—Cleaved wood is inserted into a chipping machine for further processing. Chipping is not necessary for mills that purchase wood chips and sawdust. The next step is a hammer mill. These machines break down sawdust and wood chips into regular smaller sizes, making drying and pressing through the pellet die quicker and more consistent; Adopted power consumption for this process is 19 $kWh/tonne_{pellet}$ of electricity [9].
- Dryer—When the pellet mill consumes freshly cut raw material, material exposed to the weather or with high moisture content, or a mix of raw materials that may contain moisture, crushed material has to be dried to a consistent moisture level (MC = 10%). Dryers may use chopped wood, natural gas, propane, or other fuels to heat the raw material, driving off the extra moisture; Adopted power consumption for this process is 24 $kWh/tonne_{pellet}$ of electricity and 1200 kWh per tonne of evaporated water from wood [9].
- The Pellet Mill—After drying the shredded material is pressed through dies at high pressure. This process causes the crushed wood to heat up and release the natural lignin in the wood to bind wood particles together. The mill also determines important characteristics, such as the density of the pellet, the diameter and the durability; Adopted power consumption for this process is 139 $kWh/tonne_{pellet}$ of electricity [10].
- Cooling and Storage—The pellets come out of the mill at temperatures between 200 °C and 250 °C and soft. A cooling tower is used to bring the temperature down and harden the pellets. After cooling they are usually stored in a large silo to await bagging or bulk distribution; Adopted power consumption for this process is 2 $kWh/tonne_{pellet}$ of electricity [9].
- Bagging or Bulk—The most common method for distribution is to put the pellets into plastic bags and stack them on pallets or skids. Bulk pellets are loaded from the pellet mill silo directly into trucks for delivery to bulk storage containers. The bulk trucks are more expensive than regular flatbed trucks, but are a part of a much more efficient system of processing, transfer, and delivery. Adopted power consumption for this process is 18 $kWh/tonne_{pellet}$ of electricity [9].
- Other power consumption—4 $kWh/tonne_{pellet}$ [9].

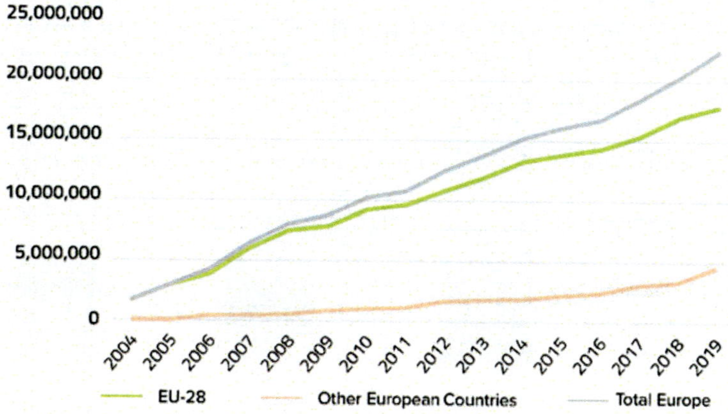

Fig. 2. Pellet production in Europe (tonnes) [15]

2.2 *Wood Pellet Production and Consumption in Europe*

World wood pellet production and consumption show massive growth after the year 2000. It is expected that this trend will continue in the coming years, regardless of the fluctuation of oil prices and climate change [15]. The largest producer, importer and consumer of wood pellets is the EU-28. In 2018, the EU-28 consumed 26 million tons of wood pellets. The production of pellets within the EU-28 in 2018 amounted to about 17 million tons [15] and the rest came from imports, mostly from North America, Russia and other European countries (Fig. 2).

A smaller portion (36%) was spent for power generation, of which most in the UK. The rest (64%) was used for heat generation (residential, commercial and CHP) [15]. The largest consumers of pellets for households were Italy (3 million tons) and Germany (1.485 million tons) [15].

2.3 *Wood Pellet Production and Consumption in Serbia*

Production and consumption of wood pellets in Serbia follows trends in the EU. The first industrial facility in the Republic of Serbia was opened in 2007 and according to data for 2018 there are 69 active pellet plants in Serbia. Their total capacity was about 541,000 t/a [15]. Production in 2018 was about 324,000 t/a [15], and domestic consumption about 241,000 t/a (75% of total production).The structure of domestic buyers is given in Fig. 3.

Households use 96% of the domestic consumption of pellets, supplied through a distributor (63%) or directly from one of the manufacturers (28%).

The dominant species for pellet production is beech (84%). It is followed by spruce and fir (13%) and pine at around 2% [16]. Serbian pellet manufacturers use

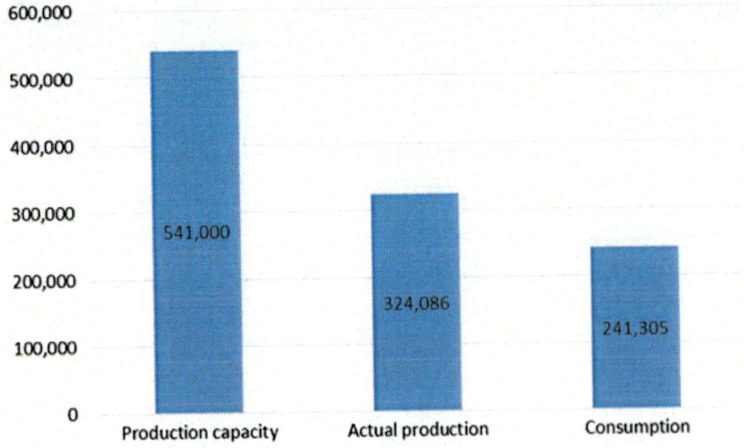

Fig. 3. Production and consumption of wood pellets in 2018 in Serbia (tonnes) [15]

most raw materials in the form of long firewood or in meter length firewood (79.6%). Solid sawmill residues (12.5%) are in the second place, followed by sawdust (7.7%). Forest residues and logs are also utilized, but their combined share is about 0.2% [16]. PE "Srbijašume", which has been entrusted with the protection, utilization and management of state and private forests, forest land and other forest resources in Serbia, fulfils 66% of the raw material requirements of pellet producers. The second largest suppliers are sawmills, with solid sawmill residues and sawdust (14%). Private forest owners share around 10% of the supply, delivering mostly meter length and long firewood. Only 4% of the supply is covered by wood residues from pellet producers own wood processing factories.

3 Material and Methods

For research purposes, the data relating to two representative pellet plants were used. The first, marked with "A", belongs to a group of high-capacity plants, and the second, marked with "B", to the medium-capacity plant group in Serbia. The collected data relates to the year 2012. The data includes the amount of raw wood and pellets produced, the consumption of all kinds of fuels, the average distance of the delivery of raw materials and finished pellets (Table 2).

From Table 2 it can be seen that in both analyzed plants firewood delivered directly from the forest dominated as raw material. This was indicated by the high moisture content of the input raw material.

Pellet production also required the consumption of different types of energy for heat, machining and internal transport. The collected data on consumed fuels for obtaining the required energy is shown in Table 3.

Table 2. Consumed raw wood and produced wood pellets in the analyzed plants in 2012

Input data	Plant "A"		Plant "B"	
	Measuring unit	Quantity	Measuring unit	Quantity
Pellet production per year	t	35,500	t	8550
Raw material consumed per year	t	79,740	t	16,852
Average moisture content of the raw material*	%, dry base	79	%, dry base	73
Structure of raw material consumed				
Beech firewood	m^3	65,000	m^3	8550
Beech sawdust	t	54	t	3852
Beech trimmings	t	–	t	3217
Conifers	m^3	1240	m^3	1011
Old pallets	Piece	–	Piece	8321

*Calculated by surplus of raw material [8, 17]

Table 3. Fuels consumed for heat and technology in the analyzed plants in 2012

Energy source	Plant "A"	Plant "B"
Woody biomass*	13,571 t/a	2951 t/a
LPG	27,222 kg/a	–
Diesel	121,828 l/a	17,983 l/a
Electricity from the grid	7,191,346 kWh/a	2,250,000 kWh/a

*Adopted caloric value is 16.6 MJ/kg (MC = 10%)

Wood biomass was used for the production of heat for the drying of input raw material and heating. In plant "A", LPG was used for the propulsion of working and transport equipment within the factory. Diesel fuel was used for the operation of mobile equipment, passenger and cargo vehicles. Consumption was not separated to the level of each department because it was assumed that all fuel consumption and electricity were tied directly or indirectly to the production of wood pellets.

The use of different types of fuel is accompanied by appropriate GHG emissions. For all necessary GHG calculations the program "GHG Balance of Pellet Production" was used. This Excel file has been created by "Bioenergy 2020+" to perform life cycle GHG calculations of wood-based pellets [18].

The program uses validated data for its work on emission factors for electricity, fuels and packaging (Table 4).

Emission factors and default values are taken, in order of preferences, from: WTT App. 1 [19], BioGrace [20].

Table 4. Emission factors—fuels and packaging

Type of fuel	GHG emission, [gCO_2eq/MJ]	Source
Woody biomass	7.94	WTT App. 1 [19]
LPG	293.13	BioGrace [20]
Diesel	248.39	BioGrace [20]
Electricity mix (Serbia)	248.39	BioGrace [20]

4 Results and Discussion

Gathered and processed data relating to production and transport of the pellets and the necessary raw materials, are used as input in the expanded program "GHG Balance of Pellet Production". In Table 5, calculated data of yearly energy consumption and GHG emission of the two selected pellet plants are presented.

The difference in the consumption of diesel fuel can be explained by the presence of the mobile chipper in the pellet plant "A" while the plant "B" had an electrical one (which then influenced the electric energy consumption levels in the plant "B). The difference would have been even greater if the forklifts in the plant "A" had not been altered to use LPG.

For the use of analysis of specific energy consumption in pellet production and the effect that pellet production has on the environment, data from both plants are calculated for 1 ton of the pellets produced (see Table 6).

Difference in energy consumption for the production of 1 ton of wood pellets of 10% of all forms of energy used in plants "A" and "B" originates from the consumption of fuels, mostly woody biomass for heat, due to the differences in the moisture content and structure of the raw material.

Additional analysis of the results showed that the amount of heat needed for the evaporation of 1 ton of water is nearly equal in both selected plants and add up to: 2040 kWh/t_{water} (for plant "A") and 2063 kWh/t_{water} (for plant "B"). The amount of heat required for evaporation of water and heat losses in boiler and dryer are both included in these numbers. When the heat loss in the boiler ($\eta_{boiler} = 0.71$) is excluded from the calculation, new calculated values (about 1450 kWh/t_{water}) are much lower

Table 5. Energy consumption and GHG emission in the selected pellet plants in Serbia (2012)

Energy source	Energy [kWh/a]		Emission [$kgCO_2e/a$]	
	Plant "A"	Plant "B"	Plant "A"	Plant "B"
Electricity from the grid	7,191,346	2,250,000	6,430,433	2,011,929
Diesel	1,213,515	179,127	382,869	55,494
LPG	352,427	0	84,625	0
Wood biomass for drying	62,577,389	13,607,389	223,279	48,987
Total	71,334,677	16,036,516	7,121,206	2,116,410

Table 6. Specific energy consumption and GHG emission for 1 ton of produced pellets in selected pellet plants in Serbia

Energy source	Energy [kWh/t_{pellet}]		Emission [$kgCO_2e/t_{pellet}$]	
	Plant "A"	Plant "B"	Plant "A"	Plant "B"
Electricity from the grid	203	264	182	236
Diesel	34	20	11	6
LPG	10	0	2	0
Wood for drying	1763	1546	6	6
Total	2010	1830	201	248

but are still standing higher than those suggested by Obernberger and Thek in "The Pellet Handbook" [9], the latter being between 1000 kWh/t_{water} and 1200 kWh/t_{water}.

All the above mentioned lead to the conclusion that in the selected plants problems are present regarding high moisture content of the raw material and low energy efficiency of the equipment. This also results in increased consumption of the wood biomass used for the heat production.

4.1 *The Possibilities for Reduction of Consumption of Energy and Raw Materials in Pellet Plants*

In Fig. 4 three different relationships between the mass of wet wood required for production of 1 ton of pellets and the input wood moisture content are shown.

Dashed line presents the case when energy for raw material drying is provided from other sources (natural gas, coal, electricity etc.). In one ton of pellets, assuming the final moisture content is 10%, there is 909 kg of absolutely dry wood and 91 kg of water. The necessary quantity of raw material for the production of 1 ton of pellets rises with the higher moisture content in the material, which means that the same quantity of absolutely dry wood (909 kg is obtained from the higher quantity of raw material), because more water is present. For example, average moisture content of raw material in plant "A" is 79%. It means that plant "A" should provide 1628 kg of raw material (which contains 909 kg of absolute dry wood and 719 kg of water) for 1 ton of pellets. In the case of plant "B" this quantity is somewhat smaller and amounts 1573 kg/t_{pellet}. The difference is 55 kg and corresponds to the difference between the quantities of water contained in the wood of 79% moisture content and in that of 73% moisture content.

Dotted lines mark the quantity of wood, with different moisture content, necessary for the production of 1 ton of pellets, if the wood is used both for pellet production and for the production of heat for drying. The calculation is made assuming that the heat for evaporation of 1 ton of water from wood is 1200 kWh/t_{water} and that the efficiency of the boiler is 0.81 [8, 21, 22]. The amount of wet wood necessary for producing 1 ton of pellet increases from 1628 to 2272 kg (plant "A") and from 1573

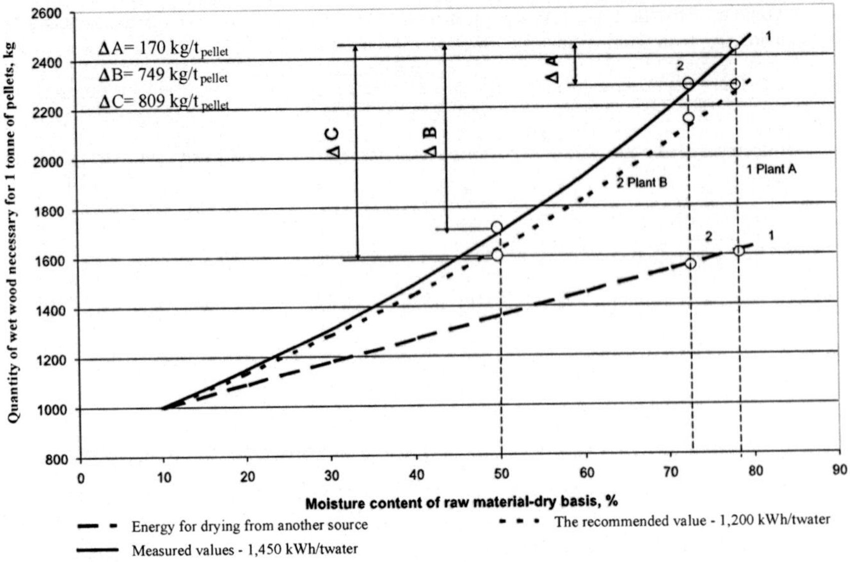

Fig. 4. Changes in the quantity of wet wood necessary for the production of 1 ton of pellets as a function of its input moisture content

to 2128 kg in plant "B". The difference in the amount of the wet wood needed for the pellet production in the plants "A" and "B" increases to 144 kg for 1 ton of pellet.

The full line shows dependence between the mass of raw wood needed for the production of 1 ton of pellets (both for production and drying) and the moisture content of the raw material. The calculation is made on the assumption that the amount of heat for evaporation of 1 ton of water from the wood equals to 1450 kWh/t_{water} and boiler efficiency is 0.81 (in both plants). The difference in the amount of wet wood needed for the pellet production in the plants "A" and "B", in this case is 170 kg. The calculated amount of wood that is used for drying raw material, from the initial moisture down to 10%, was 815 kg. (plant "A"), in the case of a plant "B" slightly less (689 kg). If the plants are successful in increasing their energy efficiency and achieve lower suggested values (1200 kWh/t_{water}), possible savings in the energy required for drying could be about 250 kWh/t_{water} or about 17%. The effect in wood consumption would not be so significant. The quantity of wood necessary for production of 1 ton of pellets would be decreased between 142 kg (plant "B") and 170 kg (plant "A"), that is between 6 and 7% (Fig. 5).

Significantly higher effects would be achieved if plants could provide raw material with lower moisture content. Considering the structure and origin of the raw material that plants are acquiring (firewood) directly from the forest, suggested possibilities for decreasing moisture content is the natural drying of the raw material in the forest or in the pellet plant itself after the delivery. By decreasing the initial moisture content of the raw material from 79 and 73%, respectively to 50% significant savings in the

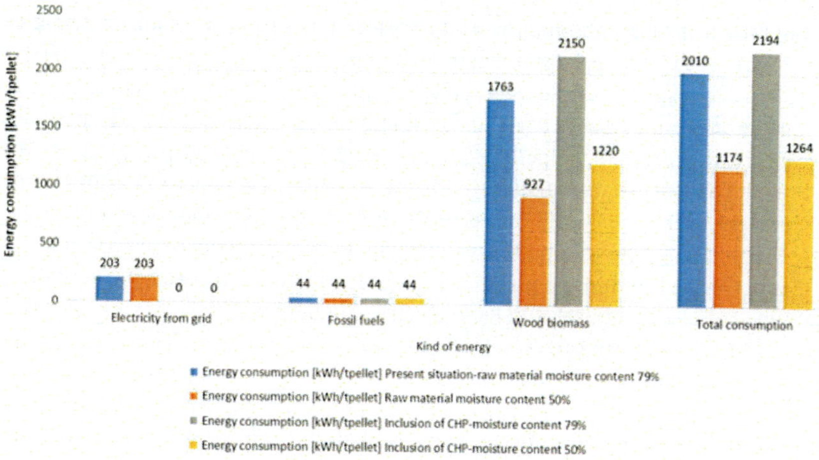

Fig. 5. General consumption of all forms of energy for pellet production—plant "A"

amount of heat required for evaporation of water and thus the amount of the raw material used could be achieved (see Fig. 6).

The problem for the suppliers of the material and the pellet manufactures would be to provide adequate conditions for the natural drying of raw wood, the necessary space, equipment and time, and, of course, providing sufficient investment and cash flow capital. The amount of time needed for the natural wood drying depends to a large extent on the weather conditions the wood stack is exposed to. These are the level of precipitation, time of the year, climate conditions on the micro-location (wind speed, insulation, air temperature and relative air humidity) [23]. According to the results of the experiments performed in several European countries, the amount

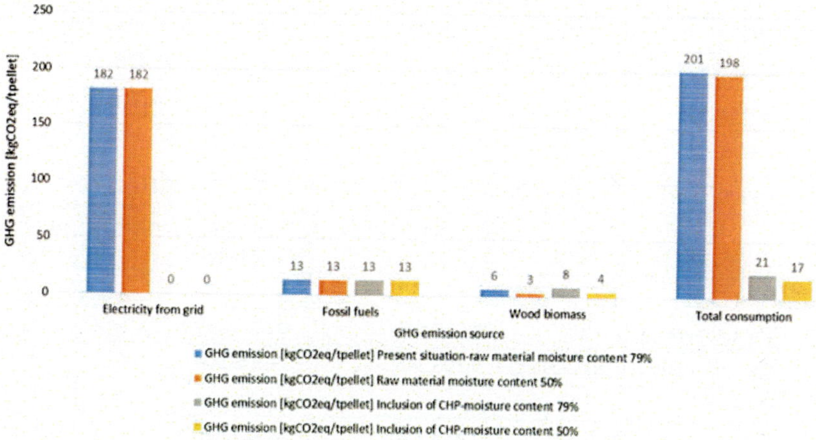

Fig. 6. General GHG emission in the pellet production—plant "A"

of time needed for chopped beech to dry to about 30% of moisture content is between 3 months in Italy and up to 6 months in Austria [24]. Natural drying to 50% should last somewhat less. Additional effect could be achieved if the natural drying of raw material were performed in the forest and the raw material, already partially dried, were transported to the pellet plants for further drying to final moisture content.

Results of the calculation, performed for the raw material of lower moisture content, using the "GHG Balance of Pellet Production" program are shown in Table 4. The calculation was made assuming that the process of pre-drying of the raw wood is performed within the plant. Also, it was assumed that the consumption of other types of energy (diesel, LPG and electricity) did not change.

When comparing the results from Tables 5 and 7, positive results of pre-drying of the raw material can be seen in the form of lower quantities both of raw wood necessary to produce the same amount of pellets and the energy needed for raw material transportation. In the case of plant "A", lower wood consumption in the dryer (7136 t/a in contrast to the previous 13,571 t/a) caused a lower raw wood requirement. The amount of raw wood with the 79% moisture according to the new calculation is 70,281 t/a instead of 79,740 t/a. Similar results were found for the plant "B".

Table 8 shows possible savings in energy as well as the reduction of emissions of GHGs if the moisture of the incoming raw materials was 50% instead of 79% (plant "A") or 73% (plant "B").

Table 7. Energy consumption and GHG emission–natural drying of raw material to moisture content of 50%

Energy source	Energy [kWh/a]		Emission [kgCO_2e/a]	
	Plant "A"	Plant "B"	Plant "A"	Plant "B"
Electricity from the grid	7,191,346	2,250,000	6,430,433	2,011,929
Diesel	1,213,515	179,127	382,869	55,494
LPG	352,427	0	84,625	0
Wood Biomass for drying	32,904,889	8,157,056	118,458	29,365
Total	41,662,177	10,586,183	7,016,385	2,096,788

Table 8. Possible savings of thermal energy in the pellet producing plants in the Serbia

	The current situation	With natural partially drying of raw material	Savings
Plant "A"			
The total energy [kWh/a]	78,260,546	48,293,217	38.29%
The total emission [kgCO_2e/a]	9,306,345	9,108,504	2.13%
Plant "B"			
The total energy [kWh/a]	17,613,157	12,123,810	31.17%
The total emission [kgCO_2e/a]	2,613,846	2,581,916	1.22%

Possible savings in energy consumption are 38.29% for plant "A" and 31.17% for plant "B". The reduction of emissions of GHG would be only symbolic, and would amount to 2.13% for plant "A" and 1.22% for plant "B".

4.2 *The Possibilities for Reduction of the GHG Emission in the Pellet Plants*

As mentioned before, the best option for decreasing GHG emission in pellet production can be found in using energy from its own renewable source. This solution requires inclusion of the CHP into the energy system. The capacity of the CHP plant should be dimensioned in accordance with the heat requirements of the pellet production plant. The electricity produced could be used both for its own use and for sale. When choosing the technology for cogeneration, priority should be given to already proven technologies that are in commercial use [25]. For the purpose of this study, the Organic Rankine Cycle (ORC) technology was selected. The main reason for such a decision is that this technology was found to be a well proven product for industrial application in pellet plants over the last 15 years [26–31].

For selection of the CHP capacity it was necessary first to determine the number of working hours throughout the year and then, using that and the data from Table 1, to calculate hourly energy and power consumption in the pellet plant. The calculations preformed for 7500 h/a are presented in Table 9. In the same table, characteristics of CHP's are given which, according to the thermal power output, correspond to the analyzed plants.

Suggested solution for plant "A" (CHP-22) has an excess of thermal power that is not too large (9630 kW > 8344 kW). This does not present a technical problem because ORC maintains 90% of cycle efficiency down to 50% of loading and functions successfully in conditions of 10% of nominal load [27]. The difference between the available and the required electric power is larger, and taking into account that the average load is 1018 kW and lower than nominal (87%), it leaves the possibility of selling the electricity to the EPS (Electric Power Industry of Serbia) at a preferential rate. A slightly more favorable situation appears with the plant "B", where available and required average heat outputs match and electric energy produced by

Table 9. Average hourly energy consumption in the pellet plants (2012) and basic characteristics of the suggested CHP ORC's

Plant	Heat	Electricity	Power requirement for CHP, kW*		Nominal power of the proposed CHP, kW [31]		
	MWh/a	MWh/a	Heat	Electricity	Designation	Heat	Electricity
"A"	62,577	7191	8344	959	CHP-22	9630	2282
"B"	13,607	2250	1814	300	CHP-4	1844	424

*Calculation is done with the assumption that the number of working hours per year was 7500 h/a

Table 10. Energy consumption and GHG emission with CHP included in the plant energy system

Energy source	Energy [kWh/a]		Emission [gCO_2e/a]	
	Plant "A"	Plant "B"	Plant "A"	Plant "B"
Electricity from the grid	0	0	0	0
Diesel	1,213,515	179,127	382,868,891	55,494,000
LPG	352,427	0	84,624,675	0
Wood Biomass for drying	76,341,556	19,090,000	274,829,600	68,724,000
Total	77,907,498	19,269,127	742,323,166	124,218,000
The surplus of electricity	7.433.522	930.000	–	–

CHP is 40% higher, also allowing the possibility of selling the electricity to EPS at the preferential rate.

Result of the GHG calculations with the "GHG Balance of Pellet Production" program are shown in Table 10.

Table 10 also shows that the calculation is done on the assumption that the electricity requirements from the net are basically reduced to zero. The amount of energy necessary to produce electricity is included when calculating the total amount of woody biomass needed. GHG emission is significantly reduced, amounting to only from 6 to 10% of calculated yearly GHG emission of the plant in 2012.

4.3 *Overall Effects of the Proposed Improvements*

Implementing both of the aforementioned suggestions at the same time in the pellet plants, significant savings in raw material and energy could be achieved, simultaneously reducing GHG emission. Three different improvement suggestions were analyzed and compared with the present situation:

1. The use of raw material with moisture content of 50% (dry basis);
2. Inclusion of CHP plants in the system for heat supply—feedstock moisture 79% (dry basis);
3. Inclusion of CHP plants in the system for heat supply—feedstock moisture 50% (dry basis).

Improvements specified under 1 and 2 have been discussed in the previous subsections. The following text presents the possible effects that could be accomplished by using a combination of raw material with lower moisture content and incorporation of CHP in the energy supply system (Table 11).

Heat requirements, when the moisture content of raw material is 50%, are significantly reduced and it is therefore necessary to re-select capacities of CHP for pellet production plants. Necessary input data and results of the selection are shown in Table 12.

Table 11. Energy consumption and GHG emission—natural drying of raw material to moisture content of 50% + CHP included in the plant energy system

Energy source	Energy [kWh/a]		Emission [gCO_2/a]	
	Plant "A"	Plant "B"	Plant "A"	Plant "B"
Electricity from the grid	0	0	0	0
Diesel	1,213,515	179,127	382,868,891	55,494,000
LPG	352,427	0	84,624,675	0
Wood biomass for drying and power	43,307,556	10,734,667	155,907,200	38,644,800
Total	44,873,498	10,913,794	623,400,766	94,138,800

Table 12. Average hourly energy consumption in the pellet plants after the improvements and basic characteristics of suggested CHP

Plant	Heat	Electricity	Power requirements for CHP, kW[a]		Nominal power of the proposed CHP, ORC, kW [31]		
	kWh/a	kWh/a	Heat	Electricity	Designation	Heat	Electricity
"A"	32,904,889	7,191,346	4387	959	CHP-14	5370	1317
"B"	8.157.056	2,250,000	1087	300	CHP-04	1844	424

[a]Calculation is done with the assumption that the number of working hours per year was 7500 h/a

The suggested solution for the plant "A" (CHP-14) has an excess of thermal power that is not too large (5370 kW > 4387 kW). Average load of CHP would be around 82%, which leaves a certain amount of power in case of overload. The lesser heat load of CHP implies a lower electrical power (1127 kW) which is still sufficient to cover the plant's needs. Surplus of the electricity (168 kW) leaves the opportunity open of selling the electrical energy to the EPS. Total consumption of the wood necessary to produce 1 ton of pellet would increase by about 6% and consumption of raw feedstock by 30%.

For plant "B" there is no solution more favorable than the former. ORC system CHP-4 has the lowest heating power in the range of "Turboden" Italia products [31].

The highest calculated energy consumption is in the "third case", and that is expected and understandable considering the fact that CHP is included in the existing heat supply system. It is higher than the consumption in the "first case" for two reasons. Firstly, energy consumption in transport is higher in this case because of the increase in the amount of the raw material needed for electricity production. The second reason is that the decreasing moisture of feedstock by natural drying reduces the energy necessary to evaporate water from it and therefore the total energy consumption is lower (case two). If CHP is integrated in the system, energy consumption rises somewhat (case 4), but remains significantly lower than in cases one and three.

Figure 6 shows that, after the inclusion of CHP, total emissions of GHG are reduced to about 8.5% of the former value, actually from 201 kg CO_2e/t_{pellet} falling to 17 kg CO_2e/t_{pellet} (comparison of the case 2 and case 4).

5 Conclusions

Serbia is currently relying almost entirely on its own forest resources. This suggests that the consumption of raw wood biomass for all purposes should be limited within current boundaries (about 9,000,000 m^3). Focus of future actions in this area should be directed on increasing energy efficiency in the production, transportation and use, especially when talking about wood pellets.

The research was carried out in selected Serbian pellet plants. The greatest attention was paid to the production and transportation, because these two segments have the greatest impact on overall energy consumption in wood pellet life cycle and on the environment. For the necessary calculations the "GHG Balance of Pellet Production" program, with some upgrades, was used.

The analysis showed that the researched Serbian pellet plants, for different reasons, are using more raw materials and energy for production per 1 ton of pellets than plants elsewhere in Europe and America. In order to correct this situation several different improvements have been suggested and analyzed:

- *Raising the technical level of equipment in the pellet plant*—raising energy efficiency in boilers and dryers would result in savings in the wood needed for fuel (around 7%);
- *Use of wood raw materials with reduced moisture content*—Calculations have proved that large energy savings could be achieved if the raw material (wood biomass) is naturally dried from fresh (79%) to about 50% of moisture content before entering the production process at the plant. In this way it would be possible to achieve savings of about 30% of wood fuel required for drying.
- The process of natural reduction of the moisture content in the raw material would take less than three months, if the appropriate conditions in the warehouse were achieved;
- *Use of electricity from renewable sources*—Reducing emissions of GHG in the process of pellet production is possible with the use of electricity from renewable sources. This could be achieved with the inclusion of the CHP in the plant energy system. In this way, GHG emissions could be reduced by almost 12 times, with minimum increase in consumption of wood for fuel;
- Consumption of wood biomass would be decreased between 7 and 10%. This means that the plant could increase the production volume by the same percentage, for the same amount of raw wood material. This increase would be lower because the inclusion of CHP requires an additional amount of wood as fuel for the boiler.

Acknowledgements This research was financed by the Ministry of Education, Sciences and Technological Development of Republic of Serbia (Project no.: 451-03-68/2020-14/200169).

References

1. BP Statistical Review of World Energy. http://www.bp.com/en/global/corporate/energy-economics/statistical-review-of-world-energy/downloads.html (2016)
2. Cornelissen, S., Koper, M., Deng, I.: The role of bioenergy in a fully sustainable global energy system. Biomass Bioenergy **41**, 21–33 (2012). https://doi.org/10.1016/j.biombioe.2011.12.049
3. Mingxin, G., Weiping, S., Jeremy, B.: Bioenergy and biofuels: history, status, and perspective. Renew. Sustain. Energy Rev. **42**, 712–725 (2015)
4. IEA World Energy Balances 2020. https://www.iea.org/subscribe-to-data-services/world-energy-balances-and-statistics (2020)
5. National Renewable Energy Action Plan of the Republic of Serbia (In accordance with the template foreseen in the Directive 2008/29/EC-Decision 2009/548/EC), Republic of Serbia, Ministry of Energy, Development and Environmental Protection. https://www.energy-community.org/pls/portal/docs/2144185.PDF (2009)
6. Paa, A., Craven, J., Xiaotao, B., Melin, S., Sokhansanj, S.: Environmental footprints of British Columbia wood pellets from a simplified life cycle analysis. Int. J. Life Cycle Assess. **17**, 220–231 (2012)
7. Katers, J., Snippen, A., Puettmann, M.: Life-cycle inventory of wood pellet manufacturing and utilization in wisconsin. For. Prod. J. **62**(4), 289–295 (2012)
8. Furtula, M.: The impact of ecological and energy factors on more efficient use of solid wood fuels in Serbia, Doctoral thesis, University of Belgrade, Faculty of Forestry, pp. 1–169 (2014)
9. Obernberger, I., Thek, G.: Pellet Handbook. Chapter: Pellet Production and Logistics. Earthscan Ltd (2010)
10. Risović, S., et al.: Analysis of pellets made of wood residues. Croat. J. For. Eng. **29**(1), 95–108 (2008)
11. Telmo, C., Lousada, J.: Heating values of wood pellets from different species. Biomass Bioenergy **35**, 2634–2639 (2011). https://doi.org/10.1016/j.biombioe.2011.02.043
12. Stevanovic Janezic, T., Danon, G., Bujanovic, B., Dedic, A.: Correlation between chemical composition and heating value of some domestic wood species. Wood Research (Drevársky Výskum) **3**, 1–7 (1993)
13. Danon G.: Valorisation of Wood Processing and Forest Residue, Enhancement of wood technologies in correlation with properties of wood chemical constituents (Ed. T. Stevanović-Janežić et al.), University of Belgrade, pp. 115–130 (1995)
14. Glavonjić, B., Vukadinović, M.: DKTI Sustainable Bioenergy Market in Serbia - Assessment of the Market of Heating and Cooking Appliances and Wood Fuels in Serbia. E4tech (UK), London (2014)
15. Calderon C., Colla, M.: European Bioenergy Outlook 2019 – Pellet, Bioenergy Europe, Brussels. http://www.bioenergyeurope.org (2019)
16. Wood Pellet Production and Market Structure in Serbia. http://www.bioenergyserbia.rs/images/documents/studies/Wood_Pellet_Production_and_Market_Structure_in_Serbia.pdf (2016)
17. Danon, G., Furtula, M., Lukačev, D.: Energy Efficiency in Wood Industry: Example of Wood Pellets Production, Maintenance Forum 2017, Budva, 24–26. June 2017, pp. 44–52. IIPP, University of Belgrade, Mechanical Faculty, www.iipp.rs (2017)
18. "GHG Balance of Pellet Production" program produced by "Bioenergy 2020+" GmbH in cooperation with "Technology & Innovation" GmbH under the commission of "Propellets" Austria. http://www.google.rs/?gws_rd=cr,ssl&ei=8hqdVuiSOIj6swH5q4agDA#q=GHG+Balance+of+Pellet+Production (2016)
19. JEC, Well-to-wheels Analysis of Future Automotive Fuels and Power trains in the European Context WTT APPENDIX 1 Description of individual processes and detailed input data (2011)
20. BioGrace List of Standard Values. http://www.biograce.net/content/ghgcalculationtools/standardvalues (2016)
21. Danon, G., Bajić, V., Furtula, M., Mandić, M.: Cradle-to-gate life-cycle fossil energy inventory and impact assessment of wood fuel manufacturing from hardwood forest biomass. In:

Proceedings of 8th Dubrovnik Conference on Sustainable Development of Energy, Water and Environment Systems, Dubrovnik, Croatia, 22–27.09.2013, pp. 1–22 (2013)
22. Furtula, M., Danon, G.: Environmental impact of the wood pellet production in Serbia. In: Proceedings of VII International Forum for Clean Energy Technologies, http://www.pkv.rs/pkv/files/VII%20FORUM%20-%20FIRST%20INFORMATION.pdf (2013)
23. Erbera, G., Routa, J., Wilhelmsson, L., Raitila, R., Toiviainen, M., Riekkinen, et al.: A prediction model prototype for estimating optimal storage duration and sorting. Finnish Forest Research Institute. http://www.metla.fi/julkaisut/workingpapers/2014/mwp297.htm (2014)
24. Erbera, G., Kanziana, C., Stampfera, K.: Modelling natural drying of European beech (Fagus sylvatica L.) logs for energy based on meteorological data. Scand. J. For. Res. **31**, 294–301 (2016)
25. Arnau, G., Jordi-Roger, R., Puig, R., Navarro, P.: Review of micro- and small-scale technologies to produce electricity and heat from Mediterranean forests' wood chips. Renew. Sustainable Energy Rev. **43**, 143–155 (2015). https://doi.org/10.1016/j.rser.2014.11.013
26. Shen, X., Kommalapati, R.R., Huque, Z.: The Comparative life cycle assessment of power generation from lignocellulosic biomass. Sustainability **7**(10), 12974–12987 (2015). https://doi.org/10.3390/su71012974
27. Kiehne, H.: Biomass fired combined heat and power plants (CHP) with organic rankine cycle (ORC) in Europe. In: Proceedings of Energy from Biomass and Waste, Pittsburgh, USA (2008)
28. Duvia, A., Tavolo, S.: Application of ORC units in the pellet production field: technical-economic considerations and overview of the operational results of an ORC plant in the Industry installed in Mudau (Germany); 2008 http://www.turboden.eu/en/public/downloads/Turboden_Paper_on_ORC_application_inthe_pellet_production.pdf (2008)
29. Pedrajas, I.: Engineering aspects of a pellet production facility - Possible association to cogeneration with ORC. http://www.novator.se/bioint/BPUA12Pres/11_BPUA12_Jose_Ignacio_Pedrajas_PRODESA.pdf
30. http://www.turboden.eu/en/public/downloads/Turboden_Paper_on_ORC_application_in_the_pellet_production.pdf (2003)
31. Campana, F.: Turboden ORC technology for the wood industry. In: Proceedings of 12th Wood – Technology Conference June 1–2, 2015, Opatija, Croatia, http://www.drvna-konferencija.hr/wpcontent/uploads/2015/06/Turboden_12_DTK.pdf (2015)

Multi-objective Optimization and Experimental Testing of a Laminated Vertical-Axis Wind Turbine Blade

Zorana Trivković, Jelena Svorcan, Marija Baltić, Nemanja Zorić, and Ognjen Peković

Abstract Vertical-axis wind turbines, despite being somewhat uncommon and less efficient renewable energy converters, still offer many advantages to small consumers and possibilities for further improvement. Their rotor is usually made up of fiberglass composite blades that can be optimized both aerodynamically and/or structurally. Although the aerodynamics of vertical-axis wind turbine rotors is unsteady, complex and challenging to simulate, it is possible to make a sufficiently accurate estimation of variable aerodynamic loads acting on the blade and use them for its structural dimensioning. This research combines a multi-objective constrained optimization procedure by particle swarm and finite element method with experimental analyses with the purpose of defining the best blade, i.e. the blade of minimal mass, least tip deflection under the loading case corresponding to the aerodynamically most demanding operational regime, greatest difference between its natural frequencies and rated rotor angular frequency, lowest manufacturing cost and complexity, etc. Imposed constraints also include acceptable failure criteria along the blade. Design variables refer to ply lay-up scheme, i.e. lamina thicknesses and orientations. Final solution, chosen from the obtained Pareto set, was manufactured and experimentally validated. The consistency of strains, measured in several different cases of bending, and corresponding numerical values was mostly below 8%. The study demonstrates the applicability of the employed multi-objective optimization methodology in wind turbine blade design. It also proposes an affordable and structurally reliable composite lay-up scheme specifically designed for small-scale vertical-axis wind turbines.

Keywords Vertical-axis wind turbine blade · Computed aerodynamic loads · Optimized laminated composite structure

Z. Trivković · J. Svorcan (✉) · M. Baltić · N. Zorić · O. Peković
Faculty of Mechanical Engineering, University of Belgrade, 11120 Belgrade, Serbia
e-mail: jsvorcan@mas.bg.ac.rs

N. Mitrovic et al. (eds.), *Current Problems in Experimental and Computational Engineering*, Lecture Notes in Networks and Systems 323,
https://doi.org/10.1007/978-3-030-86009-7_3

1 Introduction and Motivation

During the past few decades there has been a constant increase in the use of renewable wind energy, with additional 50 GW being installed in the year 2018 [1]. Large wind farms have become a common sight throughout China, USA, Europe (particularly Germany, Spain, UK and France), India, Brazil, etc. The main reasons for the growing investments in this field are: significant technological development, increased reliability and decreased initial costs of contemporary wind turbines that appear in all shapes and sizes. A comparative study of different types of existing wind turbines was performed by Eriksson et al. [2]. Along with the large-sized, horizontal-axis (HA) structures that function in clean and strong winds, particularly interesting topics are small-scale urban wind turbines designed to operate in extremely changeable, low-speed winds as well as self-sufficient machines installed in inaccessible rural environments. Here, also, there are various alternatives. One interesting invention is Magnus wind turbine investigated in [3] particularly suitable for low wind speed conditions. Another possible solution to the imposed requirements of omnidirectional functioning together with cost-effectiveness, and the main topic of present research, is a vertical-axis Darrieus wind turbine (VAWT). Bhutta et al. [4] provide a good review of various VAWT configurations and design techniques. While being simple in design, a VAWT operates in aerodynamically complex unsteady flows that involve perplexing flow phenomena such as dynamic stall and is subjected to significant cyclic loads that can be (more or less accurately) estimated by computational models of various levels of complexity [5]. Nonetheless, VAWTs also offer many possibilities of improvement of both aerodynamic performances and structural behavior which is why they present an interesting, contemporary research topic.

Today, with so many digital and manufacturing tools and technologies at our disposal, careful conceptual design and optimization are necessary initial steps in the process of engineering design. Instead of simply producing a working mechanical part, it should be made to operate in the best possible way, i.e. offer the greatest benefits at the lowest possible cost. This is also true for composite structures so much employed in aerospace [6–9] including wind energy sector [10], naval/shipbuilding and automotive industries as well as civil engineering [11]. Inspired by the huge numbers of possible design solutions, Dulcey et al. [12, 13] even developed a robust supporting platform for reliable decision making in the design of composite laminated structures that was also used for optimization studies. Nikbakt et al. [14] provide a detailed list of numerous studies performed on laminated composites and their structural behavior. While many diverse approaches (differing in number of input parameters, definition of cost functions and constraints, employed algorithms) have been implemented, some usual optimization goals include: weight reduction, cost reduction, improved performances, increased structural reliability, simplified manufacturing, reduction of material scrap/waste, improved dynamic behavior, reasonably priced maintenance, etc. Since the topic of this paper is a laminated composite wind turbine blade, an overview of the most interesting, previously performed work is given in the following paragraphs.

Wang et al. [15] considered compound layered composite VAWT blades with internal structural elements—shear webs, and performed its structural optimization using finite element analysis (FEA) model and genetic algorithm (GA). The optimization was single-objective, with the intent to minimize the composite blade mass with five imposed constraints (stress, deformation, vibration, buckling, manufacturing and laminate continuity). The design variables were: number of unidirectional plies, locations of the spar cap and shear web thicknesses. The blade model, weighing over 220 kg, corresponds to the ELECTRA 30 kW wind turbine.

Fagan et al. [16] performed both experimental (static load case) and optimization studies of a 13 m long composite HAWT blade. The optimization goal was the minimization of the used material (and consequently manufacturing costs) for what purpose the authors used GA. Considered input parameters included thickness distribution and internal structural layout of the blade. Imposed constraints referred to blade stiffness and stress distribution. The correspondence of computed to measured deflection was within 9%.

An example of a coupled aerodynamic–structural HAWT blade two-objective optimization was provided by Dal Monte et al. [17]. The study was based on coupling a simple aerodynamic model, FEM and GA. Cost-functions included maximal aerodynamic power and minimal tip displacement. As a result, improved chord and twist angle distributions along the blade were achieved.

The aims of the works of Albanesi et al. [18, 19] were again the optimization of composite HAWT blade (where the goal was to minimize its mass). However, in [18], an optimization method (GA) was coupled with a predictive tool (artificial neural networks, ANN) in order to reduce the computational cost of the optimization procedure. Based on the initial computation of representative samples, ANNs were built, trained and validated and later used for the estimation of other blades' performances. Mass reduction of 20% together with a 40% computational cost reduction was achieved. Additional examples of optimal lay-up design of geometrically simpler structures (even with ply-drops) obtained by evolutionary algorithms can be found in [20–23].

All mentioned studies demonstrate the modernity, relevance and applicability of structural optimization in wind turbine blade design. They also confirm that, due to the specifics of their working life, wind turbine blades require the use of light-weight but also reliable and chemically persistent composite materials. By adequately orienting the fibers/plies (mostly uniaxial or biaxial fiberglass) in the matrix (mostly epoxy) it is possible to achieve improved mechanical characteristics of the new laminated material designed specifically for the application on wind turbine blades. It is also feasible to reduce the blade mass and manufacturing cost as well as prolong its working life. This research paper presents one such attempt, i.e. it describes the steps of a multi-objective optimization of a laminated composite (from biaxial plies) VAWT blade. Since many contemporary VAWT designs are a product of acquired experience, such computational studies have not been performed often. The input aerodynamic loads used for structural dimensioning were obtained through flow simulations. Moreover, the computational part of the study is complemented and validated by the actual blade manufacturing and static load case experiments.

2 Numerical Calculation

2.1 *Starting Geometric Model of the Wind Turbine Rotor*

The wind turbine model used in this study corresponds to a small-scale VAWT that can be mounted in urban environment (rooftops, parks, river banks, etc.). Its geometric features have been defined in accordance with the findings of aero-structural optimizations presented in [24, 25] that can be summarized as follows:

- Since blade angles-of-attack change significantly while rotating, curved airfoils do not particularly enhance the aerodynamic performances of VAWTs. Furthermore, since thinner airfoils generally have higher lift gradients as well as maximal lift coefficients, while thicker airfoils are structurally more convenient, it is best to adopt a medium value of airfoil relative thickness.
- Lesser values of rotor solidity (achieved by shorter blade chords, larger diameters or both) imply that each blade operates in cleaner air stream and that the interaction between the blade and shed vortices is reduced resulting in higher aerodynamic efficiency of the wind turbine.
- Extracted power is proportional to the wind turbine diameter and height (i.e. length of the blades).

The chosen wind turbine rotor with the diameter $D = 3$ m (and radius $R = 1.5$ m) consists of three straight, untwisted, identical blades whose length is $L = 1$ m, chord $c = 0.15$ m and cross-section shaped like NACA 0018 airfoil. Its design angular velocity is $\Omega = 200$ rpm. Figure 1 illustrates the oncoming uniform wind velocity V_o, the wind turbine model and the simplified computational rotor model (colored in red) necessary for the estimation of its aerodynamic performances (shaft, struts, generator and other elements were not considered in flow simulations). This chubby rotor design is primarily the consequence of wind turbine height limitations and can simply be modified by increasing the rotor height.

2.2 *Flow Simulation*

Since aerodynamic loads present the prerequisite for structural analysis, it was first necessary to complete the unsteady surrounding fluid flow computations. Here, a simple 1-way fluid–structure interaction (FSI) analysis that implies computing aerodynamic loads on rigid geometry, extracting them and applying them to the subsequent structural models was performed. All flow simulations were performed by

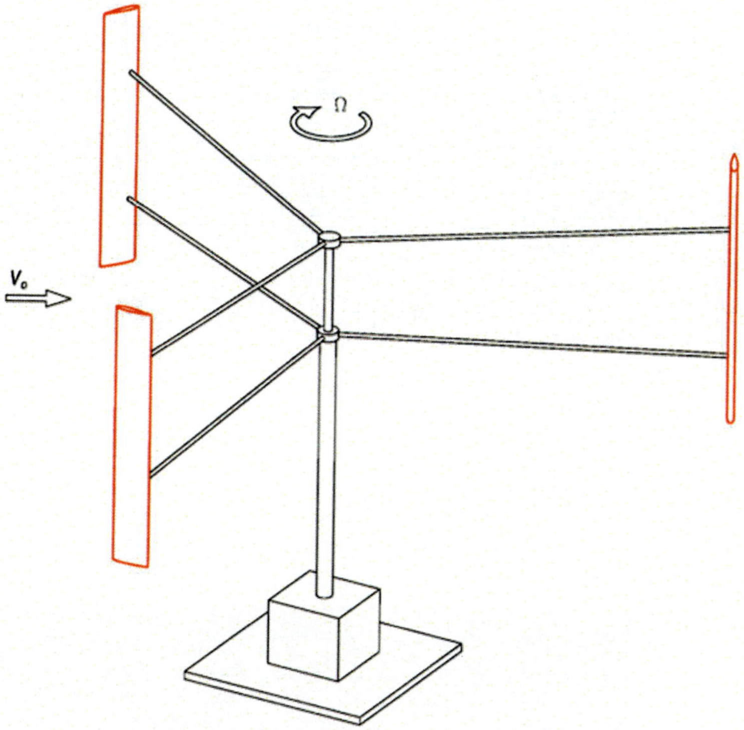

Fig. 1. Wind turbine model (blades used in flow simulations are colored in red).

finite volume method (FVM) in the commercial software package ANSYS FLUENT [26]. Compared to simpler concepts (blade element-momentum or vortex theory as mentioned by [4, 5]), this computational approach can deal with numerous flow phenomena (flow asymmetry, large angles-of-attack, flow separation) successfully. Seeing that the details of numerical set-up and numerous obtained results can be found in [27], only a short summary is given here. Other extremely useful examples of flow simulations around VAWTs can be found in [28–34].

The flow was considered as 3D, unsteady, incompressible and viscous (turbulent). Unsteady Reynolds-averaged Navier–Stoker (URANS) equations were closed by 2-equation k-ω SST turbulence model [26]. The computational domain comprises two zones: outer–stator, shaped like a prism and extending $-3R$ fore and $11R$ aft of the wind turbine model in the longitudinal direction, $\pm$ $4R$ along the lateral and $\pm$ $2.5L$ along the vertical axis, and inner cylindrical—rotor with the radius $1.33R$, as illustrated in Fig. 2a. It was discretized into approximately 3.2 million prismatic and tetrahedral elements. Computational mesh is additionally refined in the vicinity of the blades, with 20 thin layers of prismatic cells forming the boundary layer, as depicted in Fig. 2b. First layer thickness of $y_1 = 0.05$ mm results in dimensionless wall distance $y^+ < 5$ which is crucial for obtaining viable and accurate numerical results.

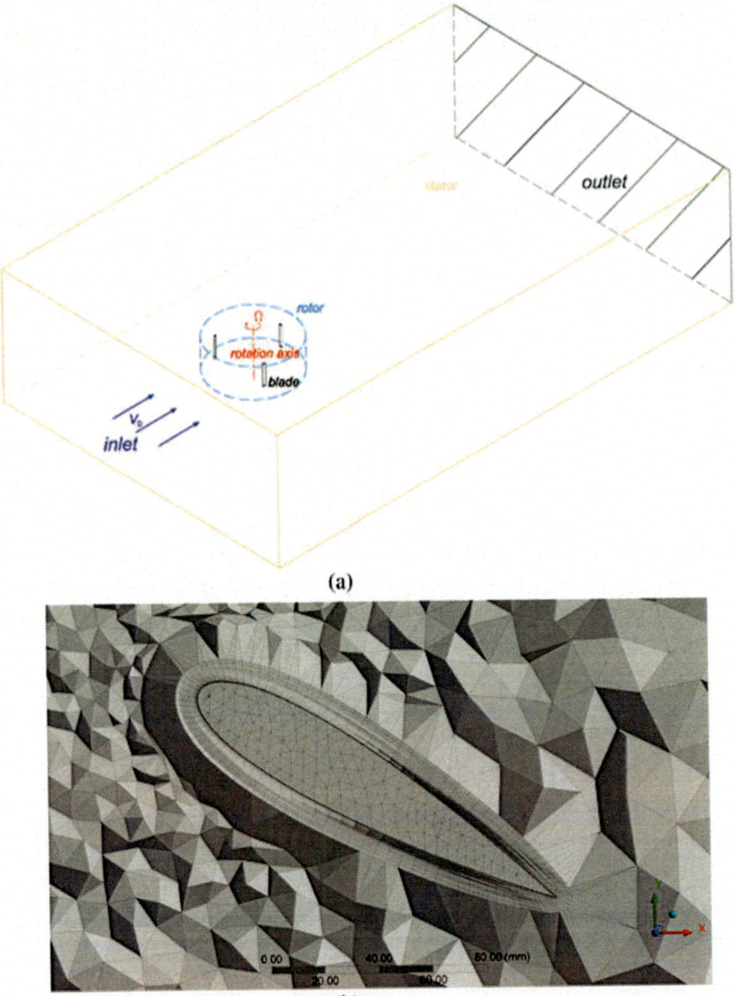

(a)

(b)

Fig. 2. **a** Computational domain, **b** Detail of the computational grid.

The rotation of the inner part of the computational domain was taken into account by the sliding mesh approach where the inner cells (and nodes) are actually moved in every time-step by angular speed Ω. The value of undisturbed wind speed V_o is defined along the inlet surface (considered wind speeds were in the range 6 m/s ≤ V_o ≤ 18 m/s), zero gauge pressure at the outlet, and no-slip boundary condition at the rotating but rigid blade walls. Segregated pressure-based solver with SIMPLEC pressure–velocity coupling scheme was used. All spatial discretizations were of 2nd order while temporal were of 1st order. The computations were performed until reaching quasi-convergence (oscillatory in character) of aerodynamic coefficients, which was usually achieved after 5 rotor rotations. The assumed time-step Δt corresponded to 5° angular increment of the inner rotational zone, while ten iterations were performed per time-step. Convergence is additionally accelerated by initial conditions that assumed the uniform velocity profile throughout the domain.

In order to ascertain the validity of the assumed computational approach (i.e. the relative extents of the rotational and stationary parts of computational domain, mesh features and its level of medium fineness, material properties, zone and boundary conditions, adopted turbulence model, numerical schemes, etc.), the complete modeling, meshing and computing processes were repeated on a VAWT of comparable dimensions, rotor solidity and employed airfoil (of slightly greater relative thickness) whose description and experimentally obtained data in a wind tunnel are available in [34, 35]. The only difference is in the smaller turbulence intensity of 1% assumed along the inlet boundary (corresponding to the flow conditions in the wind tunnel). Figure 3a illustrates both the measured and computed power coefficient curves with respect to tip-speed ratio $C_P(\lambda)$, while Fig. 3b presents the measured and computed total force coefficients $C_R(\lambda)$, obtained as the vector sum of longitudinal and lateral components. As can be seen, the differences in the two sets of results are within acceptable tolerance (in most design flow cases the relative variance is below 10%), particularly given some discrepancies in free-stream flow quantities at boundaries and inevitable complexity and uncertainty of both experiments and

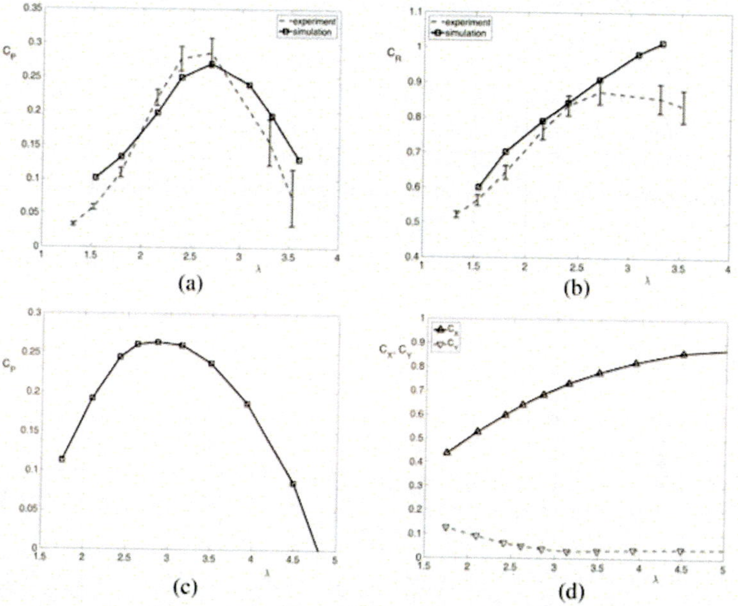

Fig. 3. Comparison of measured and computed: **a** power coefficients C_P, and **b** force (thrust) coefficients C_R of VAWT from [34, 35]; computed: **c** power coefficient C_P, and **d** longitudinal and lateral thrust coefficients C_X and C_Y of present VAWT model

computations. Same as with the experimental measurements, the accuracy of simulating these kinds of flows decreases with increasing tip-speed ratio λ (i.e. at lower wind speeds for the constant angular velocity) because the wakes travel downwards slower and their complex interaction with rotating blades is more pronounced. Fortunately, these phenomena are not so significant at higher free stream velocities which are paramount for structural dimensioning of the blades.

Although not crucial for the following structural dimensioning of the blade, it is also possible to depict some fundamental characteristics of the designed VAWT such as its power coefficient, Fig. 3c, or longitudinal thrust (i.e. axial force) and lateral thrust coefficients (C_X and C_Y, respectively) acting on the complete wind turbine at different operating conditions, Fig. 3d. As can be noted, the aerodynamic efficiency is satisfactory for this type of wind turbine and can result in maximal power of approximately 1200 W. On the other hand, the dominant component of the aerodynamic force—axial or longitudinal force F_X (parallel to the free-stream flow) is important since the greatest momentum exchange happens precisely in this direction. Although Fig. 3 presents only values averaged per rotation, i.e. global characteristics of the two VAWTs, both longitudinal and lateral thrusts have highly oscillatory characters. Their mean values and amplitudes primarily depend on the number of blades (here 3), operating conditions and blade geometry.

In the process of blade structure dimensioning, one of the most important results of the flow simulation is a portion of the cyclic aerodynamic force acting on a single blade surface that increases with wind speed Vo as opposed to the total axial and side force acting on the whole VAWT illustrated in Fig. 3d. It is usually represented in blade-fixed coordinate system through its two orthogonal components i.e. normal force F_N (acting along the radial direction and marked in blue in Fig. 4) and tangential force F_T (marked in red in Fig. 4) whose highly variable character is illustrated. The intensities of normal and tangential force presented in Fig. 4 are obtained for undisturbed wind velocity of $V_o = 18$ m/s (that corresponds to the VAWT cut-out

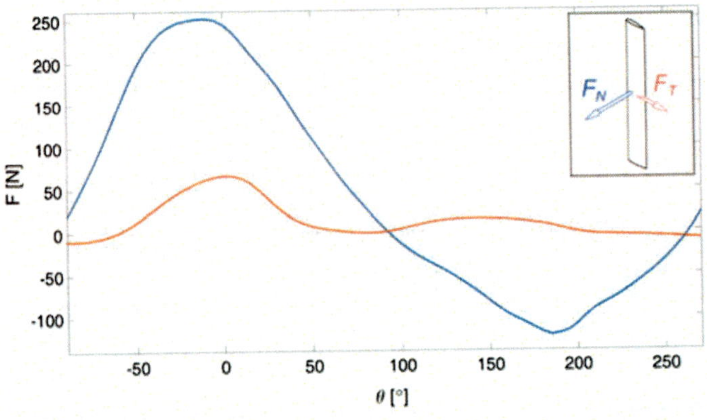

Fig. 4. Change of normal F_N and tangential force F_T acting on a single blade during one rotation at maximal wind speed $V_o = 18$ m/s

speed) at the nominal rotor angular velocity of $\Omega = 200$ rpm, i.e. for the aerodynamically most demanding operational regime (i.e. the worst case scenario). Their maximum values were later used for structural analyses of the blade. In this way, the most unfavorable results of the performed unsteady flow computations were utilized for static structural simulations of the blade behavior. As follows, it was assured that the structure of the designed blade would be able to withstand both the most as well as the less intensive aerodynamic loads and that it would remain operational in all expected working conditions.

2.3 *Description of the Structural Model and Design Variables*

While the behavior of anisotropic composite materials depends on a number of macroscopic parameters: material of fibers, matrix material, ply type, orientation and thickness, fiber volume fraction, number of layers, lay-up sequence, etc. most studies consider fiber orientation angles and number of layers as prime design variables [13, 15–23]. This assumption (and limitation) is adequate for commercially available plies with guarantied 2D mechanical properties.

Due to the small size of the considered WT, the simplified structural blade model consists of a single shell (lower and upper surfaces of the airfoil) without any additional inner structural elements. Blade is a symmetric fiberglass/epoxy laminate with the assumed lay-up sequence $[(\theta_1)_{n1}/(\theta_2)_{n2}/(\theta_1)_{n1}]$ layered towards the inside as shown in Fig. 5 so that the outer aerodynamic shape is not obstructed. There are 4 design variables: numbers of plies, n_1 and n_2, and ply orientations, θ_1 and θ_2. The mechanical characteristics of the main construction brick—a single biaxial layer, taken from [36], are listed in Table 1.

The parameterized FE models were also establishes using ANSYS software package [37] that enables the consideration of thin composite structures according to First-Order Shear Deformation Theory (FSDT) applied to each layer. The following

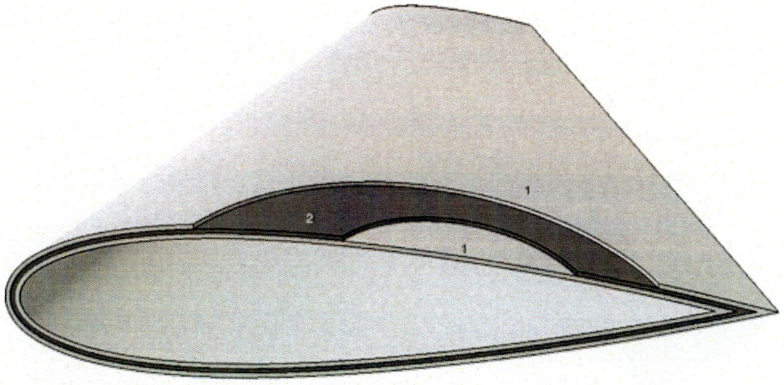

Fig. 5. Structural model of the blade

Table 1. Mechanical properties of BIAX material.

Property	Symbol	Unit	Value
Density	ρ	[kg/m^3]	1740
Axial elasticity modulus	E_1	[GPa]	17.72
Transverse elasticity modulus	E_2	[GPa]	17.72
Shear modulus	G_{12}	[GPa]	2.47
Poisson coefficient	ν_{12}	[–]	0.25
Axial tension strength	F_{1t}	[MPa]	310
Axial compression strength	F_{1c}	[MPa]	−280
Transverse tension strength	F_{2t}	[MPa]	310
Transverse compression strength	F_{2c}	[MPa]	−280
Shear strength	F_6	[MPa]	10
Thickness	dt	[mm]	0.1

computational procedure was conducted for each considered blade. After importing an external geometric file (the same for every model) and assigning the momentary (i.e. instantaneous) values of design variables and material properties, a computational mesh was created. It comprised approximately 4400 quadrilateral elements that were refined in the vicinity of the leading and trailing edges, Fig. 6. Generated grids were of reasonable density in order to be appropriate for a large number of repeated computations. Total thickness of the blade t (and consequently blade mass m) depends on the assumed numbers of plies, $t = (2n_1 + n_2)dt$. The chosen element type is SHELL181, a four-node element with six degrees of freedom at each node,

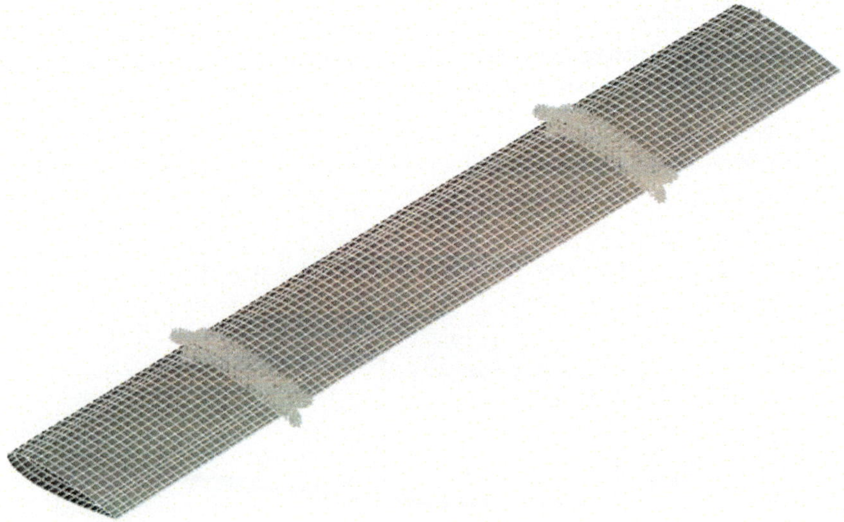

Fig. 6. Detail of the generated FE model with clamps

suitable for analyzing thin to moderately-thick shell structures of both layered or sandwich construction [37].

Next, boundary conditions were defined. In both static and modal analyses the blade was clamped along the cross-sections located at the 1st and 3rd quarter of the blade length *L*, as marked in Fig. 6, since that is how the blades are planned to be connected to the central shaft (and the complete wind turbine assembly). In the computations of static load case, the forces acting on the blade were uniformly distributed along its surface. Due to the high aspect ratio of the blade, the performed 3D flow simulations showed that this approximation can be employed without much losing on the accuracy since aerodynamic force acts almost uniformly along the blade and tip effects are negligible.

For each study case the three force components were obtained by vector summing the: maximal aerodynamic loading (estimated at undisturbed wind speed $V_o = 18$ m/s and nominal rotor angular velocity $\Omega = 200$ rpm being the same for all considered lay-ups and acting along normal and tangential directions), inertial loads (i.e. $m\Omega^2 R$, due to the blade rotation at nominal rotor angular speed $\Omega = 200$ rpm, acting along normal direction) and mass forces (i.e. mg, acting along the third, vertical direction). The last two components depend on the blade mass and were calculated anew for each study case. The estimated forces were additionally multiplied by the safety factor of 1.25. Finally, after performing the computations, it was possible to determine and post-process various characteristics of each blade model, namely its: total mass, natural frequencies, deflection of the structure, stress distribution, failure criteria (e.g. Tsai–Wu), etc.

3 Multi-Objective Particle Swarm Optimizations (PSO)

3.1 Basics of PSO

PSO, that imitates the behavior of a large group of social organisms (such as bees, ants, birds, etc.) that are in a search for food, was chosen for finding a set of optimal composite blade structures. Originally defined by Eberhart and Kennedy [38], but mostly due to its rapid convergence and ease of implementation, it experienced a real boom in engineering applications [14] and is widely used today for both single- and multi-objective, unconstrained and constrained, continuous and discrete optimizations.

In the beginning, each particle (defined by randomly initialized design variables) that can move freely across the design space is assigned with a randomly determined position and velocity vectors. In every iteration, the values of the output (cost functions, constraints) are estimated and particle's position is recomputed anew from the previous position and velocity vectors, where each velocity depends on the previous value but also cognitive (individual) and collective (social) behavior that are randomly taken into account. In this way, both local and global achievements (optima) are considered and the possibility that the algorithm will get stuck in a local extremum is diminished i.e. reduced to minimal. Additional safety measure is the possibility (usually small, <1%) that the particle might *go crazy*, when it forgets everything it learned about the local and global optima, and its position and velocity are reinitialized.

The described method must be slightly modified for multi-objective applications. Firstly, a new method for the comparison of two particles must be adopted since simple < or > are no longer applicable. Alternatively, we use the relation of particle dominance where particle dominates over the other one only if all of its output parameter values are better. Optimum is now a non-dominated particle. Secondly, instead of a singular global optimum, there is now a set of optimal, equally good (non-dominated) solutions. For that reason, global optimum in every iteration is randomly chosen from the existing Pareto set. Thirdly, in a case of constrained optimization, an additional term appears in the velocity expression in order to maintain the diversity of solutions (since the existence of constraints additionally binds/restricts the optimization procedure) [39].

3.2 *Formulation of the Optimization Problem*

Previously described optimization procedure was adapted to the matter of composite blade structure. Its flowchart is depicted in Fig. 7. Position of each particle, describing the composite lay-up scheme, is in the form of a vector $\mathbf{X} = [n_1\ \theta_1\ n_2\ \theta_2]$. As can be seen from Table 2 that describes the extensions of the design space, the minimal number of layers per blade is 3. In the beginning, the swarm of 200 particles is randomly initialized, after which an iterative computational process starts. The values of blade mass, tip deflection, maximal failure criteria, natural frequencies, etc. are computed for every combination of input parameters (ply numbers and orientations). The optimization stops after a predetermined number of computational steps, here $N = 30$.

4 Numerical Results and Discussion

The greatest advantage of multi-objective optimizations, compared to single-objective, is their capability to simultaneously consider several opposing aspects (e.g. mass, structural behavior, reliability) while increasing the objectiveness of the design process. This enables the design of a truly improved and/or customized product, even for a range of operating regimes (both optimal and non-optimal). In most cases, examining a single characteristic (cost-function) leads to insufficient comprehension of the problem physicality as well as its unfavorable performance in certain disregarded working conditions. On the other hand, two or more cost-functions and constraints should be carefully selected and analyzed in order to obtain reasonable and practical

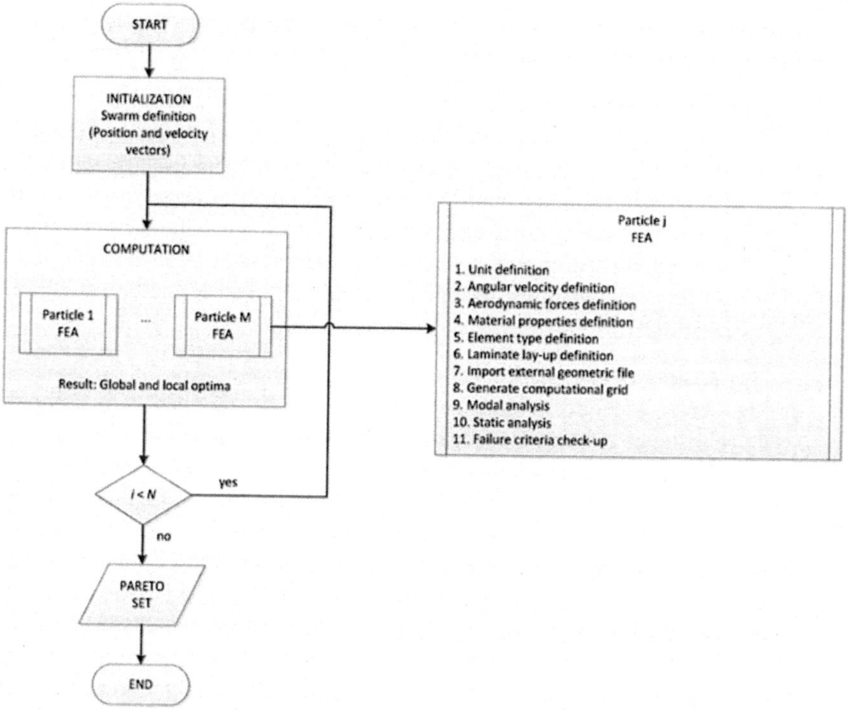

Fig. 7. Flowchart of the optimization algorithm

Table 2. Design space

Design variable	Symbol	Unit	Min	Max	Increment
Ply number	n_i	[−]	1	20	1
Ply orientation	θ_i	[°]	0	75	15

final results. If the considered goals are tightly coupled or the constraints are too harshly imposed, finding an optimal solution will not be possible. This particular aspect of optimization is still primarily based on designer's experience and therefore includes some partiality in the engineering design (as will be seen in continuation).

4.1 Two-Criteria Optimization

The main goal of the performed study was to define an optimal blade structure that is simultaneously: light, economic, easily maintained, dynamically stable and structurally reliable with respect to the expected operating conditions. This objective demanded an appropriate choice of optimization goals and constraints. Therefore, a two-objective optimization was performed with respect to two distinct goals:

- minimal blade mass $f_1 = m$, and
- minimal value of an additional function f_2 that incorporates the coupled effects of blade mass m (that is proportional to the amount of used material, i.e. blade cost), tip deflection u_{tip} under the combined effect of aerodynamic, inertial and gravitational loading (structural reliability), maximal difference of integer multiplies of rated rotational $n\nu_{\text{rated}}$ and computed natural frequencies ν_i, i.e. $|n\nu_{\text{rated}} - \nu_i|$ (dynamic stability, resonance avoidance) and a number of different ply orientations (manufacturing complexity).

An additional explanation on the auxiliary function f_2, that is meant to assist in the choice of optimal blade structure, should be provided. It is computed as a relative distance Δ from a *fictional point* whose first three coordinates correspond to the *desired* values of blade mass, tip deflection and frequency difference, while the fourth coordinate quantifies the blade manufacturing simplicity and speed by favoring the blades with higher numbers of identically oriented layers in the lay-up sequence (since that relieves the work load of the operator). The effects of blade total cost are estimated through a weighting factor of 2.0 assigned to the relative distance along the blade mass coordinate Δ_{mass} (making the heavier blades less desirable) since the amount of used material is directly proportional to the cost. The additional function f_2, whose minimal values denote the best blades, can be written down in a simplified form by Eq. (1) as follows:

$$f_2 = \sqrt{2^2 \Delta_{\text{mass}}^2 + \Delta_{\text{deflection}}^2 + \Delta_{\text{frequency}}^2 + \Delta_{\text{manufacturing}}^2} \tag{1}$$

Imposed constraints also required that blade mass, tip deflection and frequency difference be below or above predefined values. Furthermore, an additional structural constraint, that demands the satisfactory value of failure criteria I_{fail} along the whole blade shell, was imposed. A failure criterion (or failure index) is often used to predict the first occurrence of failure in any of the laminas of the laminated composite exposed to multi-axial stress. It is roughly defined as stress-to-strength ratio (with the desired value that is lower than 1) although expressions for 3D models are more complicated. Here, Tsai–Wu criterion [37] was employed. The optimization problem can be formulated by Eqs. (2–3) as:

$$\min(f_1) \wedge \min(f_2) \tag{2}$$

$$\text{subject to } (m < 1.5\text{kg}) \wedge \left(u_{\text{tip}} < 0.1L\right) \wedge \left(|n\nu_{\text{rated}} - \nu_i| > 0.5\text{Hz}\right) \wedge \left(I_{\text{fail}} \leq 1\right) \tag{3}$$

It should be mentioned here that this is not the only way to define the goal function f_2. E.g. manufacturing cost, speed, flexibility and simplicity as well as maintenance demands, although very important, were only implicitly covered by the adopted goal functions and constraints. Depending on the particular structure in question, a different quantification of initial requests during the optimization procedure might be established.

The obtained 2D Pareto front together with the output parameters of all the other considered (dominated) particles are marked by downward-pointing triangles and dots, respectively, in Fig. 8. Owing to a relatively strict definition of goal functions and constraints, the obtained Pareto set is somewhat limited and numbers only 8 non-dominated particles listed in Table 3 (sorted first according to f_1, and then according to f_2).

Still, when all computed output parameters are analyzed, several things can be noted:

- It is possible to use very light blades that weigh less than a kilogram.
- The increase of blade mass induces greater differences between rotational and natural frequencies (nearly 1 Hz) and assures greater dynamic stability of the blade.
- Expected tip deflections are quite acceptable since they are approximately 1–2 mm (for a blade whose length is 1 m) for all blade masses (ranging from less than a kilogram to 3 kg).

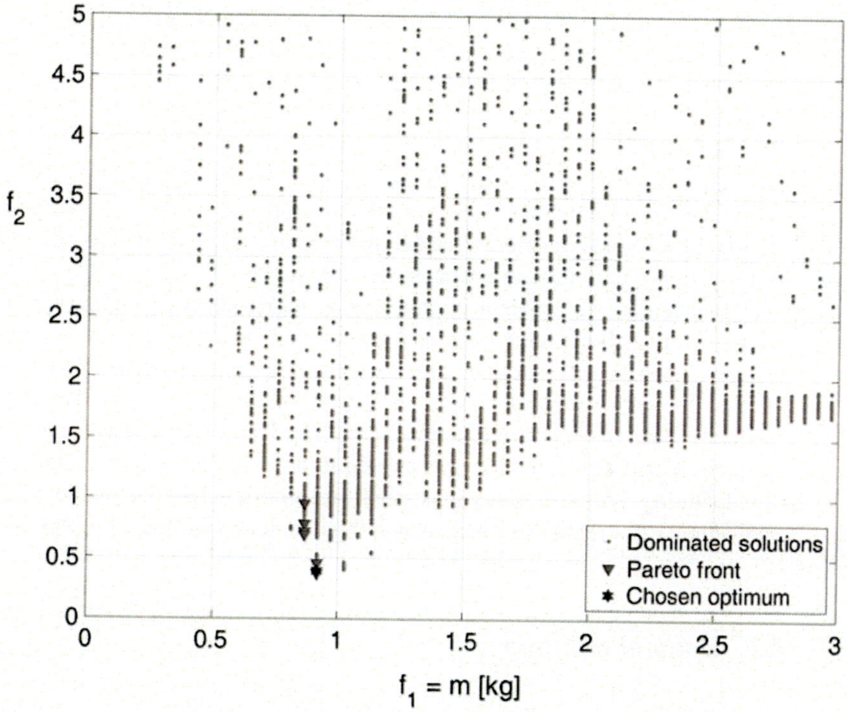

Fig. 8. Obtained 2D Pareto front together with output parameters of dominated particles and chosen optimum

Table 3. Optimal solutions

n_1	θ_1 [°]	n_2	θ_2 [°]	n	f_1 [kg]	f_2 [−]
7	0	2	45	16	0.8650	0.6930
7	0	2	30	16	0.8650	0.6984
7	0	2	0	16	0.8650	0.7238
7	75	2	45	16	0.8652	0.7865
7	15	10	0	16	0.8650	0.9535
8	0	1	45	17	0.9191	0.3689
8	**0**	**1**	**0**	**17**	**0.9191**	**0.3913**
8	75	1	30	17	0.9191	0.4566

Given that all blades in the obtained Pareto set satisfy the imposed criteria and constraints, it is reasonable to first sort them in ascending order according to the estimated mass (function f_1), and then according to the values of function f_2. Table 3 enables inspecting the optimal blades in more detail. It is immediately apparent that the lightest blades (weighing around 1 kg) should comprise no more than 17 layers in total, with the minimal total number of layers (both outer and inner) being $2n_1 + n_2 = 16$. The recommended number of outer layers is mostly 7 (and no more than 8), while the recommended number of inner layers is much smaller, mostly spanning between 1 and 2, thus implying that its contribution to overall structural behavior of the blade is less pronounced. Furthermore, it seems that the recommended value of outer orientation θ_1 is mostly 0° while the recommended value of θ_2 is mostly 45°, but 0° and 30° also appear. It can be concluded that the numbers of plies play a more important role in structural dimensioning of the small-scale VAWT blade than their respective orientations. However, although all listed solutions satisfy imposed structural goals and constraints, the fact remains that by applying lay-up schemes at 0° it is possible to obtain somewhat stiffer structures.

Consequently, the problem of choosing a single optimal solution persists with Pareto set since all particles are equally good. Two ideas that first come to mind are to select the point either closest to the theoretical optimum (denoted by the minimal values of objective functions) or farthest from the worst solution (denoted by the maximal values of objective functions) [40]. Here, however, another approach was assumed. A particular solution was chosen in order to account for some unwanted manufacturing errors or omissions (that are unfortunately inevitable in real-life applications) and reduce as much as possible the manufacturing complexity.

4.2 Optimal Solution

After a comprehensive analysis of the obtained Pareto set and imposed goal functions, a final solution characterized by $\mathbf{X}_{opt} = [8\ 0°\ 1\ 0°]$ was chosen (marked in bold

Table 4. Characteristics of the chosen optimal solution

Output parameter	Symbol	Unit	Value
Mass (1st goal function)	m	[kg]	0.9191
2nd goal function	f_2	[–]	0.3913
Tip deflection	u_{tip}	[mm]	2.191
Frequency difference	$\lvert n\nu_{rated} - \nu_i\rvert$	[Hz]	0.6867
Fail criteria	I_{fail}	[–]	0.6639
1st natural frequency	ν_1	[Hz]	4.02
2nd natural frequency	ν_2	[Hz]	4.13
3rd natural frequency	ν_3	[Hz]	5.32
4th natural frequency	ν_4	[Hz]	8.01

in Table 3). One supplementary layer (in comparison to the minimal number of layers) is included for additional safety i.e. to allow for manufacturing inconsistencies (or errors) that are unfortunately inevitable during the wet lay-up manufacturing procedure. Also, the constant value of zero ply orientation increases the stiffness of the blade while much relieving the manufacturing workload as well as manufacturing costs. The chosen optimum is denoted by a hexagram symbol in Fig. 8. Further computational check of the adopted optimal design $\mathbf{X}_{opt}$ was performed by applying the combined aerodynamic, inertial and gravitational forces to the clamped blade model. The values of obtained output parameters are listed in Table 4. It can be confirmed that the chosen optimum satisfies the imposed goals (its mass is slightly over 900 g, maximal tip deflection around 2.2 mm and frequency difference nearly 0.7 Hz) and constraints (maximal failure criterion is well below 1, thus assuring that no lamina failure should occur under expected operating conditions), as well as being light-weight, economic and the easiest to manufacture with the satisfactory value of the second objective function f_2.

Figure 9 depicts the computed displacements of the laminated structure of the blade (where red color corresponds to the maximal value of approximately 2 mm, and blue color to minimal value of 0 mm) under combined static loading corresponding

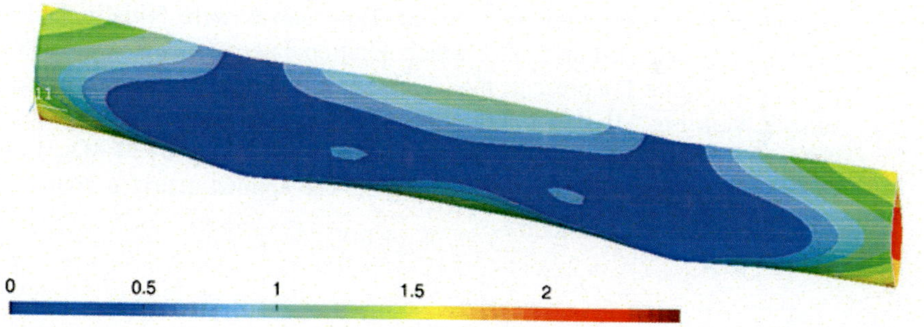

Fig. 9. Computed displacement field

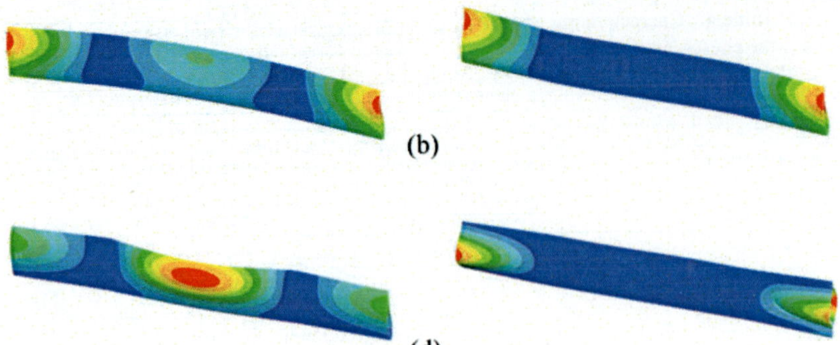
(a) (b) (c) (d)

Fig. 10. Natural modes: **a** 1st mode, **b** 2nd mode, **c** 3rd mode, and **d** 4th mode

to the highest loads that occur once per blade rotation. As expected, the highest deflections appear solely in the vicinity of blade tips, and their value is acceptable.

Figure 10 illustrates the first 4 natural modes, i.e. oscillation shapes, (whose corresponding values of natural frequencies are listed in Table 4) for the chosen way of connecting the blade to the rest of the wind turbine structure. As before, the red color denotes the highest values of deflections, and blue color the lowest, while the actual values (intensities) may vary with dynamic impetus. Although the resulting shapes are complex, it can be stated that the first 3 modes are predominantly flexural.

5 Validation of the Defined Design Methodology Through Experimental Testing

5.1 Blade Manufacturing

The process of blade manufacturing commences with the mould design and fabrication. In this case, the mould was simple, comprising two equal parts due to the symmetric airfoil and untwisted blades. Both halves were initially shaped by a numerically-controlled mill, after which additional necessary surface finishing was performed. Figure 11a presents the sweeping of the aerodynamic part of the mould surface, while Fig. 11b shows the completed blade. The blade was made manually, by wet lay-up method. Specific weight of the used BMS 9–3 Style 120—Dry Fiberglass Cloth is 107 g/m^2, with thickness 0.09 mm. The applied matrix is Henkel Loctite Hysol EA 9396 AERO Epoxy.

(a)

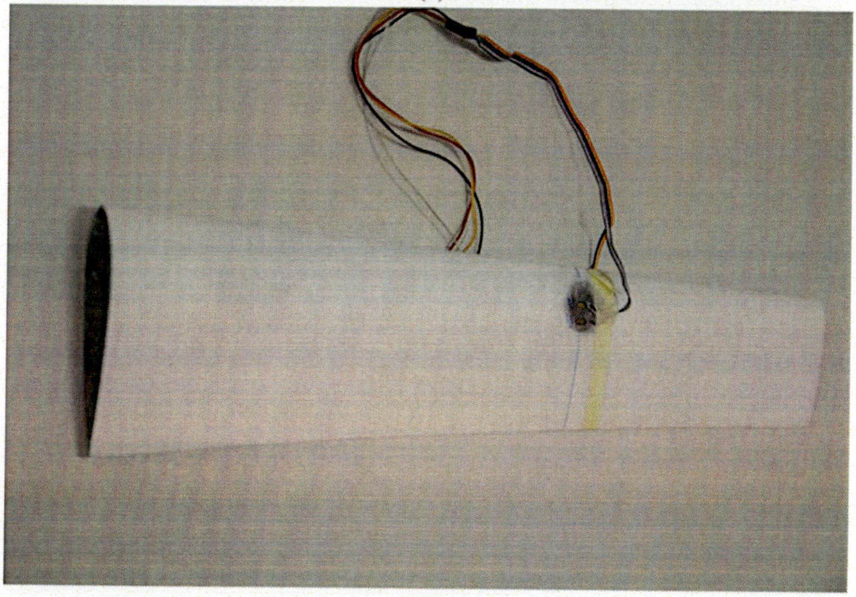

(b)

Fig. 11. **a** Sweeping of the mould surface, **b** Produced blade

5.2 Experiment

In order to determine the manner in which the blade responds to imposed static loads, strain gauges were applied to its surface. The only disadvantage of this extremely reliable, affordable and simple method is that it only returns local values (in a single point, i.e. across a small surface patch).

A full-bridge strain gauge configuration that compensates temperature effects was employed for measuring bending strains. Four strain gauges forming Wheatstone bridge (two at the upper—R_2 and R_4, two at the lower surface—R_1 and R_3) were glued in the blade middle cross-section at approximately 30% of blade chord c as shown in Fig. 11b. The measured voltage output U_o is a function of the four electrical resistances (of the four strain gauges R_1, R_2, R_3 and R_4) and input voltage U_i according to Eq. (4). Since the resistance changes with resistor length, measuring the output voltage enables the determination of the enforced blade deformation. However, in most cases, the output voltage must be amplified (here, integrated AD623 was used).

$$U_o = U_i\left(\frac{R_3}{R_2 + R_3} - \frac{R_4}{R_1 + R_4}\right) \tag{4}$$

Identically to the computational set-up, the blade was clamped along the two cross-sections located at the 1st and 3rd quarter (i.e. at 25 and 75%) of the blade length L. On the other hand, for simplicity reasons, the forces were applied only at two places between the clamps, at 37.5 and 62.5% of L., i.e. in the vicinity of the glued strain gauges. This rather simplified representation of the continuous loading that happens in reality is sufficient for the validation of the employed numerical set-up. Furthermore, it is common practice to use discretely distributed forces in static experimental investigation of aerospace structures [16]. Five experiments that differ in intensities and locations of the applied forces were performed, as listed in Table 5. Figure 12a and b display experiment no. 4 and acquisition system schematic, respectively.

Measured, amplified values of the output voltage $U_{o,\text{AMP}}$ for both unloaded and loaded blade structure in the five experimental cases are listed in Table 6. For results accuracy and reliability, each measurement was repeated several times. The last column shows values of the indirectly measured strain ε according to Eq. (5) where gauge factor $GF = 2.08$ and the fractional change in electrical resistance $\Delta R/R$ is

Table 5. The experimental arrangement

Experiment no	Mass at 37.5% L [kg]	Mass at 62.5% L [kg]
1		5
2		10
3	5	10
4	10	10
5	15	15

(a)

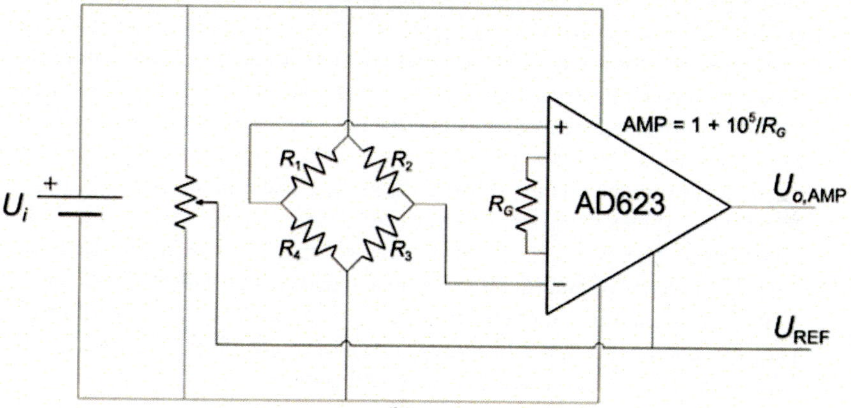

(b)

Fig. 12. **a** Application of two 10 kg weights at 37.5 and 62.5% of blade length *L*, **b** Schematic of the strain acquisition system

Table 6. Measured voltages and strains

Experiment no	Output voltage [V] (unloaded blade)	Output voltage [V] (loaded blade)	Indirectly measured strain ε
1	3.15 ± 0.01	2.87 ± 0.01	$(5.9 \pm 1.9)\cdot 10^{-5}$
2	3.35 ± 0.01	2.90 ± 0.01	$(9.5 \pm 1.9)\cdot 10^{-5}$
3	3.15 ± 0.01	2.40 ± 0.01	$(15.8 \pm 1.9)\cdot 10^{-5}$
4	3.30 ± 0.01	2.35 ± 0.01	$(20.0 \pm 1.9)\cdot 10^{-5}$
5	3.35 ± 0.01	1.80 ± 0.01	$(32.0 \pm 1.9)\cdot 10^{-5}$

Table 7. Comparison of computed and measured bending strains

Experiment no	Computed strain ε^{num}	Measured strain ε^{exp}	Relative difference [%]
1	$5.45 \cdot 10^{-5}$	$(5.9 \pm 1.9) \cdot 10^{-5}$	7.64
2	$10.9 \cdot 10^{-5}$	$(9.5 \pm 1.9) \cdot 10^{-5}$	14.73
3	$16.0 \cdot 10^{-5}$	$(15.8 \pm 1.9) \cdot 10^{-5}$	1.16
4	$21.2 \cdot 10^{-5}$	$(20.0 \pm 1.9) \cdot 10^{-5}$	6.24
5	$31.9 \cdot 10^{-5}$	$(32.0 \pm 1.9) \cdot 10^{-5}$	0.39

proportional to the ratio of output and input voltage.

$$\varepsilon = \frac{\Delta R/R}{GF} = \frac{U_o/U_i}{GF} = \frac{U_{o,\mathrm{AMP}} - U_{\mathrm{REF}}}{\left(1 + 10^5/R_G\right) U_i GF} \tag{5}$$

5.3 *Results and Discussion*

In order to perform a quantitative comparison between the two sets of results (experimental versus numerical) all computations were repeated in accordance with the conducted measurements. This time, the necessary outputs from numerical simulations are strain fields along the strain gauge placement direction, particularly in the node located in the vicinity of the 30% of the blade middle cross-section.

Table 7 shows computed and measured strains in parallel for the 5 considered load cases. The first column denotes the number of experiment, while columns 2 and 3, respectively, present the computed and measured values of the bending strain. The final column lists the relative differences between the two sets of results (mean measured values are used). In addition, for more clarity, Fig. 13 graphically illustrates the comparison between the measured and computed micro-strains.

As the first two experiments differ only in force intensity (which is doubled in the 2nd case), numerical values also exactly follow that linear trend. There is a slight deviation in the experimental results which can be explained by the low values of applied forces, i.e. small voltages and deformations i.e. increased measurement error that reduces with increased load. However, although the highest mean relative difference of approximately 15% was obtained in experiment 2, the correlation between the two sets of results can be considered satisfactory. In all other load cases, mean relative difference remains below 8% and computed results fall within the measurement error range. In the final experiment, where the greatest forces are applied, the relative error is even below 1%. Also, the character of the blade behavior as well as strain field redistributions are well captured in both types of experiments.

Force increase consequently instigates strain growth, particularly in the vicinity of the two blade cross-sections where the forces are applied and nearer to the trailing edge (where inertia moments are smaller and the two sides of the blade surface

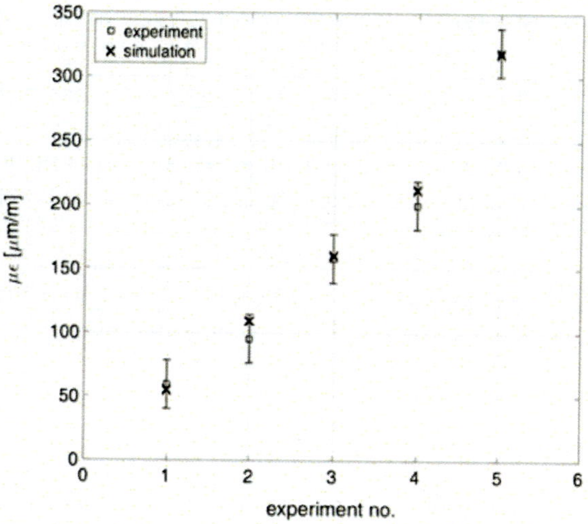

Fig. 13. Measured vs. computed micro-strains in [μm/m]

connect). As expected, the changes follow linear trend i.e. the deformations are elastic and the blade can return to its undeformed shape after the forces stop acting which is quite important for the rotor aerodynamic performances. Graphical representations of the computed strain fields from the top layer are shown in Fig. 14. Blue and red colors correspond to the values of 0 (lowest) and 0.001 (highest), respectively.

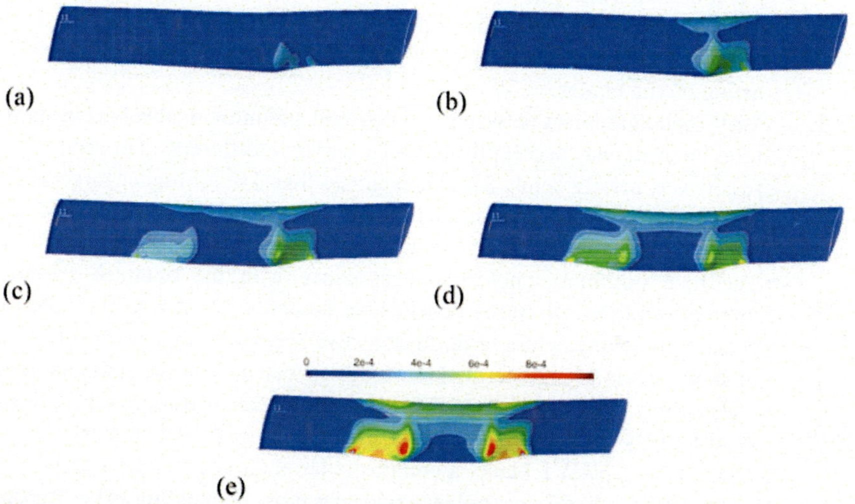

Fig. 14. Computed strain fields in: **a** experiment 1, **b** experiment 2, **c** experiment 3, **d** experiment 4, and **e** experiment 5

6 Conclusions

The research topic of the paper is the improved structural design of a laminated composite straight untwisted blade of a small-scale VAWT. This contemporary, but still insufficiently investigated subject can induce further advancements in the field of renewable energy. It requires a multi-disciplinary approach—knowledge of both aerodynamic and structural behavior of the blade in various operating conditions as well as basics of material science, electrical engineering and control. It is therefore necessary to employ all available CAD/CAE/CAM techniques, apply advanced design methods, i.e. multi-objective optimization and conduct coupled comparative numerical and experimental studies.

Here, FEA and constrained multi-objective PSO were brought together in order to define the optimal laminate structure of a wind turbine blade with respect to several different goals (lightest, least deformable, easy to manufacture blade whose operation is not likely to result in resonance). A necessary prerequisite for the optimization procedure was an accurate estimation of aerodynamic loads acting on the blade described in the Sect. 2.2 of the paper. Design variables included ply numbers and orientations. After the set of possible optimal solutions (Pareto set) has been determined, the final choice was made by the engineering team with respect to blade manufacturing cost and complexity. The designed blade was manufactured and experimentally tested. Most important conclusions and gained experience can be formulated as follows:

- Computational analyses of rotational flow fields around VAWTs in different operating conditions can be conducted with satisfactory accuracy for the estimation of its global characteristics (power and thrust coefficients) as well as cyclic loads acting on the blades.
- Very good agreement between experimental and numerical values in static structural analysis was accomplished. Mean relative difference between two sets of results was mostly below 8% and even less than 1% for higher force intensities.
- The measured values of the strain field validate the previously conducted structural analyses and demonstrate the applicability of the employed design methodology.
- The final optimum, chosen in accordance with the obtained Pareto set of optimal solutions, primarily depends on the defined cost-functions and imposed constraints. Nonetheless, a multi-objective approach applied in this study also includes previous engineering knowledge as well as some unquantifiable aspects. It also offers more diversity in design and allows consideration of some previously insufficiently known relations between the investigated parameters.
- The resulting VAWT blade, a composite laminate made from biaxial fiberglass plies, dimensioned in accordance with the most demanding expected operating conditions, weights approximately 900 g and satisfies all of the imposed structural requirements.
- Presented work can be adjusted for use on any wind turbine, helicopter or propeller blade, as well as other composite structures found in aeronautical and mechanical applications.

Acknowledgements The research work is supported by the Ministry of Education, Science, and Technological Development of the Republic of Serbia through contract no. 451-03-9/2021-14/200105.

References

1. IRENA (The International Renewable Energy Agency): Renewable Energy Statistics 2019. IRENA, Abu Dhabi (2019)
2. Eriksson, S., Bernhoff, H., Leijon, M.: Evaluation of different turbine concepts for wind power. Renew. Sustain. Energy Rev. **12**(5), 1419–1434 (2008)
3. Svorcan, J., Peković, O., Simonović, A., Tanović, D., Hasan, M.S.: Design of optimal flow concentrator for vertical-axis wind turbines using computational fluid dynamics, artificial neural networks and genetic algorithm. Adv. Mech. Eng. **13**(3), (2021)
4. Bhutta, M.M.A., Hayat, N., Farooq, A.U., Ali, Z., Jamil, Sh.R., Hussain, Z.: Vertical axis wind turbine – a review of various configurations and design techniques. Renew. Sustain. Energy Rev. **16**(4), 1926–1939 (2012)
5. Wendler, R., Calderon-Munoz, W.R., LeBoeuf, R.: Energy-based iteration scheme of the double-multiple streamtube model in vertical-axis wind turbines. Acta Mech. **227**, 3295–3303 (2016)
6. Fusi, F., Congedo, P.M., Guardone, A., Quaranta, G.: Shape optimization under uncertainty of morphing airfoils. Acta Mech. **229**, 1229–1250 (2018)
7. Duan, Z., Yan, J., Lee, I., Wang, J., Yu, T.: Integrated design optimization of composite frames and materials for maximum fundamental frequency with continuous fiber winding angles. Acta. Mech. Sin. **34**(6), 1084–1094 (2018)
8. Zorić, N.D., Tomović, A.M., Obradović, A.M., Radulović, R.D., Petrović, G.R.: Active vibration control of smart composite plates using optimized self-tuning fuzzy logic controller with optimization of placement, sizing and orientation of PFRC actuators. J. Sound Vib. **456**, 173–198 (2019)
9. Baltić, M., Svorcan, J., Perić, B., Vorkapić, M., Ivanov, T., Peković, O.: Comparative numerical and experimental investigation of static and dynamic characteristics of composite plates. J. Mech. Sci. Technol. **33**(6), 2597–2603 (2019)
10. Kim, S.H., Bang, H.J., Shin, H.K., Jang, M.S.: Composite structural analysis of flat-back shaped blade for multi-MW class wind turbine. Appl. Compos. Mater. **21**, 525–539 (2014)
11. Kaveh, A., Zolghadr, A.: Meta-heuristic methods for optimization of truss structures with vibration frequency constraints. Acta Mech. **229**, 3971–3992 (2018)
12. Dulcey, G.F., Fischer, X., Joyot, P., Fadel, G.: Support for decision making in design of composite laminated structures. Part 1: parametric knowledge model. Appl. Compos. Mater. **26**, 643–662 (2019)
13. Dulcey, G.F., Fischer, X., Joyot, P., Fadel, G.: Support for decision making in design of composite laminated structures. Part 2: reduced parametric model-based optimization. Appl. Compos. Mater. **26**, 663–681 (2019)
14. Nikbakt, S., Kamarian, S., Shakeri, M.: A review on optimization of composite structures Part I: laminated composites. Compos. Struct. **195**, 158–185 (2018)
15. Wang, L., Kolios, A., Nishino, T., Delafin, P.L., Bird, T.: Structural optimisation of vertical-axis wind turbine composite blades based on finite element analysis and genetic algorithm. Compos. Struct. **153**, 123–138 (2016)
16. Fagan, E.M., Flanagan, M., Leen, S.B., Flanagan, T., Doyle, A., Goggins, J.: Physical experimental static testing and structural design optimisation for a composite wind turbine blade. Compos. Struct. **164**, 90–103 (2017)

17. Dal Monte, A., De Betta, S., Raciti Castelli, M., Benini, E.: Proposal for a coupled aerodynamic–structural wind turbine blade optimization. Compos. Struct. **159**, 144–156 (2017)
18. Albanesi, A., Roman, N., Bre, F., Fachinotti, V.: A metamodel-based optimization approach to reduce the weight of composite laminated wind turbine blades. Compos. Struct. **194**, 345–356 (2018)
19. Albanesi, A., Bre, F., Fachinotti, V., Gebhardt, C.: Simultaneous ply-order, ply-number and ply-drop optimization of laminate wind turbine blades using the inverse finite element method. Compos. Struct. **184**, 894–903 (2018)
20. Zheng, Y., Chen, J.: A modified multi-objective particle swarm optimization approach and its application to the design of a deepwater composite riser. Acta. Mech. Sin. **34**(2), 275–284 (2018)
21. Zhong, J., Zheng, Y., Chen, J., Jing, Z.: Variable-stiffness composite cylinder design under combined loadings by using the improved Kriging model. Acta. Mech. Sin. **35**(1), 201–211 (2019)
22. Fan, H.T., Wang, H., Chen, X.H.: An optimization method for composite structures with ply-drops. Compos. Struct. **136**, 650–661 (2016)
23. Axinte, A., Taranu, N., Bejan, L., Hudisteanu, I.: Optimisation of fabric reinforced polymer composites using a variant of genetic algorithm. Appl. Compos. Mater. **24**, 1479–1491 (2017)
24. Svorcan, J., Stupar, S., Komarov, D., Peković, O., Kostić, I.: Aerodynamic design and analysis of a small-scale vertical axis wind turbine. J. Mech. Sci. Technol. **27**(8), 2367–2373 (2013)
25. Posteljnik, Z., Stupar, S., Svorcan, J., Peković, O., Ivanov, T.: Multi-objective design optimization strategies for small-scale vertical-axis wind turbines. Struct. Multidiscip. Optim. **53**(2), 277–290 (2016)
26. ANSYS: ANSYS Fluent Theory Guide, release 16.0. ANSYS Inc., Canonsburg (2015)
27. Svorcan, J., Trivković, Z., Ivanov, T., Baltić, M., Peković, O.: Multi-objective constrained optimizations of VAWT composite blades based on FEM and PSO. FME Trans. **47**(4), 887–893 (2019)
28. Howell, R., Qin, N., Edwards, J., Durrani, N.: Wind tunnel and numerical study of a small vertical axis wind turbine. Renewable Energy **35**(2), 412–422 (2010)
29. Urbina, R., Peterson, M.L., Kimball, R.W., deBree, G.S., Cameron, M.P.: Modeling and validation of a cross flow turbine using free vortex model and a modified dynamic stall model. Renewable Energy **50**, 662–669 (2013)
30. Liang, C., Li, H.: Effects of optimized airfoil on vertical axis wind turbine aerodynamic performance. J. Braz. Soc. Mech. Sci. Eng. **40**, 88 (2018)
31. Franchina, N., Persico, G., Savini, M.: 2D–3D Computations of a vertical axis wind turbine flow field: modeling issues and physical interpretations. Renewable Energy **136**, 1170–1189 (2019)
32. Guillaud, N., Balarac, G., Goncalvès, E., Zanette, J.: Large eddy simulations on vertical axis hydrokinetic turbines - power coefficient analysis for various solidities. Renewable Energy **147**, 473–486 (2020)
33. Mazarbhuiya, H.M.S.M., Biswas, A., Sharma, K.K.: Low wind speed aerodynamics of asymmetric blade H-Darrieus wind turbine-its desired blade pitch for performance improvement in the built environment. J. Braz. Soc. Mech. Sci. Eng. **42**, 326 (2020)
34. Franchina, N., Kouaissah, O., Persico, G., Savini, M.: Three-dimensional CFD simulation and experimental assessment of the performance of a H-shape vertical-axis wind turbine at design and off-design conditions. Int. J. Turbomach. Propul. Power **4**(3), ijtpp4030030 (2020)
35. Battisti, L., Persico, G., Dossena, V., Paradiso, B., Raciti Castelli, M., Brighenti, A., Benini, E.: Experimental benchmark data for H-shaped and troposkien VAWT architectures. Renewable Energy **125**, 425–444 (2018)
36. Zhong-Jia, Y., Yi-Zhuo, G., Min, L., Yan-Xiab, L., Jie, L., Zuo-Guang, Z.: Design and analysis of small-scale lift-type vertical-axis wind turbine using composite blade. Polym. Polym. Compos. **22**(4), 423–430 (2014)
37. ANSYS: ANSYS Mechanical APDL Element Reference, release 16.0. ANSYS Inc., Canonsburg (2015)

38. Eberhart, R., Kennedy, J.: New optimizer using particle swarm theory. In: Proceedings of the 6th International Symposium on Micro Machine and Human Science, 39–46. Nagoya, IEEE (1995)
39. Chowdhury, S., Tong, W., Messac, A., Zhang, J.: A mixed-discrete particle swarm optimization algorithm with explicit diversity-preservation. Struct. Multidiscip. Optim. **47**(3), 367–388 (2013)
40. Wang, X.D., Hirch, C., Kang, Sh., Lacor, C.: Multi-objective optimization of turbomachinery using improved NSGA-II and approximation model. Comput. Methods Appl. Mech. Eng. **200**, 883–895 (2011)

Analysis of the Use of Renewable Energy Sources in the Republic of Serbia

Marko Ristić, Ljiljana Radovanović, Jasmina Perisić, Ivana Vasović, and Luka Đorđević

Abstract One of the consequences of the growth of the world industry is the increased need for electricity. It is imperative for developed countries to use renewable sources, both due to the declining reserves of conventional fuels and the reduction of environmental pollution. The unstoppable growth in the use of alternative sources is more than evident in these countries, while less developed countries are trying to keep up with them. This paper analyzes the development of renewable energy sources in the Republic of Serbia and some surrounding countries and the USA, Germany, France, Japan, Australia, and the Netherlands in the previous decade. A comparative analysis of utilization trends in developed countries in the world is also given, as well as a brief overview of the development of devices used in the processes of the utilization of these energy sources.

Keywords Renewable energy · Electricity production · Solar energy · Wind energy

1 Introduction

The increase in the industry's growth is directly related to the increase in the need for electricity [31]. Since the main energy sources are conventional, they have the largest share in obtaining electricity, with the increase in the use of conventional fuels, the negative impact on the environment also increases [26, 33].

M. Ristić (✉)
Institute Mihajlo Pupin, Volgina 15, 11000 Belgrade, Serbia
e-mail: marko.ristic@pupin.rs

L. Radovanović · L. Đorđević
Technical Faculty "Mihajlo Pupin", University of Novi Sad, ĐureĐakovica bb, Zrenjanin, Serbia

J. Perisić
Faculty of Entrepreneurial Business and Real Estate Management, UNION "Nikola Tesla" University, Cara Dušana 62-64, Belgrade, Serbia

I. Vasović
Lola Insititute, KnezaVišeslava 70a, 11000 Belgrade, Serbia

N. Mitrovic et al. (eds.), *Current Problems in Experimental and Computational Engineering*, Lecture Notes in Networks and Systems 323,
https://doi.org/10.1007/978-3-030-86009-7_4

Developed countries are rapidly working to reorient themselves to obtaining electricity from renewable sources, which are incomparably less or do not pollute the environment at all. The trend of replacing the main energy sources from conventional to renewable sources is noticeable worldwide. An increasing number of countries in their strategic plans, as an imperative, emphasize switching to the highest possible percentage of renewable energy use [33]. As developed countries succeed more and more in these plans, so do less developed countries try to keep up with them and use the available capacities of their countries in using these resources.

These cases, both through the prism of increasing the use of renewable sources and through the prism of socio-economic connections, have already been described in various literature [20, 23]. The world's 20 most developed countries, the G20, account for 86% of world GDP and 85% of the world's energy-related emissions. The primacy of reducing the environmental pollution in the world certainly puts in the first place the reduction of pollution in these countries that participate most in it [4, 34].

Renewable energy sources such as solar energy, wind energy, geothermal water, biomass are available to a greater or lesser extent worldwide [2, 35]. Taking only the example of the solar potential for obtaining electricity, which is available on the entire Earth's surface, i.e., that in just one hour, the Earth is irradiated with an amount of energy that is sufficient to meet the total annual energy consumption consumed in the world, is exceptional [24]. The progress in the application of solar energy in the last decade is reflected in the growing production of equipment for solar energy, which led to an average drop in the price of equipment in the world of 82% in the period from 2009 to 2019 [35]. In the European market, the price of equipment has dropped by about 90%. At the same time, the conversion of solar radiation into electricity of average equipment in the same period increased from 14 to 18%, which indicates a constant improvement in this field [5].

According to experts, the energy potential of renewable energy sources in Serbia is significant, equal to almost half of the country's annual energy needs. In European terms, Serbia has greater potential than Malta, but on the other hand, it lags significantly behind Denmark or Spain in the field of wind use. The use of biomass shows the greatest potential, and it is estimated at 2.7 million tons of oil equivalent (ten) or 63% of the total potential [26]. Of the other renewable energy sources, 0.6 million ten is in unused hydro potential (14%), 0.2 million ten in geothermal sources (4.5%), 0.2 million ten in wind energy (4.5%), and 0.6 million ten in solar radiation (14%).

EU directives on renewable energy have obliged members to ensure that by 2020, renewable energy accounts for 20% of total energy consumption in the European Union. The National Action Plan for Renewable Energy Sources of the Republic of Serbia envisages that it is necessary to build 1,092 MW of new electricity production capacities to achieve the goal of a 27% share of renewable energy sources. The division would be as follows: 500 megawatts for wind power, 438 megawatts for mini-hydropower, 100 megawatts for biomass, 30 megawatts for biogas, 10 megawatts for landfill gas and solar energy, 3 megawatts for waste power plants, and 1 megawatt on geothermal energy.

This paper aims to consider the potentials of using renewable energy sources in the Republic of Serbia, as well as to consider the possibilities for Serbia to become energy independent when it comes to electricity. Increasing the use of renewable energy would reduce the demand for fossil fuels and the associated greenhouse gas emissions.

2 Materials and Methods

As a database used in this paper, available data from The International Renewable Energy Agency (IRENA) were used for data from developed countries and countries in the region of the Republic of Serbia. Available data from the Energy Agency of the Republic of Serbia and IRENA were used for the data for the Republic of Serbia. These data are given through the reports of these institutions and with them, comparative analysis of the development of renewable energy in certain developed countries: United States, Germany, France, Japan, Netherlands and Australia, and countries in the region of the Republic of Serbia, which represent developing countries. For the purposes of writing this paper, from the reports of these organizations, as relevant, a part of the data was taken, which performed the analysis of the mentioned countries and determined their movement in applying renewable sources.

2.1 The International Renewable Energy Agency (IRENA)

Founded as an intergovernmental organization, IRENA represents a link between countries and serves to guide countries to use their renewable energy capacities. In addition to promoting these energy sources, this organization makes detailed reports and comparative analyses of the movement of countries in the world towards the use of renewable sources [30].

2.2 Energy Agency and Electric Power Industry of Serbia

Reports of the Energy Agency of the Republic of Serbia represent, in addition to organizational and economic data, a set of data on the amount of energy produced, their structure, and plans for the development of the energy sector in the Republic of Serbia [19].

2.3 *Statistics of Countries*

In the first section of this paper, an analysis of developed countries is made individually for the countries mentioned above and their comparative analysis.

United States of America

Table 1 shows the data for the USA, taken from the official IRENA website [6]. For both the USA and other countries covered in this paper, data for TPES (Total Primary Energy Supply), Capacity in 2019, Capacity change (%) 2014–2019, 2018–2019, and Generation in 2018 are analyzed.

The table shows an increase in the capacity of renewable energy sources in the period 2014–2019 by 47%, while at the same time, the capacity of non-renewable energy sources decreased by 3%. In the available capacities for electricity generation in 2019, non-renewable sources make up 77% and renewable sources 23%. In the period from 2012–2017, there was an increase in TPES by 1.83%. The most significant increase in the capacity of renewable energy sources in the period from 2014–2019 was recorded in solar energy, 253%, while in next place is wind energy with 61%. In the total capacities of renewable energy sources in 2019, the leaders are wind energy with 9%, followed by hydro energy with 7%, and solar energy with 6%.

Germany

Table 2 shows the data for Germany, taken from the official IRENA website [7]. The table shows an increase in the capacity of renewable energy sources in the period 2014–2019 by 39%, while at the same time, the capacity of non-renewable energy sources increased by 3%. In the available capacities for electricity generation in 2019, non-renewable sources account for 47% and renewable sources for 53%. In the period from 2012–2017, there was an increase in TPES by 0.18%. The largest increase in the capacity of renewable energy sources in the period from 2014–2019 was recorded in wind energy, 58%, while in next place is geothermal energy with

Table 1. Statistics of USA

TPES	2012	2017
Non-renewable (TJ)	82 716 367	83 265 938
Renewable (TJ)	5 983 878	7 059 509
Total (TJ)	88 700 245	90 325 447
Renewable share (%)	7	8

Capacity in 2019	MW	%
Non-renewable	**863 995**	**77**
Renewable	**264 504**	**23**
Hydro/marine	83 617	7
Solar	62 298	6
Wind	103 584	9
Bioenergy	12 450	1
Geothermal	2 555	0
Total	**1 128 500**	**100**

Capacity change (%)	2014-19	2018-19
Non-renewable	**- 3**	**- 1.0**
Renewable	**+ 47**	**+ 7.3**
Hydro/marine	+ 0	- 0.1
Solar	+ 253	+ 17.1
Wind	+ 61	+ 9.7
Bioenergy	+ 0	- 2.1
Geothermal	+ 2	+ 0.6
Total	**+ 5**	**+ 0.8**

Generation in 2018	GWh	%
Non-renewable	**3 712 178**	**83**
Renewable	**743 177**	**17**
Hydro and marine	295 501	7
Solar	85 184	2
Wind	275 834	6
Bioenergy	67 885	2
Geothermal	18 773	0
Total	**4 455 355**	**100**

Table 2. Statistics of Germany

TPES	2012	2017
Non-renewable (TJ)	11 639 411	11 339 464
Renewable (TJ)	1 365 613	1 690 049
Total (TJ)	13 005 024	13 029 513
Renewable share (%)	11	13

Capacity in 2019	MW	%
Non-renewable	**110 477**	**47**
Renewable	**125 476**	**53**
Hydro/marine	5 595	2
Solar	49 018	21
Wind	60 840	26
Bioenergy	9 981	4
Geothermal	42	0
Total	**235 953**	**100**

Capacity change (%)	2014-19	2018-19
Non-renewable	**+ 3**	**+ 0.5**
Renewable	**+ 39**	**+ 5.2**
Hydro/marine	+ 0	+ 0.2
Solar	+ 29	+ 8.5
Wind	+ 58	+ 3.4
Bioenergy	+ 22	+ 3.4
Geothermal	+ 45	+ 16.7
Total	**+ 19**	**+ 2.9**

Generation in 2018	GWh	%
Non-renewable	**418 391**	**65**
Renewable	**224 768**	**35**
Hydro and marine	17 975	3
Solar	45 784	7
Wind	109 951	17
Bioenergy	50 880	8
Geothermal	178	0
Total	**643 159**	**100**

45% and solar energy with 29%. In the total capacity of renewable energy sources in 2019, the leading energy is wind energy with 26%, followed by solar energy with 21% and bioenergy with 4%.

France

Table 3 shows the data for France, taken from the official IRENA website [8]. The table shows an increase in the capacity of renewable energy sources in the period 2014–2019 by 31%, while at the same time, the capacity of non-renewable energy sources decreased by 7%. In the available capacities for electricity generation in 2019, non-renewable sources account for 61% and renewable sources for 39%. In the period from 2012–2017, a decrease in TPES by 3.81% was recorded. The largest increase in the capacity of renewable energy sources in the period from 2014–2019 was recorded in wind energy, 77%, while in the next place was solar energy with 75% and was energy with 45%. In the total capacities of renewable energy sources in 2019, the leaders are hydro energy with 18%, followed by wind energy with 12% and solar energy with 8%.

Table 3. Statistics of France

TPES	2012	2017
Non-renewable (TJ)	9 820 365	9 288 824
Renewable (TJ)	925 325	1 061 550
Total (TJ)	10 745 690	10 350 374
Renewable share (%)	9	10

Capacity in 2019	MW	%
Non-renewable	**82 888**	**61**
Renewable	**52 928**	**39**
Hydro/marine	24 304	18
Solar	10 571	8
Wind	16 260	12
Bioenergy	1 776	1
Geothermal	16	0
Total	**135 816**	**100**

Capacity change (%)	2014-19	2018-19
Non-renewable	**- 7**	**- 0.1**
Renewable	**+ 31**	**+ 4.8**
Hydro/marine	+ 1	+ 0.1
Solar	+ 75	+ 9.9
Wind	+ 77	+ 9.1
Bioenergy	+ 45	+ 3.8
Geothermal	- 2	0.0
Total	**+ 4**	**+ 1.8**

Generation in 2018	GWh	%
Non-renewable	**468 545**	**81**
Renewable	**113 397**	**19**
Hydro and marine	65 765	11
Solar	10 569	2
Wind	28 599	5
Bioenergy	8 335	1
Geothermal	130	0
Total	**581 943**	**100**

Table 4. Statistics of Japan

TPES	2012	2017
Non-renewable (TJ)	18 149 690	17 108 798
Renewable (TJ)	829 464	1 031 072
Total (TJ)	18 979 154	18 139 870
Renewable share (%)	4	6

Capacity in 2019	MW	%
Non-renewable	**241 222**	**71**
Renewable	**97 462**	**29**
Hydro/marine	28 114	8
Solar	61 840	18
Wind	3 786	1
Bioenergy	3 197	1
Geothermal	525	0
Total	**338 683**	**100**

Capacity change (%)	2014-19	2018-19
Non-renewable	**- 8**	**- 3.8**
Renewable	**+ 87**	**+ 7.6**
Hydro/marine	+ 1	- 0.1
Solar	+ 220	+ 11.4
Wind	+ 38	+ 3.2
Bioenergy	+ 96	+ 14.4
Geothermal	+ 3	+ 8.9
Total	**+ 8**	**- 0.8**

Generation in 2018	GWh	%
Non-renewable	**924 363**	**84**
Renewable	**176 651**	**16**
Hydro and marine	80 968	7
Solar	62 668	6
Wind	7 481	1
Bioenergy	23 076	2
Geothermal	2 457	0
Total	**1 101 014**	**100**

Japan

Table 4 shows the data for Japan, taken from the official IRENA website [9]. The table shows an increase in the capacity of renewable energy sources in the period 2014–2019 by 87%, while at the same time, the capacity of non-renewable energy sources decreased by 8%. In the available capacities for electricity generation in 2019, non-renewable sources make up 71% and renewable sources 29%. In the period from 2012–2017, a decrease in TPES by 4.62% was recorded. The largest increase in the capacity of renewable energy sources in the period from 2014–2019 was recorded in solar energy, 220%, followed by bioenergy with 96%, and wind energy with 38%. In the total capacities of renewable energy sources in 2019, the leaders are solar energy with 18%, followed by hydro energy with 8%, and bioenergy and geothermal energy with 1% each.

Australia

Table 5 shows the data for Australia, taken from the official IRENA website [10]. The table shows an increase in the capacity of renewable energy sources in the period

Table 5. Statistics of Australia

TPES	2012	2017
Non-renewable (TJ)	5 129 057	4 983 261
Renewable (TJ)	311 168	384 578
Total (TJ)	5 440 225	5 367 839
Renewable share (%)	6	7

Capacity in 2019	MW	%
Non-renewable	**46 427**	**62**
Renewable	**28 980**	**38**
Hydro/marine	7 719	10
Solar	13 252	18
Wind	7 133	9
Bioenergy	875	1
Geothermal	0	0
Total	**75 407**	**100**

Capacity change (%)	2014-19	2018-19
Non-renewable	**- 8**	**+ 0.1**
Renewable	**+ 67**	**+ 26.6**
Hydro/marine	+ 3	+ 0.1
Solar	+ 151	+ 53.6
Wind	+ 88	+ 25.6
Bioenergy	+ 6	+ 1.3
Geothermal	+ 210	0.0
Total	**+ 11**	**+ 8.9**

Generation in 2018	GWh	%
Non-renewable	**216 556**	**83**
Renewable	**44 443**	**17**
Hydro and marine	15 831	6
Solar	9 930	4
Wind	15 164	6
Bioenergy	3 518	1
Geothermal	0	0
Total	**260 999**	**100**

2014–2019 by 67%, while at the same time, the capacity of non-renewable energy sources decreased by 8%. In the available capacities for electricity generation in 2019, non-renewable sources make up 62% and renewable sources 38%. In the period from 2012–2017, a decrease in TPES by 1.4% was recorded. The largest increase in the capacity of renewable energy sources in the period from 2014–2019 was recorded in geothermal energy, 210%, followed by solar energy with 151%, and wind energy with 88%. In the total capacities of renewable energy sources in 2019, the leaders are solar energy with 18%, followed by hydro energy with 10%, and wind energy with 9%.

Netherlands

Table 6 shows the data for Netherlands, taken from the official IRENA website [11]. The table shows an increase in the capacity of renewable energy sources in the period 2014–2019 by 167%, while at the same time, the capacity of non-renewable energy sources decreased by 17%. In the available capacities for electricity generation in 2019, non-renewable sources make up 65% and renewable sources 35%. In the period from 2012–2017, a decrease in TPES by 4.84% was recorded. The largest increase in the capacity of renewable energy sources in the period from 2014–2019 was recorded in solar energy, 542%, followed by wind energy with 56%, and bioenergy with 13%. In the total capacities of renewable energy sources in 2019, the leaders are solar energy with 20%, followed by wind energy with 13%, and geothermal energy with 2%.

Montenegro

Table 7 shows the data for Montenegro, taken from the official IRENA website [12]. The table shows an increase in the capacity of renewable energy sources in the period 2014–2019 by 19%, while at the same time, the capacity of non-renewable energy sources increased by 3%. In the available capacities for electricity generation in 2019, non-renewable sources make up 22% and renewable sources 78%. In the period from 2012–2017, there was a decrease in TPES by 2.60%. The largest increase in the

Table 6. Statistics of Netherlands

TPES	2012	2017
Non-renewable (TJ)	3 090 882	2 914 560
Renewable (TJ)	147 530	170 456
Total (TJ)	3 238 412	3 085 016
Renewable share (%)	5	6

Capacity in 2019	MW	%
Non-renewable	**22 360**	**65**
Renewable	**12 075**	**35**
Hydro/marine	39	0
Solar	6 725	20
Wind	4 463	13
Bioenergy	849	2
Geothermal	0	0
Total	**34 435**	**100**

Capacity change (%)	2014-19	2018-19
Non-renewable	**- 17**	**- 11.4**
Renewable	**+ 157**	**+ 23.2**
Hydro/marine	+ 5	0.0
Solar	+ 542	+ 48.7
Wind	+ 56	+ 1.6
Bioenergy	+ 13	0.0
Geothermal	0	0.0
Total	**+ 8**	**- 1.7**

Generation in 2018	GWh	%
Non-renewable	**95 584**	**84**
Renewable	**18 884**	**16**
Hydro and marine	72	0
Solar	3 693	3
Wind	10 564	9
Bioenergy	4 555	4
Geothermal	0	0
Total	**114 468**	**100**

Table 7. Statistics of Montenegro

TPES	2012	2017
Non-renewable (TJ)	29 444	29 333
Renewable (TJ)	18 774	17 660
Total (TJ)	48 218	46 993
Renewable share (%)	39	38

Capacity in 2019	MW	%
Non-renewable	**225**	**22**
Renewable	**779**	**78**
Hydro/marine	658	65
Solar	3	0
Wind	118	12
Bioenergy	0	0
Geothermal	0	0
Total	**1 004**	**100**

Capacity change (%)	2014-19	2018-19
Non-renewable	**+ 3**	**0.0**
Renewable	**+ 19**	**+ 0.5**
Hydro/marine	+ 1	+ 0.6
Solar	+ 58	0.0
Wind	0	0.0
Bioenergy	0	0.0
Geothermal	0	0.0
Total	**+ 15**	**+ 0.4**

Generation in 2018	GWh	%
Non-renewable	**1 555**	**41**
Renewable	**2 260**	**59**
Hydro and marine	2 113	55
Solar	4	0
Wind	143	4
Bioenergy	0	0
Geothermal	0	0
Total	**3 815**	**100**

capacity of renewable energy sources in the period from 2014–2019 was recorded in solar energy, 58%, followed by hydropower with 1%. In the total capacities of renewable energy sources in 2019, the leading ones are hydro energy with 65%, followed by wind energy with 12%.

Croatia

Table 8 shows the data for Croatia, taken from the official IRENA website [13]. The table shows an increase in the capacity of renewable energy sources in the period 2014–2019 by 17%, while at the same time, the capacity of non-renewable energy sources increased by 11%. In the available capacities for electricity generation in 2019, non-renewable sources account for 40% and renewable sources for 60%. In the period from 2012–2017, a decrease in TPES by 0.04% was recorded. The largest increase in the capacity of renewable energy sources in the period from 2014–2019 was recorded in bioenergy, 430%, while in next place is solar energy with 108% and wind energy with 79%. In the total capacity of renewable energy sources in 2019, the leaders are hydro energy with 44%, followed by wind energy with 12%, and bioenergy with 3%.

Table 8. Statistics of Croatia

TPES	2012	2017
Non-renewable (TJ)	272 898	269 991
Renewable (TJ)	100 575	103 296
Total (TJ)	373 473	373 287
Renewable share (%)	27	28

Capacity in 2019	MW	%
Non-renewable	**2 021**	**40**
Renewable	**3 014**	**60**
Hydro/marine	2 200	44
Solar	69	1
Wind	606	12
Bioenergy	130	3
Geothermal	10	0
Total	**5 035**	**100**

Capacity change (%)	2014-19	2018-19
Non-renewable	**+ 11**	**- 0.3**
Renewable	**+ 17**	**+ 1.2**
Hydro/marine	+ 1	+ 0.0
Solar	+ 108	+ 1.5
Wind	+ 79	+ 3.4
Bioenergy	+ 430	+ 12.5
Geothermal	0	0.0
Total	**+ 14**	**+ 0.6**

Generation in 2018	GWh	%
Non-renewable	**3 850**	**28**
Renewable	**9 781**	**72**
Hydro and marine	7 701	56
Solar	75	1
Wind	1 335	10
Bioenergy	668	5
Geothermal	2	0
Total	**13 632**	**100**

Table 9. Statistics of North Macedonia

TPES	2012	2017
Non-renewable (TJ)	110 650	98 178
Renewable (TJ)	23 007	23 278
Total (TJ)	133 657	121 456
Renewable share (%)	17	19

Capacity in 2019	MW	%
Non-renewable	**1 104**	**59**
Renewable	**754**	**41**
Hydro/marine	682	37
Solar	26	1
Wind	37	2
Bioenergy	9	0
Geothermal	0	0
Total	**1 858**	**100**

Capacity change (%)	2014-19	2018-19
Non-renewable	**+ 3**	**0.0**
Renewable	**+ 11**	**+ 1.9**
Hydro/marine	+ 8	+ 1.1
Solar	+ 74	+ 27.0
Wind	0	0.0
Bioenergy	0	+ 7.3
Geothermal	0	0.0
Total	**+ 6**	**+ 0.8**

Generation in 2018	GWh	%
Non-renewable	**3 642**	**65**
Renewable	**1 966**	**35**
Hydro and marine	1 791	32
Solar	23	0
Wind	97	2
Bioenergy	54	1
Geothermal	0	0
Total	**5 607**	**100**

North Macedonia

Table 9 shows the data for North Macedonia, taken from the official IRENA website [14]. The table shows an increase in the capacity of renewable energy sources in the period 2014–2019 by 11%, while at the same time, the capacity of non-renewable energy sources increased by 3%. In the available capacities for electricity generation in 2019, non-renewable sources make up 59% and renewable sources 41%. In the period from 2012–2017, a decrease in TPES by 9.9% was recorded. The largest increase in the capacity of renewable energy sources in the period from 2014–2019 was recorded in solar energy, 74%, while in next place is hydro energy with 8%. In the total capacities of renewable energy sources in 2019, the leaders are hydro energy with 37%, followed by wind energy with 2% and solar energy with 1%.

Bosnia and Herzegovina

Table 10 shows the data for Bosnia and Herzegovina, taken from the official IRENA website [15]. The table shows an increase in the capacity of renewable energy sources in the period 2014–2019 by 12%, while at the same time, the capacity of non-renewable energy sources increased by 13%. In the available capacities for electricity generation in 2019, non-renewable sources make up 57% and renewable sources

Table 10. Statistics of Bosnia and Herzegovina

TPES	2012	2017
Non-renewable (TJ)	257 091	245 508
Renewable (TJ)	31 968	41 387
Total (TJ)	289 059	286 895
Renewable share (%)	11	14

Capacity in 2019	MW	%
Non-renewable	**2 576**	**57**
Renewable	**1 931**	**43**
Hydro/marine	1 816	40
Solar	18	0
Wind	87	2
Bioenergy	9	0
Geothermal	0	0
Total	**4 507**	**100**

Capacity change (%)	2014-19	2018-19
Non-renewable	**+ 13**	**0.0**
Renewable	**+ 12**	**+ 1.9**
Hydro/marine	+ 6	+ 0.0
Solar	+ 157	+ 1.7
Wind	+ 28 867	+ 70.7
Bioenergy	0	0.0
Geothermal	0	0.0
Total	**+ 13**	**+ 0.8**

Generation in 2018	GWh	%
Non-renewable	**12 610**	**66**
Renewable	**6 550**	**34**
Hydro and marine	6 418	33
Solar	21	0
Wind	103	1
Bioenergy	8	0
Geothermal	0	0
Total	**19 160**	**100**

Table 11. Statistics of Hungary

TPES	2012	2017
Non-renewable (TJ)	914 307	984 601
Renewable (TJ)	125 953	131 035
Total (TJ)	1 040 260	1 115 636
Renewable share (%)	12	12

Capacity in 2019	MW	%
Non-renewable	**7 581**	**78**
Renewable	**2 153**	**22**
Hydro/marine	57	1
Solar	1 277	13
Wind	329	3
Bioenergy	487	5
Geothermal	3	0
Total	**9 733**	**100**

Capacity change (%)	2014-19	2018-19
Non-renewable	**- 1**	**- 0.9**
Renewable	**+ 110**	**+ 34.7**
Hydro/marine	- 0	- 0.3
Solar	+ 1 335	+ 75.9
Wind	0	0.0
Bioenergy	- 11	+ 0.7
Geothermal	0	0.0
Total	**+ 12**	**+ 5.2**

Generation in 2018	GWh	%
Non-renewable	**28 251**	**88**
Renewable	**3 753**	**12**
Hydro and marine	222	1
Solar	620	2
Wind	607	2
Bioenergy	2 292	7
Geothermal	12	0
Total	**32 004**	**100**

43%. In the period from 2012–2017, a decrease in TPES by 0.25% was recorded. The largest increase in the capacity of renewable energy sources in the period from 2014–2019 was recorded in wind energy, 28,867%, while in next place is solar energy with 157% and hydro energy with 6%. In the total capacities of renewable energy sources in 2019, the leading ones are hydro energy with 40%, followed by wind energy with 2%.

Hungary

Table 11 shows the data for Hungary, taken from the official IRENA website [16]. The above shows an increase in the capacity of renewable energy sources in the period 2014–2019 by 110%, while at the same time, the capacity of non-renewable energy sources decreased by 1%. In the available capacities for electricity generation in 2019, non-renewable sources account for 78% and renewable sources for 22%. In the period from 2012–2017, there was an increase in TPES by 7.24%. The largest increase in the capacity of renewable energy sources in the period from 2014–2019 was recorded in solar energy, 1335%. In the total capacities of renewable energy sources in 2019, the leading ones are solar energy with 13%, followed by bioenergy with 5%.

Romania

Table 12 shows the data for Romania, taken from the official IRENA website [17]. The above shows an increase in the capacity of renewable energy sources in the period 2014–2019 by 0%, while at the same time, the capacity of non-renewable energy sources decreased by 6%. In the available capacities for electricity generation in 2019, non-renewable sources account for 51% and renewable sources for 49%. In the period from 2012–2017, there was decrease in TPES by 6.37%. The largest increase in the capacity of renewable energy sources in the period from 2014–2019 was recorded in bioenergy, 51%, while in next place were solar energy with 7% and hydro energy with 2%. In the total capacities of renewable energy sources in 2019, the leading ones are hydro energy with 29%, followed by wind energy with 13%.

Table 12. Statistics of Romania

TPES	2012	2017
Non-renewable (TJ)	1 251 316	1 151 373
Renewable (TJ)	236 422	256 592
Total (TJ)	1 487 738	1 407 965
Renewable share (%)	16	18

Capacity in 2019	MW	%
Non-renewable	**11 875**	**51**
Renewable	**11 190**	**49**
Hydro/marine	6 621	29
Solar	1 386	6
Wind	3 040	13
Bioenergy	142	1
Geothermal	0	0
Total	**23 064**	**100**

Capacity change (%)	2014-19	2018-19
Non-renewable	**- 6**	**- 3.2**
Renewable	**+ 0**	**+ 0.2**
Hydro/marine	+ 2	+ 0.2
Solar	+ 7	+ 0.0
Wind	- 6	+ 0.3
Bioenergy	+ 51	+ 0.7
Geothermal	0	0.0
Total	**- 3**	**- 1.6**

Generation in 2018	GWh	%
Non-renewable	**38 682**	**60**
Renewable	**26 195**	**40**
Hydro and marine	17 664	27
Solar	1 771	3
Wind	6 322	10
Bioenergy	437	1
Geothermal	0	0
Total	**64 876**	**100**

Serbia

Table 13 shows the data for Serbia, taken from the official IRENA website [18]. The above shows an increase in the capacity of renewable energy sources in the period 2014–2019 by 19%, while at the same time, the capacity of non-renewable energy sources increased by 3%. In the available capacities for electricity generation in 2019, non-renewable sources account for 63% and renewable sources for 32%. In the period from 2012–2017, there was an increase in TPES by 7.25%. The largest increase in the capacity of renewable energy sources in the period from 2014–2019 was recorded in wind energy, 71,200%, while in next place were bio energy with 200% and solar energy with 67%. In the total capacities of renewable energy sources in 2019, the leading ones are hydro energy with 32%, followed by wind energy with 5%.

Table 13. Statistics of Serbia

TPES	2012	2017
Non-renewable (TJ)	527 399	567 023
Renewable (TJ)	80 859	85 351
Total (TJ)	608 258	652 374
Renewable share (%)	13	13

Capacity in 2019	MW	%
Non-renewable	**4 968**	**63**
Renewable	**2 870**	**37**
Hydro/marine	2 489	32
Solar	10	0
Wind	357	5
Bioenergy	15	0
Geothermal	0	0
Total	**7 838**	**100**

Capacity change (%)	2014-19	2018-19
Non-renewable	**+ 3**	**0.0**
Renewable	**+ 19**	**+ 13.1**
Hydro/marine	+ 4	0.0
Solar	+ 67	0.0
Wind	+ 71 200	+1 326.0
Bioenergy	+ 200	0.0
Geothermal	0	0.0
Total	**+ 8**	**+ 4.4**

Generation in 2018	GWh	%
Non-renewable	**26 521**	**71**
Renewable	**10 906**	**29**
Hydro and marine	10 640	28
Solar	13	0
Wind	150	0
Bioenergy	103	0
Geothermal	0	0
Total	**37 427**	**100**

3 Results and Discussion

From the data given in Tables 1 to 13, we can see the development of renewable energy capacity. In all countries developed and developing countries, there is an increase in the capacity of renewable energy sources.

Comparing the capacities of non-renewable/renewable energy sources and generation of energy, it is noticed that in all analyzed countries, except in Croatia, the generation of energy of non-renewable energy sources is higher, i.e., the generation of energy from renewable sources is less than the percentage distribution of possible capacities. Such a difference is the impact of constant availability of production from non-renewable sources, the difference in obtaining energy from renewable sources depending on the source (change in wind speed, number of sunny days, etc.), and energy demand at a given time.

Figure 1 shows the percentage data of the analyzed countries, with the percentage distribution between non-renewable/renewable capacities in 2019.

A balance between imports and exports has been established in Serbia. EPS imported electricity in the winter but exported it in the summer, so it can be said that there is no imbalance.

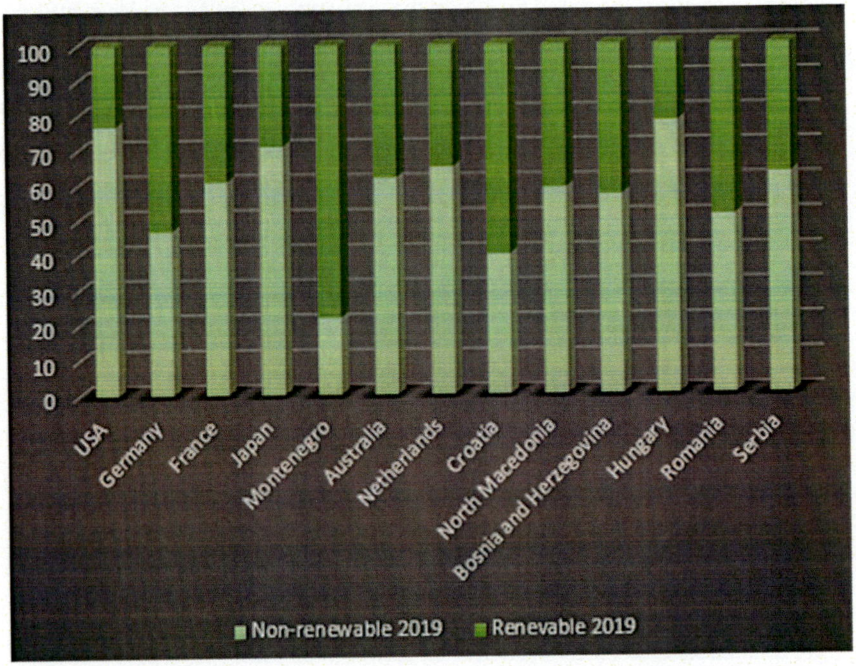

Fig. 1. Percentage distribution between non-renewable/renewable capacities in 2019

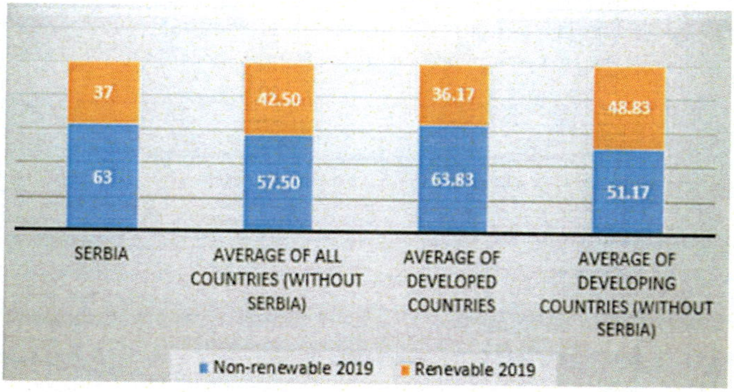

Fig. 2. Average Non-renewable/Renewable percentage

After extracting the average data of the percentage distribution of these sources, and their grouping into the total average (without Serbia), the average of developed countries, and the average of developing countries (without Serbia), we get the situation shown in Fig. 2.

In the overall average of the analyzed countries, the picture shows that Serbia has a lower average capacity of renewable energy sources, i.e., a higher average capacity of non-renewable energy sources. Comparing the percentage distribution, we can see that it is very similar to the average of the analyzed developed countries, i.e., has a deviation of only 1%.

In Fig. 3, we can see the percentage change in the capacity of renewable energy sources in relation to the total capacity in the analyzed countries in the period 2010–2019.

Given the significant differences, both in the size of the analyzed countries, population, different sources of renewable sources, economic situation, it is not considered expedient to compare data based on numerical energy capacity, but the analysis is based on the share of renewable sources in total capacity over the past decade. In all countries, there is a noticeable increase in the share of renewable sources.

4 Conclusion

From all the above, it can be concluded that the trend of increasing renewable energy sources in total electricity production in developed countries, in a similar percentage followed in developing countries.

From Fig. 2, it was concluded that Serbia has a very approximate distribution of the capacity of non-renewable and renewable energy sources as well as developed countries.

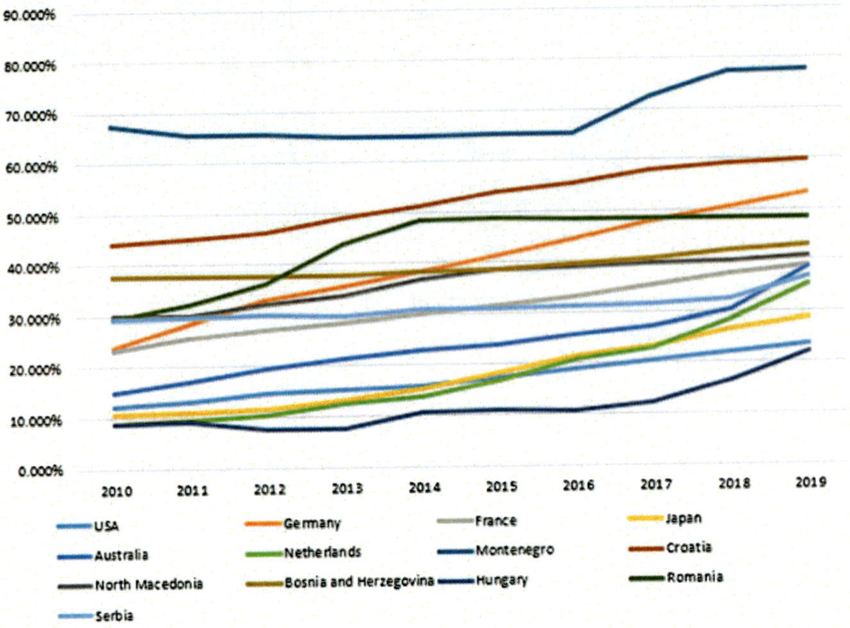

Fig. 3. Percentage change in the capacity of renewable energy sources in relation to the total capacity

Based on the analyzed countries, and according to Fig. 3, it can be seen that the Republic of Serbia follows approximately the same development of its renewable energy capacities as other countries. This way of analyzing, i.e., analyzing in percentage terms the growth of available capacities in relation to total capacities, has certain advantages, because as mentioned, many factors that are specific to each country significantly affect the true values of the analysis.

According to estimates from the energy balance of Serbia for 2020, about 20% of energy in Serbia is produced from RES, of which energy produced from water is dominant. The situation is similar in terms of consumption, where RES and electricity produced from RES participate with about 21% in final energy consumption [21, 25, 30, 37].

At first glance, according to these indicators, Serbia does not lag behind the EU average. However, when looking at the structure of RES consumption by sectors, there is a large lag given that RES consumption is mostly limited to households (firewood and electricity from large rivers), while consumption in other sectors such as transport and industry is relatively modest [3, 22].

Also, in the previous ten years, there has been a significant increase in the production of energy from RES in the world, primarily energy from wind [1, 29, 36] and solar [29] while in Serbia, these two energy sources are still at modest levels. In this regard, the RES strategy in the coming period should go in two directions. The first is to increase energy production from RES, and the second is to stimulate RES consumption in the household, industry and transport sectors [27, 28, 32].

As one example of a problematic way to compare the situation for each country, the line of Serbia in Fig. 3, which is above the line of the USA, was taken. In terms of the percentage use of renewable energy sources, Serbia is above the USA in the previous decade. Seen by the amount of W, i.e., that in the same period, the USA increased its capacity of renewable sources by 126,780 MW and Serbia by 575 MW, a different picture is obtained.

Taking into account the capacities of renewable energy sources in the Republic of Serbia, the accelerated increase in the utilization of these sources recorded in previous years, as well as the current announcements of the authorities in the country about this topic, it was concluded that Serbia does not lag behind other countries in terms of using and increasing the use of opportunities for electricity production and reducing the environmental pollution.

References

1. Benkovic S, Makojevic N, Jednak S (2013) Possibilities for development of the electric power industry of Serbia through private source financing of small hydropower plants. Renewable Energy 50:1053–1059. ISSN 0960–1481. https://doi.org/10.1016/j.renene.2012.08.042
2. Das S, Akella AK (2018) Power flow control of PV-wind-battery hybrid renewable energy systems for stand-alone application. Int J Renewable Energy Res 8:36–43 ISSN 1309–0127
3. Gigović L, Pamučar D, Božanić D, Ljubojević S (2017) Application of the GIS-DANP-MABAC multi-criteria model for selecting the location of wind farms: a case study of Vojvodina, Serbia. Renewable Energy 103:501–521. ISSN 0960–1481. https://doi.org/10.1016/j.renene.2016.11.057
4. Heryadi MD, Hartono D (2016) Energy efficiency, utilization of renewable energies, and carbon dioxide emission: case study of G20 countries. Int Energy J 16:143–152. ISSN: 1513–718X
5. IRENA (2020) Renewable power generation costs in 2019. International renewable energy agency, Abu Dhabi. ISBN 978–92–9260–244–4
6. IRENA, Energy profile United States of America, International renewable energy agency. https://www.irena.org/IRENADocuments/Statistical_Profiles/North%20America/United%20States%20of%20America_North%20America_RE_SP.pdf
7. IRENA, Energy profile Germany, International renewable energy agency. https://www.irena.org/IRENADocuments/Statistical_Profiles/Europe/Germany_Europe_RE_SP.pdf
8. IRENA, Energy profile France, International renewable energy agency. https://www.irena.org/IRENADocuments/Statistical_Profiles/Europe/France_Europe_RE_SP.pdf
9. IRENA, Energy profile Japan, International renewable energy agency. https://www.irena.org/IRENADocuments/Statistical_Profiles/Asia/Japan_Asia_RE_SP.pdf
10. IRENA, Energy profile Australia, International renewable energy agency. https://www.irena.org/IRENADocuments/Statistical_Profiles/Oceania/Australia_Oceania_RE_SP.pdf
11. IRENA, Energy profile netherlands, International renewable energy agency. https://www.irena.org/IRENADocuments/Statistical_Profiles/Europe/Netherlands_Europe_RE_SP.pdf
12. IRENA, Energy profile Montenegro, International renewable energy agency. https://www.irena.org/IRENADocuments/Statistical_Profiles/Europe/Montenegro_Europe_RE_SP.pdf
13. IRENA, Energy profile Croatia, International renewable energy agency. https://www.irena.org/IRENADocuments/Statistical_Profiles/Europe/Croatia_Europe_RE_SP.pdf

14. IRENA, Energy profile North Macedonia, International renewable energy agency. https://www.irena.org/IRENADocuments/Statistical_Profiles/Europe/North%20Macedonia_Europe_RE_SP.pdf
15. IRENA, Energy profile Bosnia and Herzegovina, International renewable energy agency. https://www.irena.org/IRENADocuments/Statistical_Profiles/Europe/Bosnia%20and%20Herzegovina_Europe_RE_SP.pdf
16. IRENA, Energy profile Hungary, International renewable energy agency. https://www.irena.org/IRENADocuments/Statistical_Profiles/Europe/Hungary_Europe_RE_SP.pdf
17. IRENA, Energy profile Romania, International renewable energy agency. https://www.irena.org/IRENADocuments/Statistical_Profiles/Europe/Romania_Europe_RE_SP.pdf
18. IRENA, Energy profile Serbia, International renewable energy agency. https://www.irena.org/IRENADocuments/Statistical_Profiles/Europe/Serbia_Europe_RE_SP.pdf
19. Izveštaj o radu agencije za energetiku za 2019. godinu. http://www.aers.rs/Index.asp?l=1&a=53. Accessed 24 May 2021
20. Perisic J, Radovanovic L, Milovanovic M, Petrovic I, Ristic M, Bugarcic M, Perisic V (2016) A brine mixing mobile unit in oil and gas industry—An example of a cost-effective, efficient and environmentally justified technical solution, Energy Sources, Part A: Recovery, Utilization, and Environmental effects, vol 38, Issue 23, pp 3470–3477, ISSN 1556–7230
21. Josimović B, Cvjetić A, Furundžić D (2021) Strategic environmental assessment and the precautionary principle in the spatial planning of wind farms—European experience in Serbia. Renew Sustain Energy Rev 136:110459. ISSN 1364–0321. https://doi.org/10.1016/j.rser.2020.110459
22. Kostić R, Mikulović J (2017) The empirical models for estimating solar insolation in Serbia by using meteorological data on cloudiness. Renewable Energy 114:1281–1293. ISSN 0960–1481. https://doi.org/10.1016/j.renene.2017.08.013
23. Late L, Phu N (2021) Institutions and geography: a "Two Sides of the Same Coin" story of primary energy use in Sub-Saharan Africa. Energy J https://doi.org/10.5547/01956574.42.2.llaw,pp.223-252
24. Lewis N, Nocera D (2006) Powering the planet: chemical challenges in solar energy utilization. PNAS 103(43):15729–15735
25. Loncar D, Milovanovic I, Rakic B, Radjenovic T (2017) Compound real options valuation of renewable energy projects: the case of a wind farm in Serbia. Renew Sustain Energy Rev 75:354–367. ISSN 1364–0321. https://doi.org/10.1016/j.rser.2016.11.001
26. Ristic M, Radovanovic L, Prokic-Cvetkovic R, Otic G, Perisic J, Vasovic I (2015) Increasing the energy efficiency of thermal power plant Kostolac B by the revitalization of ventilation mills. Energy Sources Part B: Economics Plann Policy. ISSN 1556–7257. https://doi.org/10.1080/15567249.2015.1014977
27. Ristic M, Prokic-Cvetkovic R, Kozic M, Ristic S, Katavic B (2015) Wear reducing of ventilation mill suction plates based on the multidisciplinary research. J Balkan Tribological Association vol 21. ISSN 1310–4772
28. Ristic M, Radovanovic L, Prokic-Cvetkovic R, Otic G, Perisic J, Vasovic I (2015) Increasing the energy efficiency of thermal power plant Kostolac B by the revitalization of ventilation mills. Energy Sources Part B: Economics Plann Policy ISSN 1556–7257. https://doi.org/10.1080/15567249.2015.1014977
29. Maksimovic M, Maksimovic K, Stamenkovic D, Vasovic Maksimovic I (2020) Initial fatigue life estimation of welded structural components. J Tech Gaz 28(4). https://doi.org/10.17559/TV-20200414015501
30. Müller F (2020) IRENA as a glocal actor: pathways towards energy governmentality. Innovation Eur J Soc Sci Res 30: 306–322. https://doi.org/10.1080/13511610.2017.1279538
31. Naveen R, Revankar P, Rajanna S (2020) Integration of renewable energy systems for optimal energy needs-a review. Int J Renew Energy Res 10:727-742. ISSN 1309–0127
32. Perisic J, Milovanovic M, Petrovic I, Radovanovic L, Ristic M, Perisic V, Vrbanac M (2017) Modeling and risk analysis of brine mixing mobile unit operation processes. Energy Sources Part B 12(7):646–653

33. Perišić J, Milovanović M, Petrović I, Radovanović L, Ristić M, Speight GJ, Perišić V (2018) Application of a master meter system to assure crude oil and natural gas quality during transportation, petroleum science and technology, Taylor & Francis. Print ISSN: 1091–6466 Online ISSN: 1532–2459
34. Pricewaterhouse Coopers (PwC) (2013) Busting the carbon budget: low carbon economy index 2013. http://www.pwc.co.uk/en_UK/uk/assets/pdf/low-carbon-economy-index-2013.pdf
35. Radovanovic L, Perisic J, Milovanovic M, Speight JG, Bozilović Z, Momcilovic O, Obucinski D (2018) Modeling of petroleum products sampling processes. Pet Sci Technol 36(23):2003–2010
36. Ristic M, Radovanovic L, Prokic-Cvetkovic R, Otic G, Perisic J, Vasovic I (2017) Increase energy efficiency of thermal power plant Kostolac B by revitalization ventilation mills. Energy Sources Part B: Economics Plann Policy 12(2):191–197
37. https://europeanwesternbalkans.rs/finansiranje-energetske-tranzicije-u-srbiji/

Computational Fluid Dynamics and Strength Analysis of Composite UAV Wing

Zoran Vasić, Katarina Maksimović, Ivana Vasović Maksimović, Mirko Maksimović, and Stevan Maksimović

Abstract This paper considers computational fluid dynamics (CFD) analysis and experimental test strength verification of the UAV wing made of composite materials. The CFD calculation and static strength testing results of specific UAV wing are illustrated. Specific attention was given to the definition of aerodynamics loads acting on UAV wing and experimental tests of UAV wing. For a precise definition of aerodynamic loading of UAV in this work the commercial CFD software was used. The work aims to carry out one-way fluid–solid interaction (FSI) for UAV structural design, in which aerodynamics loads obtained from CFD analysis can be applied on the vehicle structure for steady-state static FE analysis. To determine precise aerodynamic loads of UAV in this research, the complete "tail-aft on booms" shaped configuration UAV is modelled and analyzed using CFD numerical simulation.

Keywords Unmanned aerial vehicle · CFD analysis · Composite materials · Strength tests · Structural analysis · Finite element method

Z. Vasić · S. Maksimović
Military Technical Institute, Ratka Resanovica 1, Belgrade, Serbia
e-mail: aig135@mts.rs

S. Maksimović
e-mail: s.maksimovic@mts.rs

K. Maksimović
City Administration of City of Belgrade, Secretariat for Utilities and Housing Services Water Management, Kraljice Marije 1, 11120 Belgrade, Serbia
e-mail: kmaksimovic@mts.rs

I. V. Maksimović (✉)
Lola Institute, Kneza Višeslava st.70a, Belgrade, Serbia

M. Maksimović
PUC Belgrade Waterworks and Sewerage, Kneza Milosa st.27, Belgrade, Serbia

N. Mitrovic et al. (eds.), *Current Problems in Experimental and Computational Engineering*, Lecture Notes in Networks and Systems 323,
https://doi.org/10.1007/978-3-030-86009-7_5

1 Introduction

The main goal in designing unmanned aerial vehicles is to get minimal weight. Due to a very good strength to weight ratio, composite materials are often used in aircraft structures [24]. Because of their specific strength and stiffness advantages along with others, composite materials are widely used in the design of unmanned air vehicles [21, 16]. From the manufacturing point of view, composite materials also have various advantages. Integration of different parts is simple in composite structures, and molded composite construction allows for strong structures that can be built without requiring expensive equipment and highly skilled assemblers.

Versatile computational analyses have become a necessity in order to optimize the weight of UAV's in the last decades. The finite element method (FEM) has been proven a computationally efficient method to solve aerospace structure problems. CFD analysis [6, 8] nowadays is so versatile, and it is usually used to investigate the flow of almost all types of vehicles, including UAV. To determine precise aerodynamic loads of UAV in this research, the complete "tail-aft on booms" shaped configuration UAV is modelled and analyzed using CFD numerical simulation.

2 Aerodynamic Computer Fluid Solver Analysis

The different methods and flow solver software were used for UAV development in our research. However, for this paper, the results of only two used flow solver software are shown, the USSAERO flow solver software and ANSYS FLUENT [1].

Relatively new methods that have been used in industry practice for aerodynamic estimation are linearized computer codes. Usually, these linearized computer codes can provide correct results only when the airflow around the aircraft is steady and does not contain any strong vortices. This is typically true only during cruising flight. For the purpose of this research software USSAERO is used for analysis of pressure distribution around 3D aerodynamic UAV shaped bodies and lifting surfaces. The software uses mathematical model of fluid flow based on linearized equation of potential, which is solved by panel method—WOODWARD method [20].

Computational fluid dynamics (CFD) has also rapidly become a key part of the aircraft design process. CFD is a phrase for a number of new computational techniques for aerodynamic analysis. It differs from traditional aerodynamic solvers by solving for the complete properties of the flowfield around the aircraft, rather than only on the surface of the aircraft. CFD codes are based upon the Navier–Stokes equations [5].

The validity of results solved in this way can be checked comparing with results solved by other numerical methods, self-empirical methods, or results obtained in wind tunnels and/or in flight (Fig. 1).

ANSYS FLUENT [1] flow solver, based on finite volume method, was also used in this study. A pressure-based type solver with the coupled scheme was used to

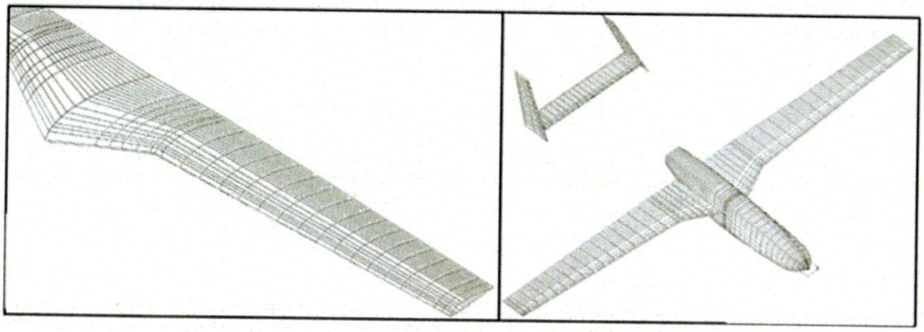

Fig. 1. The finite volume models; isolated UAV wing (left) with 960 panels; and UAV without tail booms (right) with 3754 panels [7]

compute the flow field. The implicit formulation (coupled scheme) with AUSM flux type was selected for the solution method. The Least Square Cell Based for gradient, Second Order for pressure, and the Second Order Upwind scheme for density, momentum, turbulent kinetic energy, specific dissipation rate, and energy were selected for spatial discretization [2, 3, 22]. Static pressure was calculated with respect to their appropriate total values according to flow field Mach number. All surfaces of the models were defined as a stationary no-slip adiabatic wall conditions. Menter's [17, 18] SST k-ω model was selected for the numerical calculation of the turbulent flow in the computational domain. For the air flow prediction, the Navier–Stokes equations have to be solved. The three-dimensional, time-dependent, RANS equations are discretized using a cell-centered finite volume approach [22]. The entire system of governing equations in conservation form, [2] can be given by Eq. (1):

$$\frac{\partial}{\partial t}\mathbf{U} + \frac{\partial}{\partial x}\mathbf{F} + \frac{\partial}{\partial y}\mathbf{G} + \frac{\partial}{\partial z}\mathbf{H} = 0 \tag{1}$$

where the column vectors **U**, **F**, **G** and **H** are defined as:

$$\mathbf{U} = \begin{Bmatrix} \rho \\ \rho u \\ \rho v \\ \rho w \\ \rho E \end{Bmatrix}$$

$$\mathbf{F} = \begin{Bmatrix} \rho u \\ \rho u^2 + p - \tau_{xx} \\ \rho v u - \tau_{xy} \\ \rho w u - \tau_{xz} \\ \rho u E + p u - q_x - u\tau_{xx} - v\tau_{xy} - w\tau_{xz} \end{Bmatrix}$$

$$\mathbf{G} = \begin{Bmatrix} \rho v \\ \rho u v - \tau_{yx} \\ \rho v^2 + p - \tau_{yy} \\ \rho w u - \tau_{yz} \\ \rho u E + p v - q_y - u\tau_{yx} - v\tau_{yy} - w\tau_{yz} \end{Bmatrix}$$

$$\mathbf{H} = \begin{Bmatrix} \rho w \\ \rho u w - \tau_{zx} \\ \rho v w - \tau_{zy} \\ \rho w^2 + p - \tau_{zz} \\ \rho u E + p w - q_z - u\tau_{zx} - v\tau_{zy} - w\tau_{zz} \end{Bmatrix} \quad (2)$$

where the column vectors F, G, and H are called the flux terms, and column vector U is called the solution vector. Here ρ, E and p are the density, total energy, and pressure, and u, v, w are velocity components of the fluid, respectively. In addition, τ_{ij} is the viscous stress tensor components, and q_i is the heat flux vector components.

The main control over the time-stepping scheme is the Courant number which was defined to be up to 200. The hybrid initialization method is used for each aerodynamic simulation initialization and represents the collection of instructions and boundary interpolation methods [1].

A fully transient flow solver has to be used in analyzing interference among the aircraft parts. The surface mesh for the UAV configuration is shown in Fig. 2.

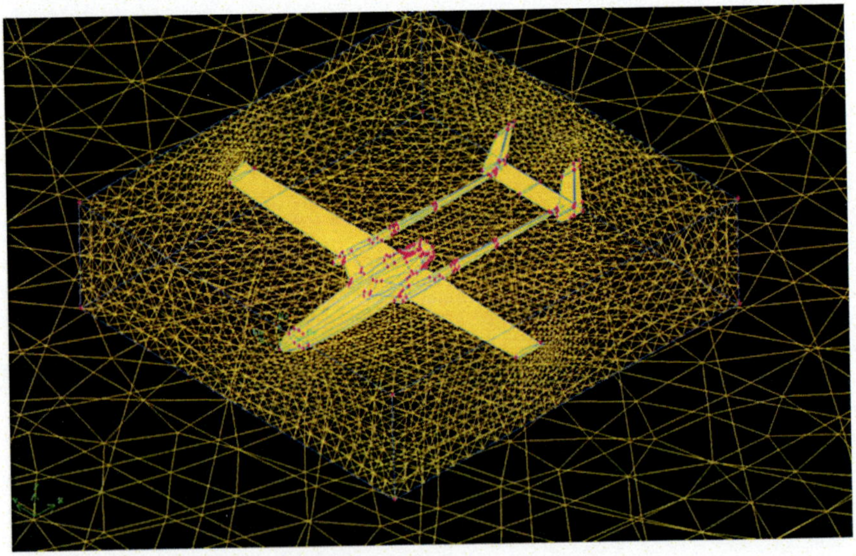

Fig. 2. Surface mesh on the whole UAV configuration

For the purpose of the analysis for this research, the standard no-slip boundary condition is applied on the solid walls. The Pressure-far field condition is set at the outer boundaries of the numerical domain. The unsteady coupled solver with second order implicit time stepping is used. It involves dual-time, i.e., physical-time step and pseudo-time-step. The implicit scheme is chosen for pseudo-time-marching with CFL equal to 5 and 30 inner iterations.

The S-A turbulence model is chosen because it is less sensitive to numerical errors when non-layered meshes and the near-wall gradients of the transported variables are much smaller than in the k-ε or k-ω models. Also, this model is the least expensive turbulence model explicitly designed for aerospace applications and is applicable for boundary layers with adverse pressure gradient appear. Although effectively a low-Reynolds-number model, the S-A enables resolving the viscous sublayer on fine enough meshes near the solid surfaces. In the simulation, the modified S-A model with wall functions is used because the centroids of the cells next to the surface are located within the log layer of the turbulent boundary layer.

2.1 Results of CFD Numerical Simulation

The main goal of the simulation is to determine precise aerodynamic loads acting on UAV for the purpose of stress and strength analysis. The instantaneous static pressure distribution on the whole configuration and details on the UAV and wing are shown in Figs. 3 and 4. Airfoil of the UAV outer part of the wing is Sd 7062 (max thickness 14%) along the wingspan with the chord length $L_{wr} = 754$ mm at the wing root and

Fig. 3. Static pressure distribution; whole UAV configuration

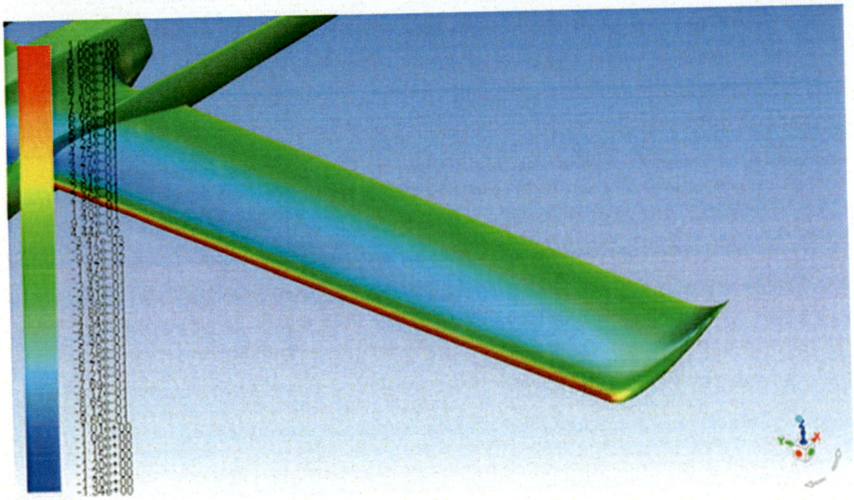

Fig. 4. Static pressure distribution on the wing

chord length $L_{wt} = 490$ mm at the wingtip. Angle of attack is $\alpha = 0°$ and Mach number is $M = 0.1301$.

In Fig. 5 the static pressure distributions on the upper part of the wing along the wingspan are presented successively for 1 m, 2 m, and 3 m from the UAV axis of symmetry.

According to calculated pressure distribution along wingspan, two different load cases were given and shown in Table 1, for the symmetrical "D" (*Dive speed*) load case with maximal UAV allowed speed given for outer part of UAV wing ($j = 1.00$), and in Table 2, for the symmetrical "C_{bmax}" load case in cruising speed and vertical gust (*Gust loads*—M_{max}), given for outer part of UAV wing ($j = 1.00$) (Fig. 6).

3 Experimental Static Testing and Comparison with Numerical Results

Structural elements of the UAV wing were tested for proof and ultimate loading conditions [4]. Deflections were measured at representative locations. Measurement of deformations and stresses is performed through a system of strain gauges. Special attention was dedicated to the deformation measurement of this structure. Figure 7 shows the composition of cloth stickers for introducing loads and displacement sensors for static strength testing of UAV wing structure.

Testing for one case load is done according to the procedure in the internal MTI Report where load factor is given on diagram ordinate. The first step in all tests is to introduce a small load into the structure to "settle" (annulment of gaps between

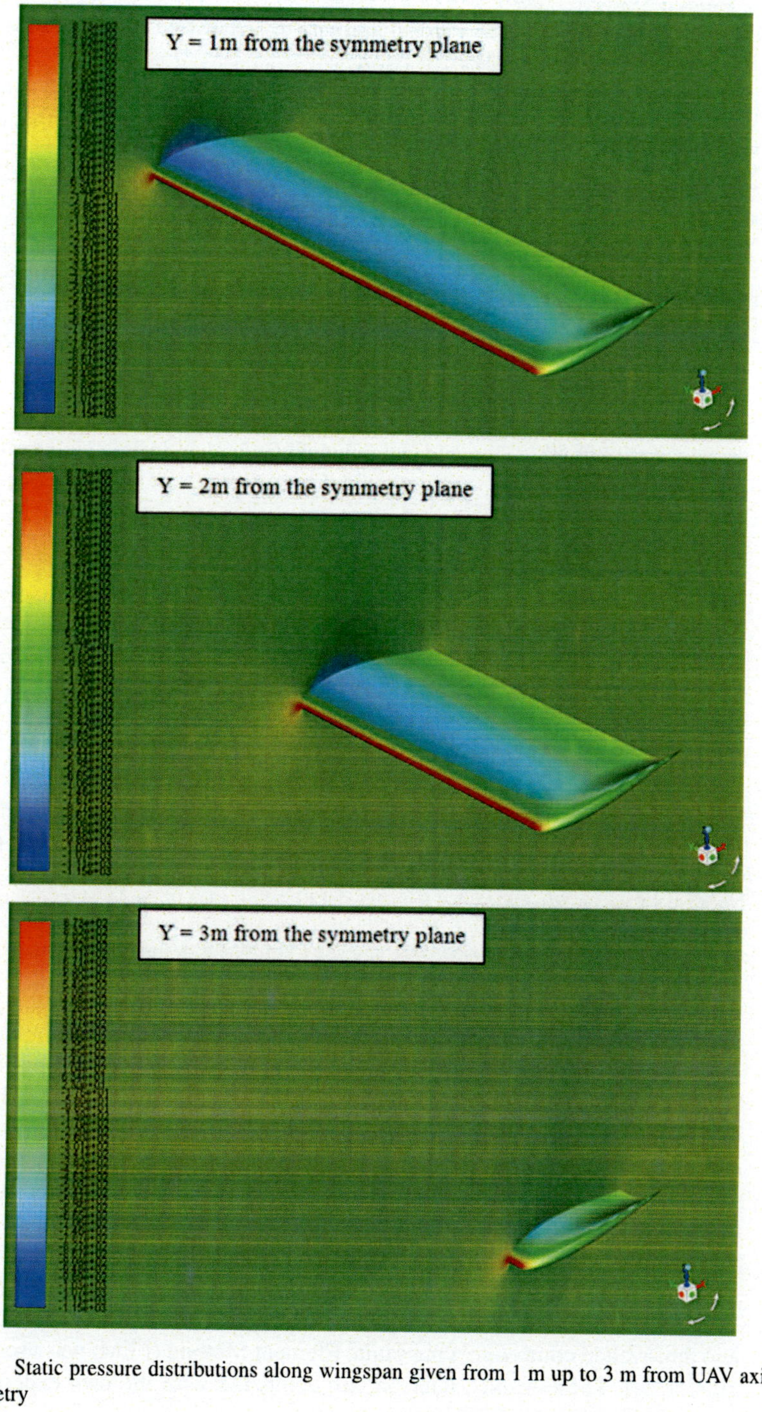

Fig. 5. Static pressure distributions along wingspan given from 1 m up to 3 m from UAV axis of symmetry

Table 1. Symmetrical "D" (*Dive speed*) load case with maximal UAV allowed speed given for outer part of the wing (j = 1.00)

Section	y	y	F_z	M_x	M_y
	% y_{max}	[m]	[daN]	[daNm]	[daNm]
1	100.00	3.170	0.0	0.0	0.0
2	97.91	3.104	2.0	0.0	−1.0
3	95.82	3.037	6.0	0.0	−1.0
4	91.26	2.893	15.0	2.0	−3.0
5	86.71	2.749	26.0	5.0	−6.0
6	82.16	2.604	39.0	9.0	−9.0
7	77.61	2.460	53.0	16.0	−12.0
8	73.05	2.316	68.0	25.0	−16.0
9	68.50	2.171	85.0	36.0	−20.0
10	63.94	2.027	102.0	49.0	−24.0
11	58.72	1.861	124.0	68.0	−29.0
12	53.50	1.696	147.0	90.0	−35.0
13	48.28	1.530	171.0	117.0	−41.0
14	43.05	1.365	197.0	147.0	−47.0
15	37.83	1.199	224.0	182.0	−54.0
16	32.61	1.034	252.0	221.0	−62.0
17	27.39	0.868	280.0	265.0	−69.0
18	22.16	0.703	310.0	314.0	−77.0
19	20.50	0.650	320.0	331.0	−80.0

Note Twisting moment is related to spar axis, which is in this case defined as 28% of local airfoil chord,
Negative values for $\mathbf{M_y}$ means that moment sinks leading edge of the local section relative to the twisting axis,
Dive speed is the maximum airspeed the airplane's airframe is designed to resist.

system elements). After settling, we begin proof load testing for a load factor of j = 1.1 and record stresses and displacements on previously determined positions on the structure. For a structure to pass the proof load test, it must have no residual stresses or deformations after load relaxation (no plastic deformations). Ultimate load test is performed after the proof load test. The load factor for the ultimate load test is j = 1.5. For the structure to pass the ultimate load test, it is necessary that it has no failure (plastic deformations are allowed). If structure passes both tests load factor is raised until effective failure is reached in order to determine margine of safety.

To validate the computation procedure in stress/strength analysis of composite structures the UAV composite wing was analyzed. For structural analysis of sandwich and thin-walled composite wing Finite Element Method (FEM) was used [9, 11–14, 19, 25, 26]. These results, later on, are compared with the own experimental test results [10, 15, 23, 27, 28].

Table 2. Symmetrical "C_{bmax}" load case in cruising speed and vertical gust (*Gust loads*—M_{avmax}), given for outer part of the wing (j = 1.00)

Section	y	y	F_z	M_x	M_y
	% y_{max}	[m]	[daN]	[daNm]	[daNm]
1	100.00	3.170	0.0	0.0	0.0
2	97.91	3.104	3.0	0.0	−0.0
3	95.82	3.037	7.0	0.0	−1.0
4	91.26	2.893	19.0	2.0	−3.0
5	86.71	2.749	33.0	6.0	−5.0
6	82.16	2.604	49.0	12.0	−8.0
7	77.61	2.460	66.0	20.0	−11.0
8	73.05	2.316	85.0	31.0	−14.0
9	68.50	2.171	105.0	45.0	−17.0
10	63.94	2.027	126.0	62.0	−21.0
11	58.72	1.861	152.0	85.0	−26.0
12	53.50	1.696	179.0	112.0	−31.0
13	48.28	1.530	207.0	144.0	−36.0
14	43.05	1.365	236.0	181.0	−42.0
15	37.83	1.199	266.0	222.0	−48.0
16	32.61	1.034	298.0	269.0	−55.0
17	27.39	0.868	330.0	321.0	−62.0
18	22.16	0.703	363.0	378.0	−69.0
19	20.50	0.650	375.0	398.0	−71.0

After the numerical stress analysis of the UAV composite structure, usually, experimental tests are undertaken. In this paper, the results of experimental tests of UAV wing structure were accomplished in order to validate the procedure of numerical stress analysis. It was carried out using a servo-hydraulic MTS system. Several test strain gauges and other sensors were placed on a composite wing. A three-dimensional CAD model of composite UAV wing with provided locations for sensors is shown in Fig. 8.

The manufactured composite wing structure with installed test strain gauges and displacement sensors (on the exact locations as in the CAD model) is shown in Fig. 9. Areas, where the test gauges are installed were of particular interest and attention during detailed numerical stress analysis where the results were outcome as critical. Data registered by test gauges are used for the final comparison of outcome results of stress analysis and experimental tests.

The aim of experimental tests was to check the static strength of the composite wing and other local structures for the pre-defined load case. Requested proof and ultimate load factors are $j_p = 1.15$ and $j_u = 1.5$, respectively. Load forces are introduced by means of a complex structure and eight cloth stickers, and it was carried

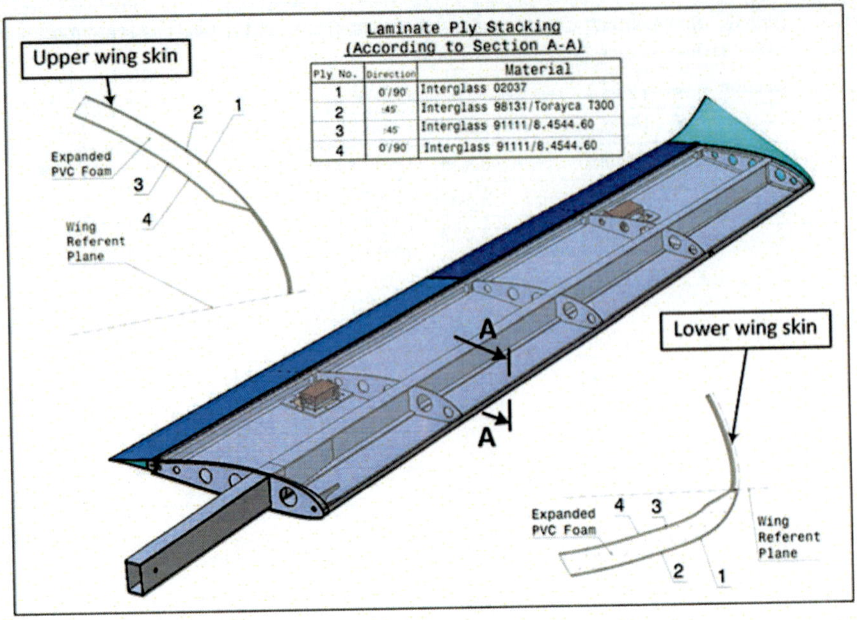

Fig. 6. Wing construction of UAVWing construction of UAV

out using a servo-hydraulic MTS system. Control measuring of the resulting force is made by an additional dynamometer. The deformations and stresses are measured through the system of strain gauges. Special attention was dedicated to the deformation measurement of this structure. Stresses were measured at 11 positions with eight strain gauges and nine rosettes (Triple strain gauges) (0°, 45°, 90°), and displacements at 12 positions with electronic displacement sensors (DS). Measuring points 11 and 12 were control points on j_{ig}. The Hottinger Baldwin Messtechnick type UGR 100 with actuator rated to 25 kN was used.

For the experimental test, the combined load cases were chosen as "C_{bmax+}" and "D". These two cases were defined as "C_{bmax+}" as the case with the highest bending moment and the highest transversal force, and case "D" as the case with the highest values of torque moment. Combined load case is with "C_{bmax+}" and "D" with the highest transversal force (T_z) and bending moment (M_x) from the "C_{bmax+}" case and the highest twist moment (M_y) from the "D" case (see two light grey columns in Table 1 and one light grey column in Table 2). Transversal force (T_z), bending moment (M_x), and twist moment (M_y) are shown together in Fig. 10.

During the determination of location for load acting points the special attention was paid to achieve the possible simulation of previously calculated load distribution as well as to be technologically rational. Figure 11 represents the (T_z) calculated load distribution and (T_{zu}) experimentally induced load distribution. Figure 12 illustrates the (M_y) calculated load distribution and (M_{yu}) experimentally introduced load

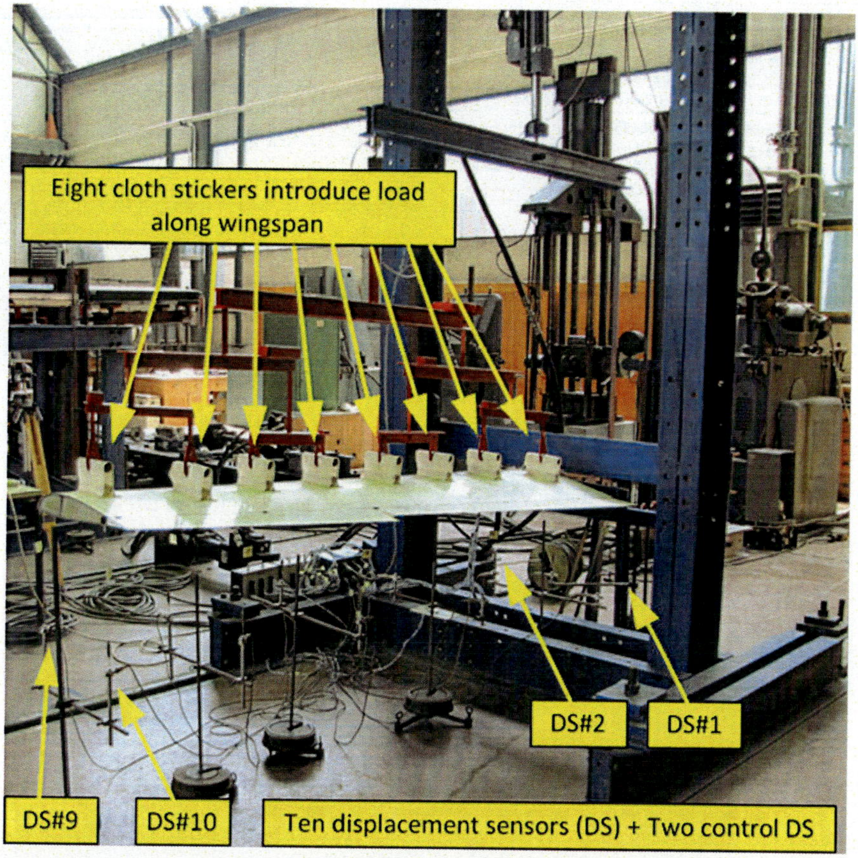

Fig. 7. Static strength testing of UAV wing with load forces introduced by means of eight cloth stickers; ten displacement sensors (DS) register deformation displacements

distribution for torque around the y-axis. Figures 11 and 12 also show values and locations of experimentally introduced loads along the wingspan.

Table 3 shows recorded micro dilatations measured by individual strain gauges for proof tests $j_p = 1.15$.

Recorded micro dilatations measured by triple strain gauges #9, #10, #11 and #15, #16, #17 as is shown in Figs. 9 and 10 are shown in Table 4.

Based on recorded micro dilatations for individual strain gauges, the values of tensile strength was calculated, and for rosettes, triple strain gauges, the values and directions of main tensile strength and shear strength, as well as the angles of main strength on all measured locations were also calculated. During calculation, the Young's Modulus E = 20,000 MPa, and Poisson's Ratio $\nu = 0.25$ were used.

Longitudinal tensile/compressive strength measured by individual stress gauges were presented in Table 5.

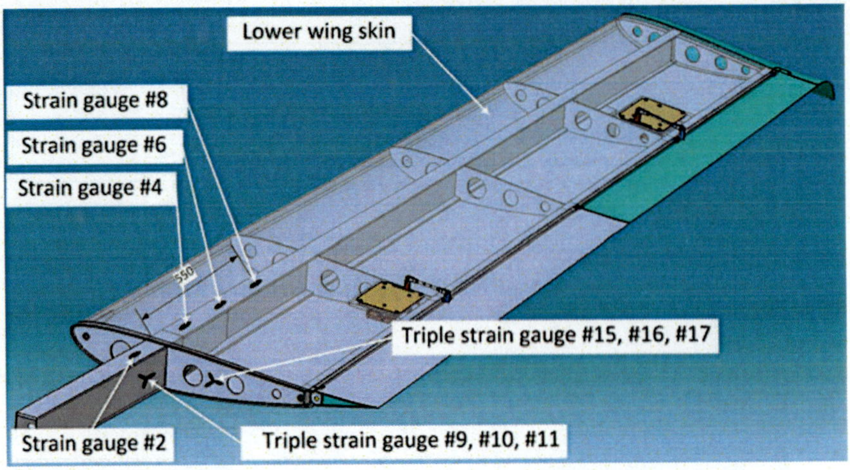

Fig. 8. CAD model of UAV composite wing with provided locations of test strain gauges

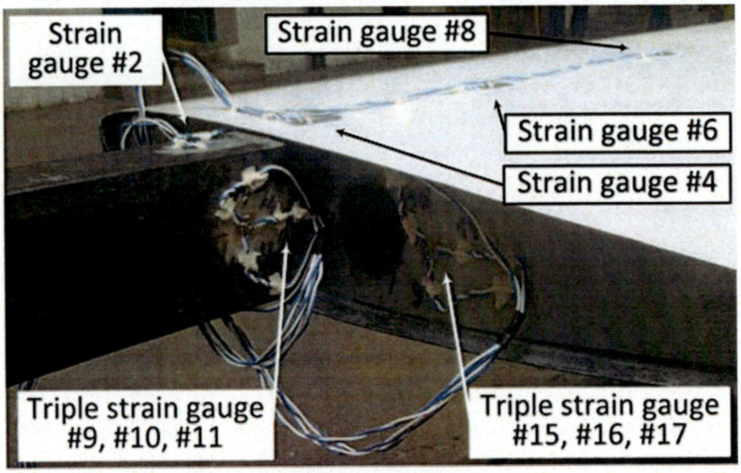

Fig. 9. Strain gauges installed on real wing structure for load tests

The bending line of the wing subjected to the load, as average values of recorded displacement on discrete wing sections is shown in Fig. 13 (cycle $j_p = 1.15$).

Angles of wing twist under the acting load (cycle $j_p = 1.15$) are shown in Fig. 14.

These two figures show that the structure well carries the acting loads, and there are no signs of structure failures. In addition, a good correlation of the results is achieved between the structural analysis and the experimental testing.

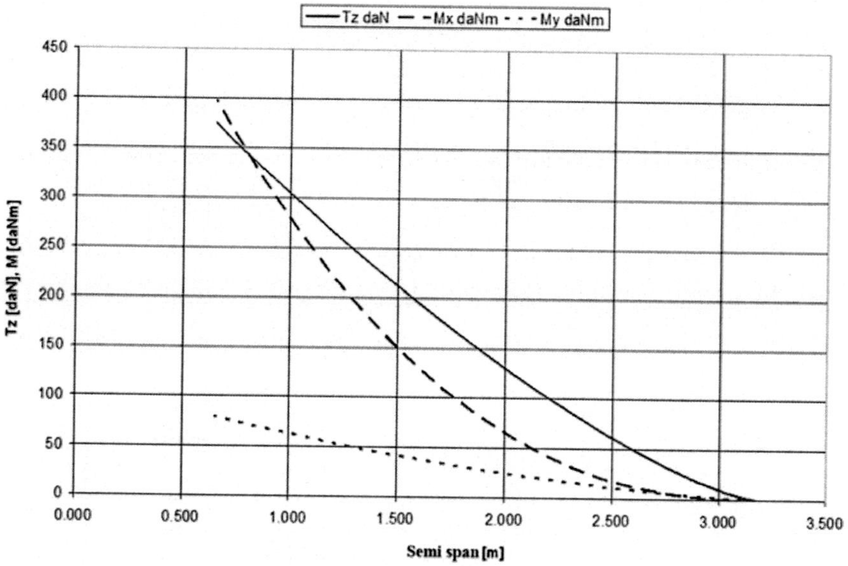

Fig. 10. Numerically calculated wing load (T_z, M_x, M_y) along wingspan

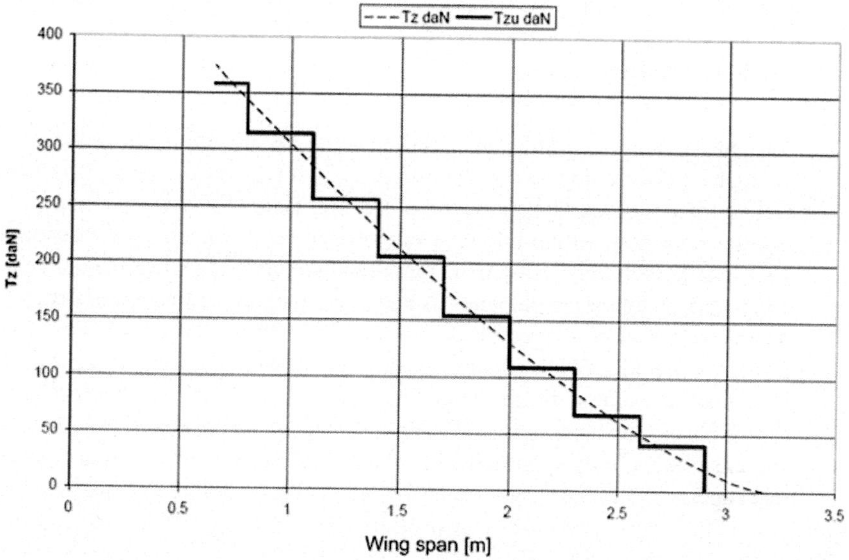

Fig. 11. Calculated load distribution (dotted line) and experimentally introduced load distribution (scattered line) for transversal wing force by means of eight cloth stickers

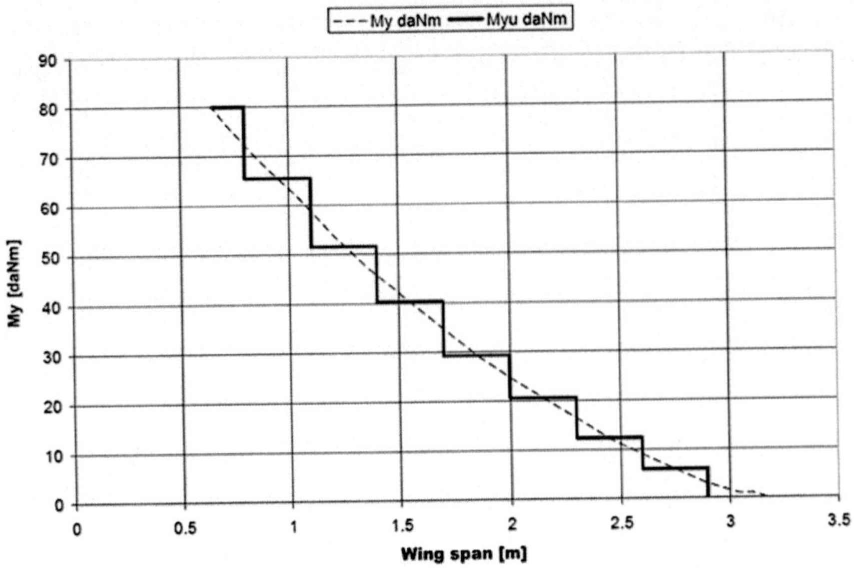

Fig. 12. Calculated load distribution (dotted line) and experimentally introduced load distribution (scattered line) for twist load around wing y-axis

4 Conclusions

This paper presents results of CFD analysis, and experimental testing of static strength of UAV composite wing structure. Two type problems are considered: (i) Determination of aerodynamic load of the complete UAV structure and (ii) Experimental static tests of tail booms. Computation results of pressure distributions on UAV that are obtained using CFD numerical simulation are used to define loads in computation stress/strength analysis and experimental verification of composite tail booms with respect to strength behavior. CFD analysis of the UAV was performed using ANSYS FLUENT software. Complete tests were performed for proof and ultimate load cases for symmetrical and nonsymmetrical load case, but for the purpose of this paper, only symmetrical load case is shown. To compare structural analysis and experiment, only symmetrical load case and proof load case was considered in this paper.

Table 3. Recorded micro dilatations measured by individual strain gauges (#2, #4, #6, #8 shown in Figs. 9 and 10) for proof tests $j_p = 1.15$

j	SG #2	SG #4	SG #6	SG #8
	mm/m	mm/m	mm/m	mm/m
0	6.6	9.9	12.0	13.3
0.21	166.9	132.8	166.1	133.4
0.41	319.1	239.5	310.4	244.6
0.62	467.0	339.5	451.0	353.5
0.72	539.5	387.0	519.6	406.6
0.82	610.9	433.5	586.2	458.1
0.92	685.5	481.8	655.6	512.2
1.02	758.3	526.3	723.2	563.8
1.07	792.7	548.7	755.2	588.6
1.12	828.2	570.9	787.9	614.0
1.17	872.2	596.5	824.6	642.0
1.02	762.3	520.7	717.3	557.9
0.82	629.7	435.1	588.4	460.1
0.61	484.8	337.2	449.6	352.4
0.41	331.9	235.0	306.2	241.0
0.20	174.7	128.9	161.8	129.2
0.00	12.5	12.7	17.1	18.3

Note Requested proof load factor usually is $j_p = 1.15$, but the construction was tested up to 1.17, which means 17% above working loads to be sure the construction does not have damages. All bolded data are maximal measured data

Table 4. Recorded micro dilatations measured by triple strain gauges #9, #10, #11 and #15, #16, #17

j	Triple strain gauge			Triple strain gauge		
	#9	#10	#11	#15	#16	#17
	μm/m	μm/m	μm/m	μm/m	μm/m	μm/m
0	14.5	14.6	1.9	3.7	3.9	4.2
0.21	−20.6	−43.9	−6.8	25.3	−183.9	−66.0
0.41	−56.5	−105.9	−17.9	48.1	−384.4	−111.5
0.62	−90.1	−150.6	−31.7	63.0	−522.0	−145.4
0.72	−105.5	−169.2	−38.5	69.7	−584.0	−160.9
0.82	−120.5	−187.0	−45.5	77.0	−649.4	−180.8
0.92	−135.7	−205.4	−53.0	84.8	−716.8	−201.4
1.02	−150.5	−221.1	−59.6	92.3	−787.4	−218.8
1.07	−156.4	−228.2	−63.4	95.9	−820.4	−231.3
1.12	−163.3	−236.4	−67.9	100.5	−858.7	−246.8
1.17	−171.5	−245.0	−71.6	108.7	−932.3	−272.7
1.02	−158.2	−233.4	−57.7	109.9	−886.5	−278.7
0.82	−117.6	−193.6	−42.5	107.4	−798.6	−272.5
0.61	−85.5	−156.9	−27.5	97.9	−688.6	−252.9
0.41	−50.1	−112.3	−12.0	81.8	−540.0	−210.5
0.20	−12.0	−57.7	1.4	55.0	−330.6	−133.3
0.00	31.4	27.3	−2.2	14.0	−5.7	2.7

Table 5. Longitudinal tensile/compressive strength (for individual strain gauges) (#2, #4, #6, #8 already shown in Figs. 5 and 6) for proof tests $j_p = 1.15$ (exactly 1.17)

j	σ_2	σ_4	σ_6	σ_8
	MPa	MPa	MPa	MPa
0	0.13	0.20	0.24	0.27
0.21	3.34	2.66	3.32	2.67
0.41	6.38	4.79	6.21	4.89
0.62	9.34	6.79	9.02	7.07
0.72	10.79	7.74	10.39	8.13
0.82	12.22	8.67	11.72	9.16
0.92	13.71	9.64	13.11	10.24
1.02	15.17	10.53	14.46	11.28
1.07	15.85	10.97	15.10	11.77
1.12	16.56	11.42	15.76	12.28
1.17	17.44	11.93	16.49	12.84
1.02	15.25	10.41	14.35	11.16
0.82	12.59	8.70	11.77	9.20
0.61	9.70	6.74	8.99	7.05
0.41	6.64	4.70	6.12	4.82
0.20	3.49	2.58	3.24	2.58
0.00	0.25	0.25	0.34	0.37

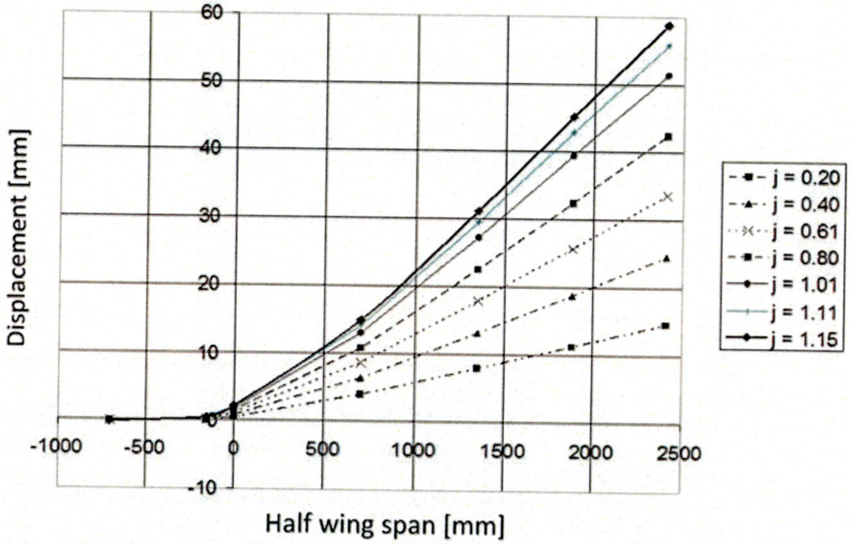

Fig. 13. Wing bending line under the acting load (cycle $j_p = 1.15$) (arithmetic means of recorded displacements values at discrete wing chords along wingspan)

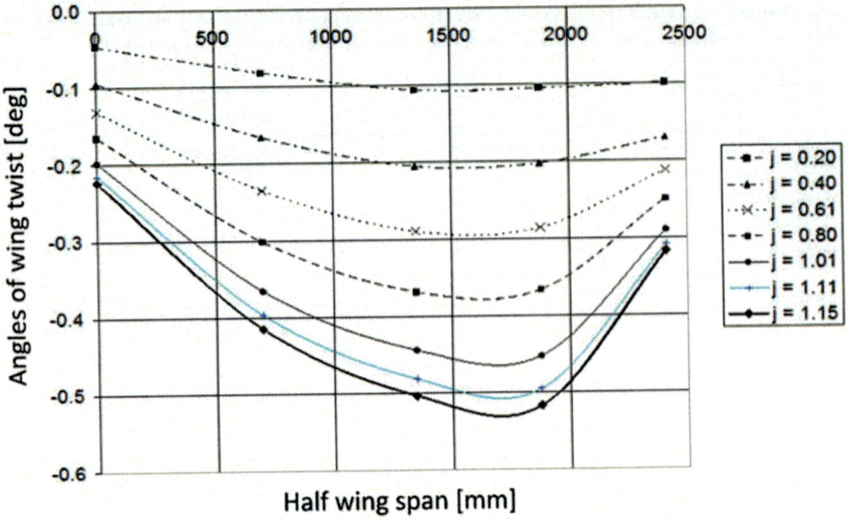

Fig. 14. Angles of wing twist under the acting load (cycle $j_p = 1.15$)

Acknowledgements This work was financially realized by the Ministry of Defense of the Republic of Serbia and has been supported by the research grants No. 451-03-9/2021-14/200066, of the Serbian Ministry of Education, Science and Technological Development.

References

1. ANSYS FLUENT User's Guide (2013) Release 15.0, ANSYS, Inc
2. Anderson JD (1995) Computational fluid dynamics. McGraw-Hill Inc., The Basics with Applications
3. Chima RV, Liou MS (2003) Comparison of the AUSM$^+$ and H-CUSP Schemes for Turbomachinery Applications. In: AIAA Paper 2003–4120, Proceedings of the AIAA 16th computational fluid dynamics conference and the 33rd fluid dynamics conference and exhibit, Orlando, Florida
4. Georgijević D et al (2014) Strength static testing of tactical UAV wing, (in Serbian), Internal MTI report 6597717, V4–3490-I-06, Belgrade
5. Gudmundsson S (2014) General aviation aircraft design: applied methods and procedures, Elsevier, ISBN: 978–0–12–397308–5
6. Jankovic D, Maksimovic S, Kozic M, Stupar S, Maksimovic K, Vasovic I, Maksimovic M (2017) CFD calculation of helicopter tail rotor airloads for fatigue strength experiments. J Aerosp Eng 30(5)
7. Kosanović N (2011) A summary of the results of the load calculation of the UAV wing, (in Serbian), Internal MTI report 6596033, V3–0558-II-06, Belgrade
8. Maksimovic S, Kozic M, Stetic-Kozic S, Maksimovic K, Vasovic I, Maksimovic M (2014) Determination of load distributions on main helicopter rotor blades and strength analysis of its structural components. J Aerosp Eng 27(6)

9. Maksimovic S (1994) Improved geometrically nonlinear finite element analysis and failure of fibre reinforced composite laminates. In: Brandt AM, Li VC, Marshall IH (eds) Proceedings of the International Symposium brittle matrix composites 4, Warsaw, Woodhead Publ
10. Maksimovic M, Maksimovic K, Stamenkovic D, Vasovic Maksimovic I (2020) Initial fatigue life estimation of welded structural components. J Tech Gaz 28(4) (Print: ISSN 1330–3651, Online: ISSN 1848–6339). https://doi.org/10.17559/TV-20200414015501
11. Maksimović S, Maksimović K, Vasović I, Maksimović M, Stamenković D (2019) Strength analysis of helicopter main rotor blade made from composite materials. In: 14th International Conference on accomplishments in mechanical and industrial engineering, 24.-25. maj. 2019., pp. 403–408, Banja Luka, Republic of Srpska, Bosnia and Herzegovina, ISBN 978–99938–39–85–9, COBISS.RS-ID 8166456
12. Maksimović M, Vasović I, Maksimović K, Maksimović S, Stamenković D (2018) Crack growth analysis and residual life estimation of structural elements under mixed modes. In: 22nd European Conference on fracture - ECF 22, Belgrade, Serbia
13. Maksimović S, Maksimović K, Vasović I, Đurić M, Maksimović M (2018) Residual life estimation of aircraft structural components under load spectrum. In: 8th International Scientific Conference on defensive technologies, OTEH 2018, Belgrade, Serbia
14. Maksimović M, Nikolić-Stanojević V, Maksimović K, Slobodan Stupar S (2012) Damage tolerance analysis of structural components under general load spectrum. Tehnicki Vjesnik-Technical Gazette 19(4):931–938
15. Maksimović S, Vasić Z, Georgijevićć D (2017) Strength analysis of wing construction of tactical unmanned vehicle made from composite materials. In: Proceedings of the materials and posibility of its applications, Požarevac
16. Maksimović S (2006) Failure analysis of layered composite structures: computation and experimental investigation. In: 16th European Conference of fracture, ECF 16, Alexandroupolis, Greece
17. Menter FR (2009) Review of the shear-stress transport turbulence model experience from an industrial perspective. Int J Comput Fluid D 23(4):305–316
18. Menter JC, Ferreira T, Konno EB (2003) The SST turbulence model with improved wall treatment for heat transfer predictions in gas turbines. In: Proceedings of the international gas turbine congress, Tokyo, Japan.
19. Msc/NASTRAN Theoretical Manuals
20. Raymer D (1999) Aircraft design: a conceptual Approach, 3rd edn. American Institute of Aeronautics and Astronautics Inc., Reston, Virginia
21. Tsai SW, Wu EM (1980) A general theory of strength for anisotropic materials. J Comp Mater 5:58–80
22. Tu J, Yeoh GH, Liu C (2008) Computational fluid dynamics. Elsevier, A Practical Approach
23. Vasic Z, Maksimovic K, Maksimovic M, Vasovic I, Vidanovic N, Simonovic A (2021) Buckling and postbuckling behavior of shell type structures under thermo-mechanical loads. Therm Sci ISSN 0354–9836; eISSN 2334–7163. https://doi.org/10.2298/TSCI201129079V
24. Vasić Z, Maksimović S, Georgijević D (2017) Applied integrated design in composite UAV development. Appl Compos Mater ISSN 0929–189X. https://doi.org/10.1007/s10443-017-9611-y
25. Vasovic I, Maksimovic S, Maksimovic K, Stupar S, Bakic G, Maksimovic M (2014) Determination of stress intensity factors in low pressure turbine rotor discs. Math Probl Eng 304638:9. https://doi.org/10.1155/2014/304638
26. Vasovic I, Maksimovic S, Maksimovic K, Stupar S, Maksimović M, Bakic G (2014) Fracture mechanics analysis of damaged turbine discs using finite element method. Therm Sci 18(1):S107-S112
27. Vasovic I, Maksimovic M, Maksimovic K (2020) Residual fatigue life estimation of structural components under mode-I and mixed mode crack problems. In: Mitrovic N et al (eds) Computational and experimental approaches in materials science and engineering, Springer Nature Switzerland AG 2020; CNN Tech 2019, LNNS 90, pp 3–21. https://doi.org/10.1007/978-3-030-30853-7_1

28. Vidanović NN, Rašuo BB, Kastratović G, Grbović A, Puharić M, Maksimović K (2019) Multidisciplinary shape optimization of missile fin configuration subject to aerodynamic heating. J Spacecraft Rockets. https://doi.org/10.2514/1.A34575

Investigations and Results Analysis of Key Parameters of Vehicle Tracking System

Djordje Dihovicni, Nada Ratković Kovačević, Zoran Lalić, and Dragan Kreculj

Abstract In this research the logistic data obtained from four companies from Eastern Serbia are analyzed, and the decision model is created considering different aggregating operators. One point of view has included application of GPS system for monitoring and measuring important parameters for vehicles. Companies from Western Serbia have implemented this kind of approach to the asset monitoring. They have applied a GPS, vehicle and machine tracking system, and thus had gained a competitive advantage. Using the GPS system, the parameters for the vehicles could be processed, such as: time interval, mileage for a given interval, fuel consumption per probe, and fuel consumption per computer board, as well as average fuel consumption, and amount of refueling—filling, amount of drained fuel and number of discharges, maximum travel speed, average speed, average engine speed, and the effective operating hours. Another approach includes creation of decision model which depends on various aggregated operators. For the chosen criteria such as quantity of gasoline drained monthly from the vehicle tank, quantity of gas drained monthly from the vehicle tank, and number of kilometers travelled monthly. The resulting weight is calculated applying SWARA method. The degree of importance is presented in the appropriate tables, taking into account different aggregation operators, such as OWA, IOWA and OWAWAIMAM. The optimal results for each company are presented and SWARA method and different aggregate functions are implemented. This approach has shown that decision making strategies depend on adoption of adequate method.

Keywords Decision making · Artificial Intelligence · System for asset tracking · Fuzzy logic · Logistics

---

D. Dihovicni (✉) · N. R. Kovačević · Z. Lalić · D. Kreculj
The Academy of Applied Technical Studies Belgrade, Blv. Zorana Djindjica 152 a, 11070 New Belgrade, Serbia

N. Mitrovic et al. (eds.), *Current Problems in Experimental and Computational Engineering*, Lecture Notes in Networks and Systems 323,
https://doi.org/10.1007/978-3-030-86009-7_6

1 Introduction

Due to the need of the companies for modernization, reduction of costs, increase of productivity, improvement of working conditions and the other needs, it is necessary to apply methods for solving this kind of problems [1, 2]. One point of view has included application of *GPS* system for monitoring and measuring important parameters for vehicles [3–5].

Some companies from Western Serbia have implemented this kind of approach to above mentioned problem. They applied *GPS*, vehicle and machine tracking system, and thus has gained a competitive advantage. Good organization of transport implies precise planning and good exploitation of means of transport [6–8].

Precise planning requires a thorough study of transport requirements and conditions, under which transport should be organized in the next period of time with an analysis of the achieved results of the vehicle in the previous period. For the analysis of the achieved results of work of the vehicle fleet, a system of indicators of the work of transport means is used, which define all the elements in the process of the work of the vehicle fleet. Installation, monitoring and maintenance of the *GPS* systems require large investments, so this capital investment should justify the invested financial resources.

The data in this paper are obtained by reviewing the professional literature, scientific journals, analyzing and developing various mathematical models [9–11], through personal experience and many years of practice, all in order to achieve optimal decision [12]. The methods used in making the paper rely on reports and diagrams without which this system would not be able to function. The basis of this paper is the analysis which provides information that is a strategic resource in a modern company [13, 14]. The quality of decisions made, depends on the quality of the available information and system stability, [15–17].

The other point of view has included creation of decision model based on aggregation functions and relative weight, [18–20]. The most frequent operator used in the literature is the weighting average, and it is connected with different engineering, mathematical and economical problems. [21, 22].

On contrary of *GPS* approach, in this paper the data obtained from several Eastern Serbian companies are analyzed, such as drained gasoline and gas from the vehicle tank, as well as number of travelled kilometers, and fuzzy logic approach is implemented [23].

2 GPS Vehicle Tracking System

The implementation of *GPS* system for tracking vehicles from companies from Western Serbia is presented.

Using the *GPS* system, the parameters for the vehicles could be processed, such as: time interval, mileage for a given interval, fuel consumption per probe, and fuel consumption per board of the computer, and as well as average fuel consumption, amount of refueling—filling, amount of drained fuel and number of discharges, maximum travel speed, average speed, average engine speed, and the effective operating hours. All these parameters are processed in diagrams and tables.

With this data, the operator daily controls the operation of the vehicle, the operation of the driver, and monitors the production process [24].

The software part of this system is represented by 2 types of applications depending on the purpose of the user: client desktop application, and WEB application. A hardware part is installed in the vehicle, which is programmed to adapt to the user. The system constantly records the position of the vehicle and its operation, its speed, the number of kilometers traveled, and the condition of the sensor. All data are sent to the center, via *GSM* network, via *GPRS* data transfer, where this data is analyzed separately for each vehicle. Such monitoring of vehicles and drivers is very useful for logisticians, because constant monitoring of vehicles allows them to react immediately to events on the streets/routes. This system helps drivers to choose the fastest, shortest route, and offers them faster assistance in the field, and on the other hand, through supervision, enables the fleet operators to introduce programmed driving that leads to lower fuel consumption and better vehicle use.

The *GPS* logger collects the location of the device at specific time intervals in the internal memory. Modern *GPS* loggers have either a memory card slot or an internal memory card and a usb port. It is possible to download data from the records for further computer analysis. Movement history or points of interest can be in *GPH*, *KML*, *NMEA* or other similar format. Many digital cameras capture the time when a photo is taken. As the camera clock is accurate, it can be connected to *GPS* data, to provide an accurate location. This can be added to the metadata in the image file. Cameras with a built-in *GPS* receiver produce directly tagged photos.

The *GPS* navigation device and the mobile phone use the same types of batteries. At certain time intervals, the phone sends a message via *SMS*, which contains data from the *GPS* receiver. Newer GPS integrated mobile phones that use *GPS* tracking software can be used in a device called a "Data pusher" or "Data logger", or a data logger or data sender.

The *GPS* tracking devices enable the so-called. "Push" technology, allowing sophisticated *GPS* tracking in business environments, specialized organizations such as commercial, vehicle and machine parks, and GPS "live tracking" are used for commercial services, which generally refers to systems that are refreshed at certain intervals every 5, 10 or 15 min. Some systems combine time-defined updates with major updates prompted by certain changes.

In contrast to the data transmission system, which sends device positions at defined time intervals, *GPS* data retrieval devices are always functional and used wherever required. This technology is not in mass use, but one type is, a computer connected to the Internet and running *GPSD*. This method can be used when the location of the vehicle or machine being monitored is needed only occasionally.

The way the vehicle tracking system works is presented in Fig. 1.

From the appropriate report, data are obtained, and they help the operational work organization, because before this monitoring of the system, there is no possibility to

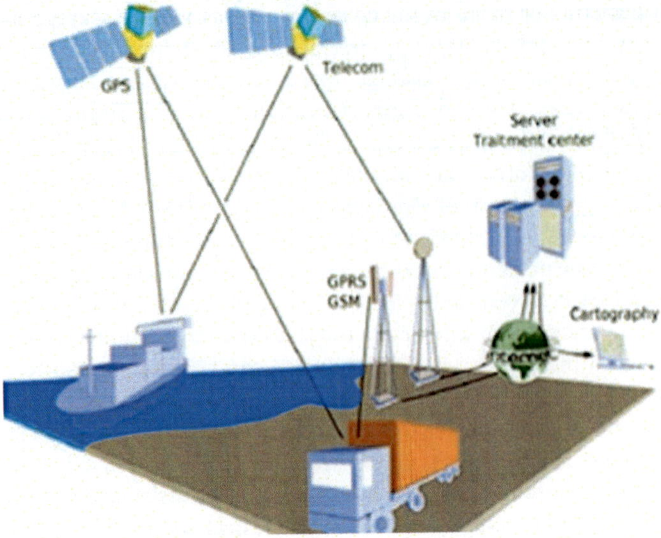

Fig. 1. The way the vehicle tracking system works

get an accurate insight into this data. On the basis of effective hours, the personal income of workers in the work organization is calculated, and with these data the number of accounting workers has decreased and thus the savings for the company has increased. Another important parameter refers to the hours of work in the place, because with this information, it should be possible to get multiple savings in fuel consumption and greater work efficiency. The report provides a display of mileage by position (constant driving without breaks) and total mileage [25].

The road report provides information such as: starting location, duration of effective driving, final location as well as road name and mileage. The parking report helps the operators in the company to be able to predict how many detentions there are on construction sites, and thus to predict and on the basis of these data to plan the work process and achieve higher productivity [26]. From the parking report, it can be seen which vehicle must perform parking after a certain number of working hours in a day of parking, which is subject to a legal obligation.

The stop report aims to show all stops during a given interval, recording stops for a maximum duration of five minutes [27]. This indicator is widely used in the control of the work of vehicles that perform work in a place without movement, such as trucks that work on excavations, and are related to construction machinery, so it should be seen the work of both vehicles and machines.

The refueling report provides the following information: fuel injection time, location, initial level and amount of fuel filled into the tanks. In order for the data on the amount of refueled fuel to be accurate, all tanks must go through a calibration process. With this report, workers which work at pumping stations would have a great relief, and there is not necessary to keep manual records, because all the data are visible in the table of refueling, which are summarized at the end of the working month.

The system enables control of fuel consumption by measuring the fuel level in the tank, by installing a fuel level sensor in the vehicle tank. In this way, the current fuel level in the tank is measured and monitored at all times. Each sudden drop in fuel level is recorded in the system, after which an additional analysis of the appropriate diagrams of fuel levels and vehicle operation in the system is executed. It is possible to determine whether the recorded drop in fuel level is due to changes in slope and terrain configuration, fuel fluctuations in the tank during the movement of the vehicle, or misuse or leakage of fuel from the tank.

The speed diagram shows the mode under which the vehicle is operating, and it is determined whether the speed and engine speed are matched. This diagram follows the data relating to altitude, starting point and executive point of vehicle movement. This data is obtained from the trip computer.

There is a serial port on the device through which this function can be realized, this information helps a lot to monitor the operating mode and the load of the vehicle. Through the number of revolutions, it can be seen in what conditions the vehicle operates, taking into account that due the mismatch of the number of revolutions leads to increased fuel consumption. When the vehicle is running at high speed, the vehicle engine and other supporting elements on the vehicle wear out. When the worker in charge of monitoring vehicles and machines notices some irregularities related to the increased number of revolutions, he immediately informs the driver to correct the driving mode.

Board computer, is a trip computer located in the vehicle and gives the driver all the necessary information related to engine operating temperature, engine speed, speed, operating hours, battery voltage, fuel level in the tank and fuel consumption.

The system is connected to the board computer so that data can be seen at any time, as well as monitoring of all events in the vehicle. Altitude data is very important in the analysis of fuel consumption, vehicle load and route planning.

Vehicle trajectory report allows the crew to run a vehicle movement simulation for a given interval, monitor vehicle operation, stop it and park it. The use of this simulation has reduced the misuse of work equipment as well as the cargo being transported. When starting the movement simulation, the speed at which the vehicle would move can be determined, which helps if some controversial situations occur, and some omissions can also be noticed in choosing the route itself, determining a new route and thus increasing productivity.

The engine temperature report, is essential for monitoring the condition of the propulsion unit on the vehicle. If it is established that the limit values deviate from the allowed parameters, then the operators react in a timely manner, the vehicle is shut down and sent to a repair shop. This report is linked to the vehicle speed and speed report. Before shutting down the vehicle, it must be established whether the increased temperature is due to uncoordinated driving, or whether it is a technical problem on the vehicle.

The system provides workers who run the fleet and machinery fleet with technical support, in terms of sending messages related to control against fire extinguishers, regular vehicle and machine services, registration and six-month inspections. This data greatly helps the operational work, speeds up the work process and reduces costs in the company itself.

The control against fire extinguishers is performed in the company by a fire protection officer who has the task of sending the extinguishers to the service every six months. All appliances in vehicles and machines, as well as appliances in buildings, must go to an authorized service center. In case this is not done, the penalty for the company is huge as well as for the person in charge. The report of the *PP* device gives the time of the last service, and the report service interval gives information on how much time is left until the device is sent for service. With the introduction of this system, the possibility of failure is minimized.

The report regular service has the task to indicate when it is necessary to do regular service on the vehicle. The mileage depends on when the regular service will be performed, and the mileage interval is determined depending on: type of the vehicle, year of production, quality of spare parts, type of oil and lubricants. It is possible to see at any time when it is necessary to perform the service, since the system records the mileage when the last service is performed and shows the current mileage of the vehicle. With this report, the number of operators in charge of monitoring the company's fleet has been reduced as well as the administration that monitors this work. It is also important that it is possible to plan when and in what period the product will be shut down, so there are no unforeseen situations, and with this precise data the efficiency of the work itself is greatly increased.

The vehicle registration report aims to warn when it is necessary to register for a given vehicle. If accurate records are not kept, and there is a failure in the registration period, the penalties for the company are high. The vehicle and machine registration officer received messages on a daily basis, thus eliminating any possibility of a failure.

Special purpose vehicles are subject to a six-month technical inspection. In addition to regular registrations, the vehicle registration officer is obliged to send special purpose for six-month inspections, in order to avoid failures in the operation of the system, and informs when it is necessary to perform an inspection in a particular case, which is 121 days.

There is a possibility that the *GPS* system sends messages about the loss of the satellite, the overspending of the vehicle in place and the expiration of fuel. Messages are sent via Gmail and *SMS*, and these notifications enable faster response and resolution of new problems in the shortest possible time.

The loss of the satellite represents the inability to track vehicles or machines (current position and all other parameters provided by the system). The message is important because if the connection to the satellite is not established for a long period of time, then there is a malfunction with the monitoring device and in that case the service is called to repair the system and thus the time for which the device will not work is reduced to a minimum. This message is activated in the notification report, and must always be active upon the recommendation of the authorized service.

Modern *GPS* systems have the ability to track the engine speed, and thus register the operation of the vehicle in place. This report enables the correction of a bad driver's habit by specifying in the program the time interval of vehicle operation in a place, that is approximately 10 min, and it is notified with a message. By sending these messages, multiple savings are achieved for the company, in the practice so far, and it has been shown that drivers depending on the job position, work with the vehicle on a daily basis and up to two hours in place more than they should. In this way, large costs are made for the company on a monthly basis, especially when there is a larger number of working means, and it is calculated on average two liters of fuel consumed per hour of vehicle operation in the place.

The GPS system allows operators to track every fuel leak from the tank. The minimum amount of drained fuel that the system will register depends on the volume of the tank, the type of vehicle, the machine and the conditions in which the vehicle will operate.

A message received via e-mail or text message, allows the operator to react quickly and determine whether there is abuse or a technical problem on the tool, and this check is determined from fuel discharge diagram. The time interval for sending messages should be as short as possible, so the operators would be able to react in a timely manner and if possible in the field where the vehicle is operating, and immediately determine the type of fuel leaking from the tank. In the practice so far, it is most often an abuse that is severely punished.

The tank lid opening report gives the exact time and location when the tank lid is opened. This message provides information that can be determined, whether it is an abuse of the means of work or the cover is removed due to the fuel being poured into the tank.

The report shows that the tank lid is opened for refueling, so that it can be determined exactly why the tank lid is opened, and the refueling report is compared in parallel. If there is a discrepancy in the reports then it is known that there has been an abuse. Also, all these changes are monitored on the fuel leak report. Workers are forbidden to open the tank lids, they can only do that if refueling is performed.

The report on nearest vehicles aims to show the nearest vehicle and machine on the construction site. It is possible to determine the number of vehicles shown in the report, and it is usually from five to ten vehicles and machines. This report provides a quick overview of where the vehicles and machines are located, how many kilometers they are from the target of the search, and the time determined by the search in this particular case is one hour, so it is possible to move from construction sites to other locations, and thus the time for relocations is greatly shortened so that there are no delays, and the employment of vehicles and machines is increased. The report on the nearest vehicles increases the profit of the entire company for the above reasons.

The distance determination report has the task of making the work easier for the operatives in the company, in the sense of determining the distance that is calculated in kilometers, and in that way it is possible to quickly determine the mileage that the vehicle has to go to the given point.

Distance reports presented the given route with the calculated mileage, and this report is very important in situations when the vehicle has more than one given location where it should work, so with this report it can be determined the nearest route.

When it is necessary to determine the area of the construction site on which the works are performed, the report of determining the area is used, which gives and calculates the area in meters and square kilometers. This report largely gives the company's operations a realistic and true picture of the size and scope of work. When this data exists, it is much easier to determine which machine should be employed and how many means of work are needed to do the given job, while respecting the set time limit.

The area determination report is widely used in quarry work, where it is widely used in determining the amount of material, determining the area to be exploited, as well as controlling and assessing the state of stocks.

Of great importance for work and reducing transport costs, and improving the work process has a trajectory report. This report works by setting a certain number of points (construction sites specifically) that the vehicle should cross, and the report finds the optimal path between the given points, and more precisely optimize routing. This optimal trajectory helps fleet and machine park managers determine the most cost-effective trajectory.

This program avoids paying tolls by marking in the report for example, and determines the routes in which there is no driving on toll roads.

The given path is used to determine whether they are cost-effective for the company in terms of saving all costs as well as reducing the time set for the execution of certain tasks.

In addition to the standard equipment used for installation in vehicles and machines, flow meters are also used, devices that are installed in case there is no technical possibility of installing probes in the tanks of vehicles and machines. It happens that due to the complexity of the work, there is no possibility to physically approach the tank, as well as due to the length of the probe, installation is not possible, and in order to do that, it is necessary to disassemble the vehicle. With machines, it is a much more complex job that takes a lot of time, and in a large number of cases, that job must be performed by an authorized service. This business requires a lot of investment and the company should allocate large financial resources, also the vehicle and machines would not work for a long period of time, and that also makes a big expense for the company. To avoid all these problems, the installation of a flow meter is performed. This installation is less complicated and is done by cutting the fuel supply and discharge pipes and placing them in the flow meter. This device is not as precise as a probe, and it is impossible to react quickly if there is any abuse regarding the discharge of fuel from the tank, but on a monthly basis the total amount of fuel consumed by the engine can be determined. On the basis of these data and the data on the amount of refueled fuel from the pump, valid data can be obtained. The difference between the two reports provides information on whether there is any misuse of funds, but on a monthly basis, the total amount of fuel consumed by the engine and which must pass through the flow meter can be determined, as well as the return fuel. Based on these data and data of fuel pumped from the pump, valid data can be obtained. The difference between the two reports provides information on whether there is any misuse of funds.

Before the installation of these devices, there is no possibility of monitoring, and if it is used, then the higher productivity is achieved, because frequent stops are an indicator that something needs to change in the process of work, and thus the number of effective hours increases.

3 Implementing of the SWARA Method

In this study, our objective is to calculate relative weights of the mean criteria for drained gasoline and gas from the vehicle tank, as well as travelled kilometers, in the first step by implementing *SWARA* method (Step-wise Weight Assessment Ratio Analysis) [28].

The *SWARA* method is often applied for two reasons. Firstly, *SWARA*'s perspective is different from other similar methods—this method gives the chance to decision makers to select their priority based on the current situation of environment and economy, [29].

Secondly, the role of the experts is very important in *SWARA* method. Experts have a key role in process of decision making on very important projects. Finally, SWARA has the advantage of more logical calculation of weights and relative importance of criteria [30].

The research is conducted for four companies in Eastern Serbia. The average monthly quantity of gasoline and gas drained from vehicle tank is shown in the Table 1, and in the Table 2, and average of travelled kilometers is shown in the Table 3.

In the next tables, the mean criteria for average monthly quantity of gasoline and gas drained from the vehicle tank, and number of traveled kilometers for four different companies in region of eastern Serbia, is presented in descending order (Tables 4, 5 and 6).

Table 1. The average monthly quantity of gasoline

Company	A_1	A_2	A_3	A_4
Average monthly quantity(l) of gasoline	40.41	25.83	49.51	27.25

Table 2. The average monthly quantity of gas

Company	A_1	A_2	A_3	A_4
Average monthly quantity(l) of gas	55.88	65.89	61.00	68.36

Table 3. The average monthly number of travelled kilometers

Company	A_1	A_2	A_3	A_4
Average monthly number of travelled kilometers	1397.25	1245.00	1008.83	1210.56

Table 4. The mean criteria for average monthly quantity of gasoline

Criteria	Mean
A_3	49.51
A_1	40.41
A_4	27.25
A_2	25.83

Table 5. The mean criteria for average monthly quantity of gas

Criteria	Mean
A_4	68.36
A_2	65.89
A_3	61.00
A_1	55.88

Table 6. The mean criteria for average monthly number of travelled kilometers

Criteria	Mean
A_1	1397.25
A_2	1245.00
A_4	1210.56
A_3	1008.83

The relative importance of the mean criteria for average monthly quantity of gasoline and gas drained from the vehicle tank, and number of traveled kilometers for four different companies in region of eastern Serbia, is presented in descending order, in Tables 7, 8 and 9.

The determination of coefficient kj is presented in the next equation:

Table 7. The relative importance of the mean criteria for average monthly quantity of gasoline

Criteria	$\tilde{s}_j$
A_3	
A_1	0.816
A_4	0.674
A_2	0.947

Table 8. The relative importance of the mean criteria for average monthly quantity of gas

Criteria	$\tilde{s}_j$
A_4	
A_2	0.963
A_3	0.925
A_1	0.916

Table 9. The relative importance of the mean criteria for average monthly quantity of travelled kilometers

Criteria	$\tilde{s}_j$
A_1	
A_2	0.891
A_4	0.972
A_3	0.833

$$OWA\left(a_1a_2\ldots\ldots\ldots a_n\right) = \sum\nolimits_{j=1}^{n} b_j\Big) \tag{1}$$

The weight q is resolved with the Eq. 2:

$$\hat{q}_j = \begin{cases} \tilde{1} & j = 1 \\ \frac{\hat{x}_{j-1}}{\hat{k}} & j > 1 \end{cases} \tag{2}$$

The Eq. 3 has solved relative weights of the evaluation criteria:

$$\hat{w}_j = \frac{\hat{q}_j}{\sum_{k=1}^{n} \hat{q}_k} \tag{3}$$

In the Tables 10, 11 and 12 it is given the summary of the represented weight criteria for monthly quantity of gasoline drained from the vehicle tank, and as well monthly number of travelled kilometers.

Table 10. The summary of the mean criteria for monthly quantity of gasoline

Criteria	$\tilde{s}_j$	$\tilde{k}_j$	$\tilde{q}_j$	$\tilde{w}_j$
A_3		1	1	0.488
A_1	0.816	1.816	0.550	0.268
A_4	0.674	1.674	0.328	0.160
A_2	0.947	1.947	0.168	0.084
			2.046	1.000

Table 11. The summary of the mean criteria for monthly quantity of gas drained from the vehicle tank

Criteria	$\tilde{s}_j$	$\tilde{k}_j$	$\tilde{q}_j$	$\tilde{w}_j$
A_4		1	1	0.523
A_2	0.963	1.963	0.509	0.266
A_3	0.925	1.925	0.264	0.138
A_1	0.916	1.916	0.137	0.073
			1.910	1.000

Table 12. The summary of the mean criteria for monthly number of travelled kilometers

Criteria	$\tilde{s}_j$	$\tilde{k}_j$	$\tilde{q}_j$	$\tilde{w}_j$
A_4		1	1	0.515
A_2	0.891	1.891	0.528	0.272
A_3	0.972	1.972	0.267	0.137
A_1	0.833	1.833	0.145	0.076
			1.940	1.000

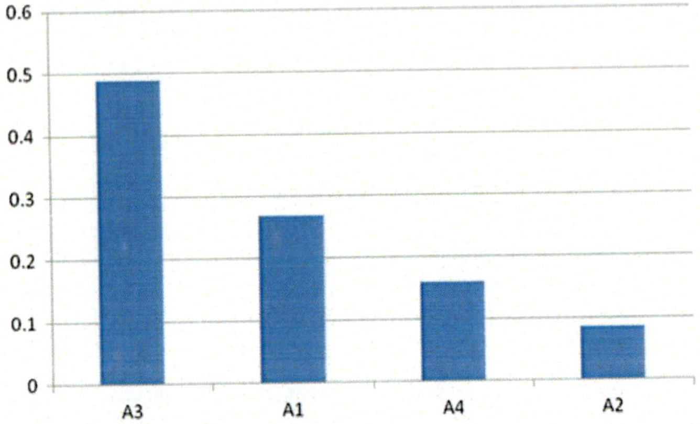

Fig. 2. Graphical representation of resulting weight of the mean criteria for average monthly quantity of gasoline drained from the vehicle tank

The summary of resulting weight of the mean criteria for monthly quantity of gasoline and gas drained from the vehicle tank, and as well the traveled kilometers, is shown in the Figs. 1, 2, 3 and 4 in descending order, as graphical representation.

4 The Aggregating Techniques and Decision Making

One of the most important tasks in engineering lies in the construction of a knowledge database of decision support, and in that way to ensure optimal conditions, improve quality and boost efficiency. The main idea to do so, is that the quality service is maintained and controlled. Applying the Fuzzy theory in decision making has given very good results, and provided a flexible framework and over the years numerous mathematical models have been developed, [31].

There are few well known stages in developing computer decision support systems based on knowledge which include choosing suitable mathematical tools, formalization of the subject area, and development of the corresponding software. In the first phase the problem lies in making the right diagnosis and in analyses of the

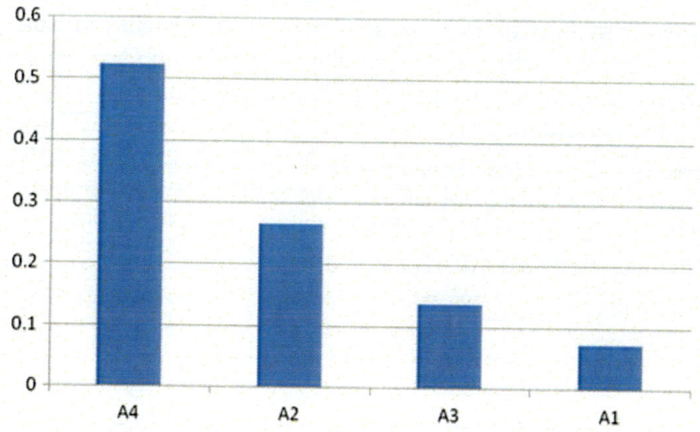

Fig. 3. Graphical representation of resulting weight of the mean criteria for average monthly quantity of gas drained from the vehicle tank

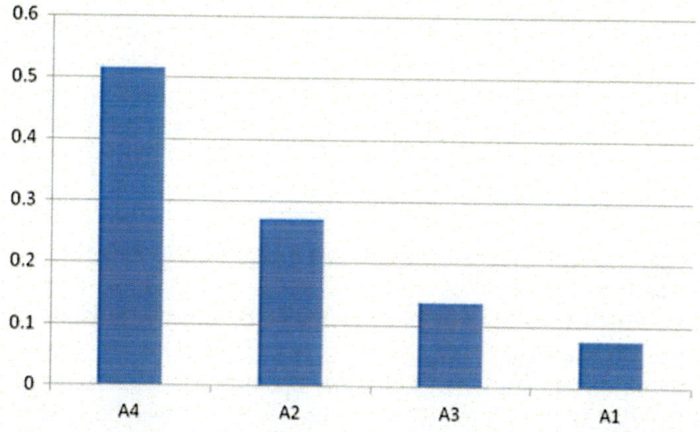

Fig. 4. Graphical representation of resulting weight of the mean criteria for average monthly number of traveled kilometers

requirements and as well the analyses of the system incidents caused by specification, design and the implementation of the project. The problem of diagnostics may be stated such as finite number of subsets, or classical investigation methods should be applied [32].

The OWA (Order Weighted Averaging) provides a parameterized group of mean type aggregation operators, and it is introduced by Jagerr [33].

Definition 1. An OWA operator of dimension n is a mapping OWA: $R^n \rightarrow R$ defined by an associated weighting vector W of dimension n, such that the sum of the weights is 1 and wj ϵ [0,1], according to the formula (1).

The *IOWA* operator is an extension of *OWA* operator, and the crucial difference represents the fact that the reordering step is not developed with the values of arguments *a*, as presented in the following definition.

Definition 2. An *IOWA* operator of dimension n is a mapping *IOWA*: $R^n \to R$ defined by an associated weighting vector *W* of dimension n, such that the sum of the weights is 1 and wj ϵ [0,1], according to the following formula:

$$IOWA((u_1a_1),(u_2a_2),(u_3a_3)\ldots(u_na_n)) = \sum\nolimits_{j=1}^{n} w_jb_j \tag{4}$$

where b_j is the j-th largest value of the pair (u_ia_i).

The *GOWA* is also introduced by Jager, it includes aggregation operators such as ordered weighted harmonic averaging operator, order weighted geometric operator, ordered weighted quadratic averaging operator and the others and it is given with the following definition, [33, 34].

Definition 3. An *GOWA* operator of dimension n is a mapping *GOWA*: $R^n \to R$ defined by an associated weighting vector *W* of dimension n, such that the sum of the weights is 1 and wj ϵ [0,1], satisfying the down below formula:

$$GOWA(a_1,a_2,\ldots a_n) = \left(\sum_{j=1}^{n} w_jb_j{}^2\right)^{\frac{1}{\lambda}} \tag{5}$$

where b_j is the j-th largest of the a_i, and λ is parameter with the characteristic that $\lambda \in (-\infty, +\infty)$.

The *OWAWAIMAM* operator combines in the same formulation, the *OWA* operator and weighted average.

Definition 4. An *OWAWAIMAM* operator of dimension n is a mapping *OWAWAIMAM*: $R^n \to R$ defined by an associated weighting vector *W* of dimension n, such that the sum of the weights is 1 and wj ϵ [0,1], such that exist the following formula, [33]:

$$OWAWAIMAM(\mu_i,\mu_i{}^b,\ldots\ldots(\mu_n,\mu_n{}^b)) = \sum_{j=1}^{n} \gamma_jK_j \tag{6}$$

where K_j represents the jth largest of all of the $\mu_i, \mu_i{}^b, \ldots \mu_n, \mu_n{}^b$.

In the down below Tables 13, 14 and 15, it is presented numerical example, taking into account following decision making criteria for obtained data for 4 companies from Eastern Serbia denoted as A_1, A_2, A_3, A_4:

(a) C_1—monthly quantity of gasoline drained from the vehicle tank;
(b) C_2—monthly quantity of gas drained from the vehicle tank;
(c) C_3—monthly number of travelled kilometers.

Table 13. The aggregated result for monthly quantity of gasoline

	OWA	IOWA	GOWA	OWAWAIMAM
A_1	0.816	0.218	0.0475	0.223
A_2	0.947	0.079	0.006	0.258
A_3	0.988	0.478	0.224	0.238
A_4	0.674	0.107	0.011	0.222

Table 14. The aggregated result for monthly quantity of gas

	OWA	IOWA	GOWA	OWAWAIMAM
A_1	0.916	0.066	0.004	0.250
A_2	0.963	0.256	0.016	0.263
A_3	0.925	0.127	0.002	0.314
A_4	0.972	0.508	0.258	0.234

Table 15. The aggregated result for monthly number of travelled kilometers

	OWA	IOWA	GOWA	OWAWAIMAM
A_1	0.978	0.074	0.001	0.267
A_2	0.891	0.202	0.040	0.243
A_3	0.833	0.114	0.012	0.264
A_4	0.973	0.5080	0.125	0.235

5 Conclusion

In this study for given data obtained from four logistic companies from Eastern Serbia, decision model is applied by using various aggregated operators. For chosen criteria such as monthly quantity of gasoline drained from the vehicle tank, monthly quantity of gas drained from the vehicle tank, and monthly number of travelled kilometers the resulting weight is calculated applying the *SWARA* method. The degree of importance is presented in the appropriate tables, taking into account different aggregation operators.

If it is analyzed monthly quantity of gasoline drained from the vehicle tank, the best results are obtained for company A_2 for *OWA* and *OWAIMAM* operators, and for company A_3 considering *IOWA* and *GOWA* operators. For monthly quantity of gas drained from the vehicle tank criteria, company A_4 has optimal results considering *OWA*, *IOWA* and *GOWA* approach and company A_3 if we take into account *OWAIMAM* approach. The best results considering monthly number of travelled kilometers are presented for company A_1 if we analyze *OWA* and *OWAWAIMAM* operators, and A_4 has optimal results considering *IOWA* and *GOWA* operators.

In this research it is shown casual connection between decision strategies and applying adequate aggregate operators.

In the further development we expect new approaches and using more factors and operators and thus getting new decision models which would optimize important parameters such as: inspection of rolling stock and machine park, reduced costs, better trained drivers, better service and satisfied customers, increased productivity and efficiency, fewer overtime hours, less idle periods, easy vehicle and machine monitoring, reduced fuel costs, and reduced maintenance and service costs.

References

1. Bagchi P, Larsen S, Soerensen L (2005) Supply chain integration a European survey. Int J Logist Manag 16(2):275–294
2. Gunter H, Grote G, Thees O (2006) Information technology in supply networks. Does it lead to better collaborative planning?. J Enterp Inf Manage 19(5):540–550
3. Kumar S, Moore K (2002) The evolution of Global Positioning System (GPS) technology. J Sci Educ Technol 19(5):59–80
4. Woznica J, Healy K, Thees O (2009) The level of information systems integration in SMEs in Irish manufacturing sector. J Small Bus Enterp Dev 16(1):115–130
5. Reusen E, Stouthuysen K, Roothoft F, Slabinnck H (2019) Imitation of management practices in supply networks: relational and environment effects. J Supply Chain Manage https://doi.org/10.1111/jscm.12216
6. Dichiara R (2003) Relationships between small and big enterprises in suppliers networks, The case of mass production industry. In: Alciatore DG, Histand MB (eds) CSB 48th World Conference, Belfase. Introduction to Mechatronics and Measurement Systems, vol 1, 3rd edn. McGraw Hill, International edition, pp 211–216
7. Ristovska N, Kozuharov S, Petkovski V (2017) The impact of logistics management practices. J Acad Res Acc Finance Manage Sci 7(1):245–252
8. Dihovicni D (201) Application of finite spectrum assignment method on distillation column. Modeling and simulation in engineering, economics and management, lecture notes in business information processing, vol 145, Springer-Verlag, pp 232–238
9. Dihovicni D, Nedic N (2008) Simulation, animation and program support for a high performance pneumatic force actuator system. Mathematical and computer modelling, vol 48, Elsevier, Washington, pp 761–768
10. Ratković Kovačević N, Škatarić D (2010) Multimodeling control via system balancing. In: Mathematical problems in engineering, vol 2010, p 20. https://doi.org/10.1155/2010/841830, Hindawi (2010). http://www.hindawi.com/journals/mpe/2010/841830.html. Accessed 30 May 2020
11. Dihovicni D, Medenica M (2011) Mathematical modelling and simulation of pneumatic systems. Adv Comput Sci Eng India 161–186. https://doi.org/10.5772/15313
12. Ćoćkalo D, Vorkapić M, Kreculj D, Đorđević D, Frantlović M (2020) Using QFD and AHP tools in the case of industrial transmitters manufacturing. FME 48(1)
13. Stock J, Boyer T, Harmon-Kitzer T (2010) Research opportunities in supply chain management. J Acad Mark Sci 38(1):32–41. https://doi.org/10.1111/s11747–009–0136–2
14. Muysinaliyev A, Aktamov S (2014) Supply chain management concepts: literature review. IOSR J Bus Manage 15(6):60–66
15. Armayor D, Batista J (2010) Decision support model for information systems integration in supply chains: a fuzzy approach. In: World Scientific Proceedings on computer engineering and information science 3, World Scientific, pp 281–28.
16. Keropyan A, Gil-Lafuente M (2010) Decision making in strategic business management. In: World Scientific Proceedings on computer engineering and information science 3, World Scientific, pp 297–304

17. Dihovicni D, Asonja A, Radivojevic N, Cvijanovic D, Skrbic S (2020) Stability issues and program support for time delay systems in state over finite time interval. Physica A: Statistical Mechanics and its Applications, vol 538, 122815, Elsevier, Washington, p 2020
18. Raman B, Monga H (2016) Method & implementation of data flow. Appl Eng Lett 1(4):105–110
19. Merigo J, Engemann K (2010) Fuzzy decision making with probabilities and induced aggregation operators. In: World Scientific Proceedings on computer engineering and information science 3, World Scientific, pp 323–332
20. Dihovicni D (2013) Decision making and fundamental matrix approach in process safety. Int J Comput Intell Syst 6(4):658–668
21. Perez D, Batista J (2010) Decision support model for information systems integration in supply chains: a fuzzy approach. In: Word Scientific Proceedings series on computer engineering and information science, pp 281–288
22. Sharma A, Goswami A, Gupta D (2009) Fts-Sql: a query language for fuzzy multidatabases. In: Proceedings of the world congress on engineering vol 1, WCE, London UK, pp 69–74. ISBN: 978–988–17012–5–1
23. Dihovicni D (2014) Fuzzy logic approach in oil treatment. J Balcan Tribological Assoc 20(4):606–614
24. Milaković A (2018) GPS system for tracking vehicles. Master theses in vocational studies (in Serbian). The Academy of Applied Technical Studies Belgrade, Technical College Beograd, Belgrade
25. Soonhong M, Zach Z (2019) Defining supply chain management: in the past, present and future. J Bus Logist 40(1):44–55. https://doi.org/10.1111/jbl.12201
26. Nicolas R (2004) Knowledge management impacts on decision making process. J Knowl Manag 8(1):20–31. https://doi.org/10.1108/13673270410523880
27. Krunić V, Ašonja A, Krunić M, Mikić D, Četić N (2013) Software for automatic control of laboratory analysis. In: 13th International Conference "Research and Development in Mechanical Industry" RaDMI 2013, vol 2, Kopaonik, Serbia, pp 688–694
28. Kaufmann A, Gilaluja J (1986) Introduction to the theory of fuzzy subsets in business management, Milladoiro, Santiago di Campostela, Spain: Technologies in human related Sciences. Leon, Spain **2**(1996), 217–223 (1986)
29. Zolfani H, Saparauskas J (2013) New application of SWARA method in prioritizing sustanability assesment indicators of energy system. Eng Econ 24(5):408–414. https://doi.org/10.5755/j01.ee.24.5.4526
30. Herrera F, Herrera-Viedma E, Lopez E (1996) Computer-base support to decision-making in orthopaedics. In: International Conference on intelligent technologies in human related sciences. vol 2, Leon, Spain, pp 205–213
31. Llopis-Albert C, Palacios-Marques D (2016) Applications of ordered weighted averaging (OWA) operators in environmental problems, multidisciplinary. J Educ Soc Technol Sci 4(1):52–63. EISSN: 2341–2593. https://doi.org/10.4995/muse.2017.700
32. Dihovicni D (2010) Constructing knowledge database in decision making, Communications, Zelina University, Slovaki, vol 3, World Scientific, pp 74–77
33. Safaraz H, Saparauskas J (2013) New application of SWARA method in prioritizing sustanability assesment indicators of energy system. Eng Econ 24(5):408–414
34. Xiuzhi L, Yaping D (2016) Fault tolerant control for fixed set-point control nonlinear networked control systems. Tech Gaz 23(1):35–41. https://doi.org/10.17559/TV-20160108182240

Morphology and Nanomechanical Properties of Ultrafine-Grained Ti-13Nb-13Zr Alloy Surface Obtained Using Electrochemical Anodization

Dragana Barjaktarević, Bojan Međo, Veljko Đokić, and Marko Rakin

Abstract Lower value of modulus of elasticity, closer to that of a bone, is one of the crucial surface properties in accepting the implant material from the surrounding tissue, and reduces the possibility of slow disappearance of bone in contact with the implant. In the present study, ultrafine-grained (UFG) Ti-13Nb-13Zr alloy was obtained using high pressure torsion process (HPT) under a pressure of 4.1 GPa with a rotational speed of 0.2 rpm up to 5 rotations at room temperature. Nanostructured surface on coarse-grained (CG) and UFG Ti-13Nb-13Zr alloy was formed using electrochemical anodization in the 1 M $H_3PO_4 + NaF$ electrolyte, during 60 and 90 min. Scanning electron microscopy (SEM) was used to characterise the morphology of the surface, while nanomechanical properties of the surface, modulus of elasticity and nanohardness were determined using the nanoindentation test. Also, in order to characterise deformation of the nanotubes after nanoindentation test, SEM was done. It was shown that the nanotubular oxide layer was formed as result of the electrochemical anodization process during both anodizing times. The surface of anodized alloys has lower modulus of elasticity than surface of non-anodized ones. Anodized UFG alloy had the lowest modulus of elasticity of the surface when compared to other tested samples, which makes it more acceptable for biomedical usage.

Keywords Electrochemical anodization · High-pressure torsion · Nanoindentation · Mechanical surface properties · Ultrafine-grained titanium alloy

D. Barjaktarević (✉) · B. Međo · M. Rakin
Faculty of Technology and Metallurgy, University of Belgrade, Karnegijeva 4, 11120 Belgrade, Serbia
e-mail: dbarjaktarevic@tmf.bg.ac.rs

V. Đokić
Innovation Centre of the Faculty of Technology and Metallurgy in Belgrade, Karnegijeva 4, 11120 Belgrade, Serbia

N. Mitrovic et al. (eds.), *Current Problems in Experimental and Computational Engineering*, Lecture Notes in Networks and Systems 323,
https://doi.org/10.1007/978-3-030-86009-7_7

1 Introduction

1.1 Titanium Based Materials for Medical Application

$\alpha + \beta$ titanium alloys, among which the most famous is Ti-6Al-4 V alloy, showed good clinical outcomes and due to their adequate properties such as specific strength, corrosion resistance and biocompatibility are used for the production of orthopaedic and dental implants. However, research has shown that the release of vanadium ions into surrounding tissues can lead to poisoning, while aluminium ions increase the chances of developing Alzheimer's disease [1]. In order to make an implant of good biomechanical compatibility, development of titanium alloys takes into account the safety of the use of alloying elements. In addition, the main factor of biomechanical compatibility and one of the most important properties that limits usage of titanium and its alloys as implants is the value of modulus of elasticity [2]. With the advancement of science and technology, the new generation of titanium alloys, $\alpha + \beta$-type and β-type, has been developed primarily for dental and orthopaedic applications.

The new generation of titanium alloys offered a unique combination of properties such as good corrosion resistance, better mechanical properties, non-toxic alloying elements and excellent biocompatibility [2, 3]. However, they attracted the most attention for medical application due to the value of the modulus of elasticity, which is much lower than the value of α-type and $\alpha + \beta$-type alloys, all thanks to alloying with β stabilizers that significantly reduce the modulus of elasticity. Bertrand et al. [4] have pointed to a very low value of the modulus of elasticity of Ti-25Ta-25Nb alloy (55 GPa), which is one of the lowest for a β-type alloy. Braille et al. [5] found that lower value of the modulus of elasticity in the Ti-Nb-Zr (Ta) β–type alloy is a consequence of martensitic transformation. They also confirmed the importance of the β phase in reducing of value of the modulus of elasticity at room temperature. The new generation of $\alpha + \beta$-type and β-type titanium alloys have excellent corrosion resistance in the human body thanks to the spontaneous formation of a protective thin oxide layer on the surface. In order to expand the application of titanium alloys as implant materials, today more and more attention is paid to the selection of alloying elements, which is one of the most important factors for improving tensile strength, reducing the modulus of elasticity and minimizing the toxic effect of released ions. Some of modern β titanium alloys are Ti-13Nb-13Zr, Ti-12Mo-6Zr-2Fe, Ti-12Mo-5Zr-5Sn, Ti-15Mo, Ti-30Ta, Ti-45Nb, Ti-35Zr-10Nb, Ti-35Nb-7Zr-5Ta, etc. [6, 7]. However, some newly developed alloys also have low osseointegration and bioactive characteristics, which limits their use in medicine. So far, various surface modifications of metallic biomaterials have been shown to improve corrosion resistance and surface bioactivity [8]. Modified surface leads to cellular reactions such as proliferation and differentiation, while preventing harmful effects such as blood coagulation and bacterial adhesion [9]. However, the greatest challenge is still to find a biomaterial that meets all the requirements for application, such as adequate biocompatibility and modulus of elasticity and high resistance to wear and corrosion.

1.2 Application of Severe Plastic Deformation in Materials Processing

Obtaining an ultrafine-grained structure can be achieved in two ways. The first way, so-called bottom up, represents making of a material by building it atom by atom or from groups of atoms [10]. Such methods include inert gas condensation and other methods of obtaining nanopowder. Another approach is so-called top down, which is based on the transformation of coarse-grained structure into ultrafine-grained one or nanostructure using severe plastic deformation (SPD) procedures [10]. SPD processes are methods of refining the structure by reducing the grain size [11–13]. Previous research conducted in the field of SPD shows that grain size can be considered a key microstructural factor that affects almost all aspects of physical and mechanical behaviour, as well as their chemical and biochemical response to body fluids [10, 14–16], which is a good starting point for a detailed analysis of the application in production of titanium-based materials for medical use. The conditions that need to be achieved for the creation of ultrafine-grained structure using SPD procedures are high hydrostatic pressure, high values of shear deformation, while avoiding damage to the structure of the material [11, 17]. It is necessary that the formed ultrafine-grained structure has high-angle grain boundaries and must be homogeneous in sample volume [12, 13]. Plastic deformations, which occur due to high external loads, lead to the refinement of grains to submicron levels. Materials obtained by SPD procedures are divided into ultrafine-grained (UFG) materials, with grain sizes 100–1000 nm and nanostructured (NS), with grain size below 100 nm [18].

There are a large number of different SPD procedures [10, 18, 19]:

- *Equal channel angular pressing, ECAP,*
- *High pressure torsion, HPT*
- *Accumulative roll-bonding, ARB,*
- *Repetitive corrugation and straightening, RCS,*
- *Multi-directional forging, MDF,*
- *Twist extrusion, TE,*
- *Caliber rolling, CAROL,*
- *Cyclic extrusion compression, CEC,*
- *Cyclic closed-die forging, CCDF* and others.

Of all the above mentioned procedures, HPT, ECAP and ARB are the most widely used. In this study, HPT was used to refine the course-grained structure, Fig. 1.

High pressure torsion processing. The thin disk is pressed between two anvils, under high pressure, and by rotating the two anvils, large shear deformations of the material occur. Both anvils have cylindrical and slightly conical holes. The diameter of the cylindrical holes is identical to the diameter of the sample, while the depth of both holes is slightly less than the sample height. In order to perform the HPT procedure correctly, it is necessary to apply a high hydrostatic pressure, which is generally three times higher than the yield stress of the material in the undeformed

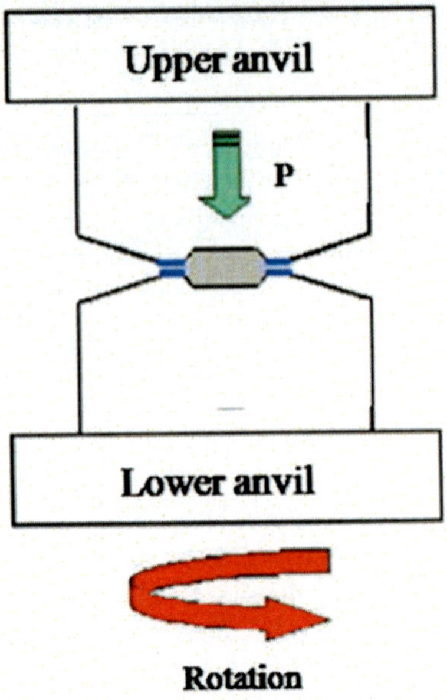

Fig. 1. High pressure torsion process.

state [17]. The number of rotations during the HPT procedure has a dominant role in terms of the impact on the physical and mechanical properties. In order to determine the influence of rotations on the material properties, in a large number of scientific papers, the HPT procedure was applied to metallic materials with the number of rotations varied. Increasing the number of rotations significantly contributed to increase of microhardness [20–22]. Vickers microhardness measurement showed a significant increase in cpTi microhardness after HPT; the microhardness increased with increasing number of rotations, slightly after one rotation and sharply after 5 and 10 rotations [23]. In addition, the microhardness value has been shown to increase from the centre of the disk toward the edges, with the lowest value at the very centre. Also, it has been shown that with increasing number of rotations during the HPT, the tensile and yield strength increase, while deformation at fracture and modulus of elasticity decrease; the largest changes are observed for 5 rotations [11].

The normal effective plastic deformation, ε_e, which occurs in the sample after the HPT, is determined using von Mises criterion based on the equation [10, 18]:

$$\varepsilon_e = \frac{\gamma_e}{\sqrt{3}} = \frac{2\pi nr}{\sqrt{3}\cdot h} \quad (1)$$

where γ_e is the effective shear deformation, n is the number of rotations, r is the distance from the centre of the sample [m] and h is the thickness of the sample [m].

Equation 1 indicates that there is no deformation in the centre of the sample, while the increase in deformation from the centre to the edge of the sample is linear [10, 18]. This distribution of deformations can affect the homogeneity of the microstructure, in such a way that larger grains will form in the centre of the sample where there is no deformation, and their size will decrease towards the edges of the sample. However, the available literature data indicate that the application of HPT under constant high pressure and a certain number of rotations can still enable formation of a homogeneous microstructure [24, 25].

1.3 Nanostructured Surface Modification of Titanium Based Materials

The behaviour of biomaterials in the human body depends on their biocompatibility and surface properties. For this reason, biomaterials often require surface modification in order to optimize implant properties and increase their bioactivity when binding to the surrounding tissue. There are various procedures that can be used to modify/improve the surface of metallic biomaterials. Nanostructured surface modification of titanium based material leads to the formation of surface morphology of several nanometers, increase of roughness and change of surface topography from micro to nano level. The methods for nanostructured surface modification can be divided into four categories [26]: mechanical, physical, biochemical and chemical.

Primary need for nanostructured surface modifications of titanium based materials is better and faster binding to bone tissue [27], improvement of the degree of osseointegration, biomechanical compatibility, surface physical and mechanical properties, corrosion and wear resistance and removal of contamination [28].

The most commonly used methods for surface modification are chemical and electrochemical treatments, i.e. electrochemical anodization (anodic oxidation), sol–gel process and chemical vapour deposition [9, 29–31]. Chemical treatments of titanium-based materials are reactions of the surface of the material and the solution to which the material is exposed. In the last decade, electrochemical methods that enable surface nanostructural modification are increasingly used in the production of implants, and one of them is electrochemical anodization (anodic oxidation). On the surface of metal material, it forms a nanostructured oxide layer composed of nanotubes. The advantage when compared to other methods of surface nanostructured modifications is the ability to control the morphology of layers and nanotubes size by careful selection of the electrolyte, pH value, voltage and/or anodizing time. Electrochemical anodization of the titanium based materials leads to creation of oxide layer composed of TiO_2-based nanotubes, with length from 10 nm to 40 μm.

The minimum time required for the formation of a nanotubular oxide layer on the surface of a titanium-based material is between 30 and 120 min [32]. The duration of anodization also affects the oxide layer thickness, i.e. the length of the formed nanotubes. Increasing anodizing time leads to increase of the layer thickness and diameter of the nanotubes [22]. Figure 2 presents the morphology of the nanostructured surface formed on Ti-45Nb alloy during different duration of electrochemical anodization. Formation of specific surface topography improves the

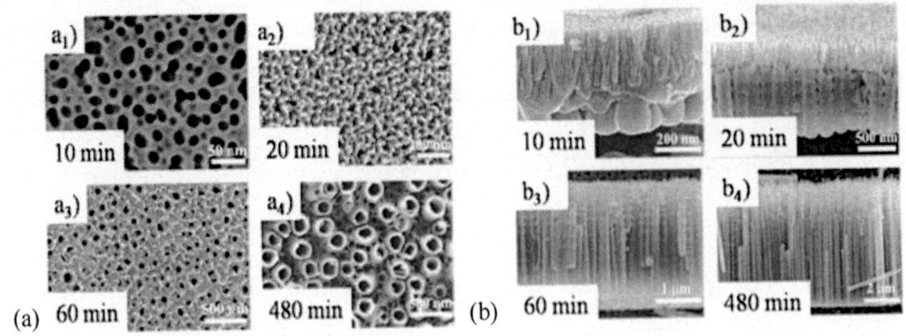

Fig. 2. **a** Nanostructured surface morphology and **b** nanotube length after electrochemical anodization of 10, 20, 60 and 480 min [33].

corrosion resistance, biocompatibility and bioactivity, reducing the surface modulus of elasticity.

2 Experimental

2.1 *Materials*

Material obtained by conventional manufacturing process. For the purposes of testing, the Ti-13Nb-13Zr alloy (coarse-grained TNZ, CG TNZ) was selected. The alloy is intended for biomedical application and made according to the ASTM F1713 standard [228]. The material was fabricated by conventional methods in as-cast condition, in the form of a rod with a diameter of 28 mm.

Material obtained by the high pressure torsion process. In order to obtain an ultrafine-grained structure, CG TNZ alloy was subjected to severe plastic deformation (SPD) using the HPT process. The HPT was performed at a temperature of 24 ± 1 °C, under a pressure of 4.1 GPa, at a speed of 0.2 rpm and with 5 rotations, *at the Erich Schmid Institute of Materials Science, Austria.*

The applied HPT process caused a change in the dimensions of the samples, in such a way that the disk thickness decreased while the diameter increased, Fig. 3.

2.2 *Electrochemical Anodization*

Electrochemical anodization was performed using a system of two electrodes: platinum, as a cathode, and a sample of CG or UFG TNZ alloy as working electrode, i.e. anode. Samples, with dimensions 10 × 10 × 1 mm, were cut from the discs by

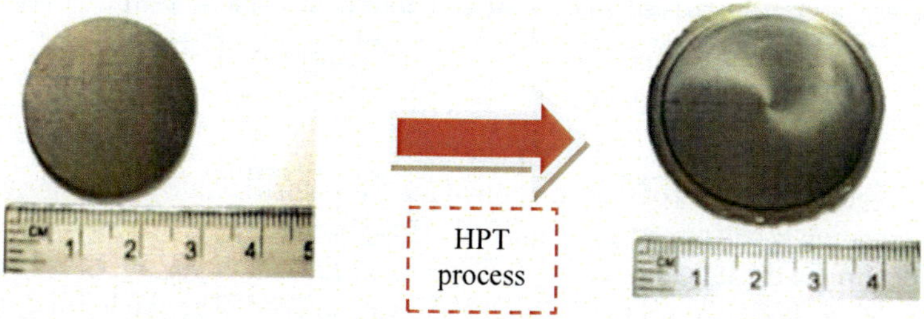

Fig. 3. Appearance of tested samples before and after HPT process.

electrical discharge machining (EDM). The area of the sample exposed to the electrolyte was 1 cm^2. The samples were sanded with SiC papers with fineness ranging from 100 μm to 4000 μm, then polished with alumina (1 μm). After polishing, the samples were cleaned with acetone and ethanol in an ultrasonic bath and washed with distilled water. Electrochemical anodization was performed at room temperature, at a voltage of 25 V. 1 M H_3PO_4 + 0.5 wt. % NaF was chosen as the electrolyte, and the duration of anodization was 60 and 90 min. After the anodization, the samples were washed with distilled water and air-dried for 24 h. The *PEQLAB EV 231* device was used as a power supply for electrochemical anodization.

Morphology characterization. As a result of anodization, a modified nanostructured surface was obtained; its characterization was done using the scanning electron microscopy (SEM)—TESCAN MIRA3 XMU microscope, voltage 20 keV.

2.3 Nanoindentation Test

The surface mechanical and physical properties were examined by nanoindentation test. Alloy made by conventional methods (CG) was examined, as well as alloy after HPT process (UFG); both of them with and without electrochemical anodization. The control of nanoindentation was done by total displacement. The displacements were 2000 nm for non-anodized samples and 10% of the thickness of nanotubular oxide layer for anodized samples, as defined by ISO 14577–4 [34–36]. The displacements were 160 nm and 275 nm for CG TNZ alloy, and 160 nm and 265 nm for UFG TNZ alloy anodized for 60 and 90 min, respectively. The test was performed on nanoindenter *G200, Agilent Technologies*, using as an indenter Berkovich-type diamond tip. Loading-displacement curves were obtained, and as a final result, the mean value of ten measurements of the surface modulus of elasticity and nanohardness. In order to determine the level and manner of deformation and damage of the nanostructured layer, SEM analysis of the surface of UFG TNZ alloy anodized after 90 min was performed.

2.4 Numerical Analysis of Influence of Nanotube Dimensions on Resistance to Loading During Nanoindentation

Numerical analysis of deformation of nanostructured oxide layer exposed to nanoindentation is performed on simplified 2D finite element models, with the main aim to determine the influence of the dimensions of the nanotubes on resistance to external loading. Software package Simulia Abaqus is used. The models are based on the layers formed after anodization during 60 and 90 min.

In the model, conical shape of the indenter is applied; it has been shown in the literature [37–39] that it provides the same results as Berkovich indenter. Half of the geometry is analysed in plane strain conditions, Fig. 4. Therefore, appropriate symmetry boundary conditions are applied; also, lower edge of the model is fully constrained. Displacement of the indenter in vertical direction is prescribed. Finite elements with linear interpolation and full integration are used.

As mentioned previously, duration of anodization affects the dimensions of the nanotubes; length and diameter tend to increase, while the wall thickness decreases. For the numerical models, mean values of dimensions measured on SEM microphotographs of samples after 90 and 60 min anodization are considered. After 90 min, length was 2740 nm, diameter 90 nm and wall thickness 20 nm. After 60 min, these values are 1630 nm, 56 nm and 20 nm, respectively. The distance between the nanotubes is 5 nm in both cases.

As for the material properties, the modulus of elasticity of the alloy is 82.9 GPa. For nanotubes, the value of modulus is obtained by application of the mixture rule, having in mind that they consist of Ti, Nb and Zr oxide; their ratio approximately corresponds to the alloy composition. The same value of Poisson ratio is used for both alloy and nanotubes: 0.33. The indentation depth is 10% of the nanotube length, i.e. 274 and 160 nm.

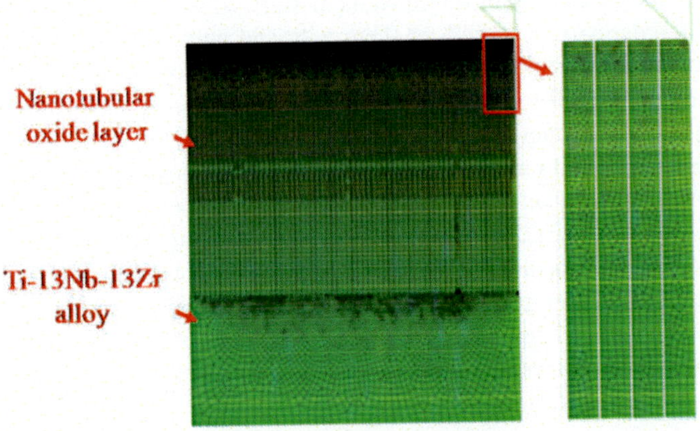

Fig. 4. Finite element mesh—2D model of nanotubes on UFG alloy.

Table 1. Dimensions of nanotubes in FE models.

Model No	Nanotube length (nm)	Nanotube diameter (nm)	Nanotube distance (nm)	Nanotube wall thickness (nm)	2D model thickness (nm)
1	1630	56	5	20	32.9
2	1630	90	5	20	47.9
3	2740	60	5	20	34.8
4	2740	90	5	20	47.9
5	2740	90	5	15	42.5
6	2740	90	5	25	50.9
7	2740	90	10	20	47.9
8	2740	90	15	20	47.9
9	2740	117	5	20	55.9

In order to represent the geometry of the layer as 2D numerical model, it is required that the nanotubes in plane strain condition have the same bending resistance, expressed through axial moment of inertia, as the real cylindrical ones. Practically, this means that the cylindrical shape is replaced by equivalent rectangle in numerical models. Since axial moment of inertia for the tube cross section is $I_X = \frac{\pi}{64}\left(D^4 - d^4\right)$, while the expression for rectangle is $I_X = \frac{ab^3}{12}$, equality of these moments is used to calculate the plane strain thickness of the models. Dimensions of all considered models are shown in Table 1. Models 1 and 4 correspond to actual layers (dimensions measured by SEM), while others are obtained by varying some of the dimensions.

3 Results and Discussion

3.1 *Morphology of the Nanotubular Oxide Layer Obtained Using Electrochemical Anodization*

Figure 5 shows the morphology of nanostructured surfaces formed on the CG and UFG TNZ alloy during the anodizing times of 60 and 90 min. On the surface of the CG TNZ alloy, the nanotubular oxide layer was formed after 90, while random oxide layer was formed after 60 min. On the surface of the UFG TNZ alloy, the nanotubular oxide layer was formed after 90, while nanoporous oxide layer was formed after 60 min. During the shorter anodizing time, nanotubes formed on the surface of UFG TNZ alloy were connected, while for longer anodizing time nanotubes were separated due to dissolution, (Fig. 5c, d). The nanotubular layer on the surface of the UFG TNZ alloy has homogeneous morphology, while the layer on the surface of the CG TNZ alloy has inhomogeneous morphology. The latter is characterised by thicker walls and smaller

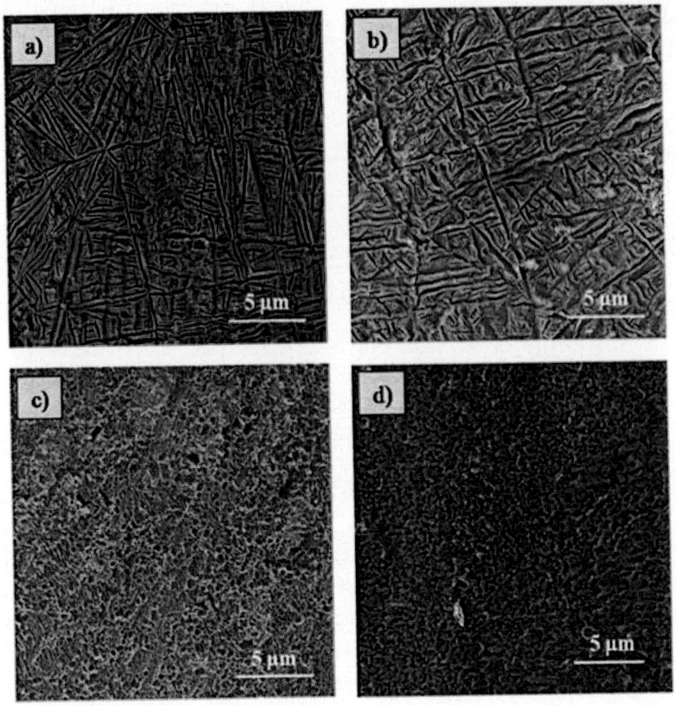

Fig. 5. Morphology of the nanostructured surface of CG and UFG TNZ alloy after electrochemical anodization for (**a**, **c**) 60 and (**b**, **d**) 90 min, respectively.

diameter, between which the furrows are formed. The achieved results show that Ti-13Nb-13Zr alloy after HPT process needs a shorter time during electrochemical anodization in order to form a homogeneous nanoporous structure on the surface. Also, as the duration of anodization increases, the diameter of nanotubes increases, while the wall thickness decreases, and the final formation of the nanotubular oxide layer is obtained.

3.2 *Influence of Electrochemical Anodization on Physical and Mechanical Properties of the Surface of Ti-13Nb-13Zr Alloy*

Examination of surface physical and mechanical properties was done using the nanoindentation test. For each non-anodized sample, the indentation depth was 2000 nm, and the maximum mean load values were approximately 200 and 250 mN for CG and UFG TNZ alloy, respectively. On the other hand, for anodized samples, the indentation depth was 10% of the thickness of the nanostructured oxide layer,

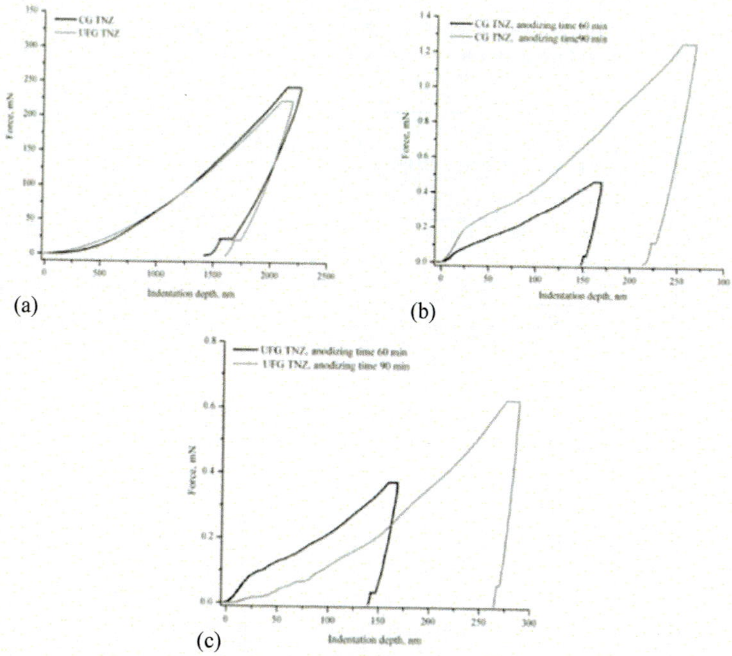

Fig. 6. Loading-displacement curves for **a** non-anodized materials, **b** anodized CG alloy, **c** anodized UFG alloy.

and the maximum mean load values were approximately 0.467 mN and 0.373 mN for CG and UFG TNZ alloy anodized after 60 min and 1.259 mN and 0.628 mN for CG and UFG TNZ alloy anodized after 90 min, respectively.

Figure 6 shows load–displacement curves obtained during the nanoindentation. Each curve consists of the loading part, the dwell period at the maximum load and the unloading part. The diagram shows the change in depth with the dwell period at the maximum load. The difference between the loading part and the unloading part on the diagram indicates the presence of the permanent deformation. From Fig. 6 it can be concluded that as the indentation depth increases, the force exerted on the material during testing increases too.

The mean values of the surface modulus of elasticity and nanohardness, obtained from the nanoindentation test, are shown in Figs. 7 and 8. The nanostructured surface modification leads to a decrease in the values of the surface modulus of elasticity, whereby they approach the values of the modulus of elasticity of the bones (10–30 GPa). A lower value of the modulus of elasticity, and closer to the value characteristic for bone tissue, is one of the key factors in the acceptance of the implant by the surrounding tissue, which leads to a reduction in the possibility of structural damage and bone decay in contact with the implant. Also, nanostructured surface modification leads to a decrease in nanohardness values. With increasing anodizing time, the values of both properties decrease. Based on Figs. 7 and 8, it can be concluded that UFG

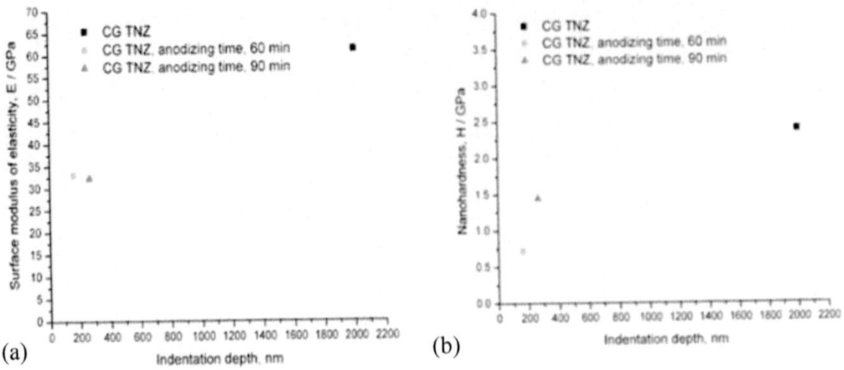

Fig. 7. Mean values of **a** surface modulus of elasticity and **b** nanohardness of CG TNZ alloy before and after electrochemical anodization.

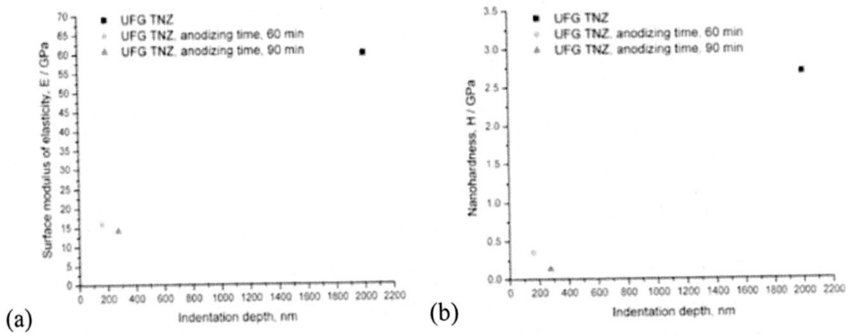

Fig. 8. Mean values of **a** surface modulus of elasticity and **b** nanohardness of UFG TNZ alloy before and after electrochemical anodization.

TNZ alloy anodized after 60 min has a mean value of surface modulus of elasticity of 16.11 GPa, while UFG TNZ alloy anodized after 90 min has a mean value of surface modulus of elasticity of 14.24 GPa. Also, the mean value of nanohardness decreased by 0.22 GPa with increasing anodization time in case of UFG TNZ alloy. On the other hand, CG TNZ alloy anodized after 60 min has a mean value of surface modulus of elasticity of 33.14 GPa, while with increasing anodization time to 90 min the mean value of surface modulus of elasticity decreases to 32.23 GPa. UFG TNZ alloy has a lower mean value of the surface modulus of elasticity than CG TNZ alloy after the same conditions of electrochemical anodization. Based on the above mentioned results, it can be concluded that anodized UFG TNZ alloy is more acceptable for implant production than the other considered materials.

However, given that the mean values of the surface modulus of elasticity of the anodized CG TNZ alloy is in the range acceptable for implant fabrication, this material cannot be completely discarded in use. Based on the literature review, it can be

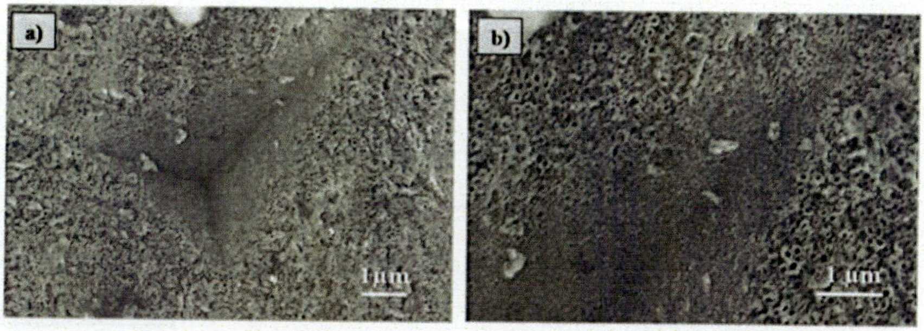

Fig. 9. Imprint of nanoindenter and damage to the nanostructured modified surface of UFG TNZ alloy anodized after 90 min.

established that the values of surface properties during nanoindentation test are influenced by the duration of electrochemical anodization, as well as the depth of indentation. The modulus of elasticity decreases with the prolongation of the anodizing process, while with the increase of the indentation depth it increases and approaches the value of the basic material. Also, the hardness decreases with the prolongation of the anodizing process, while the hardness values increase with the increase of the indentation depth [34, 40].

In addition to optimizing the chemical composition of the implant material, the porosity of the material in the surface layer also enables the reduction of the value of the modulus of elasticity. It has been shown that a porosity of about 30% leads to a value of modulus of elasticity that is almost equal to the value of the surrounding bone tissue. For example, when nanotubular oxide layer on the surface of CG cpTi was formed, the volume fraction of nanotubes in the layer was 72%, while the modulus of elasticity of dense amorphous TiO_2 is about 130–150 GPa (for one nanotube). This gave a value of modulus of elasticity for the nanotubular oxide layer from 36 to 43 GPa [26]. Also, the decrease in the value of the surface modulus of elasticity and nanohardness after the nanostructured surface modification of titanium-based materials is already known in the literature [28, 34, 40–43]. It was shown that the values of the surface modulus of elasticity decrease with increasing length of the nanotubes in the nanotubular oxide layer and it depends on the morphology of this layer [40]. Crawford et al. [34] indicated that by increasing anodizing time from 15 to 240 min, the thickness of the nanotubular oxide layer formed on the commercially pure titanium increased, while the elastic modulus decreased from 7.2 GPa to 4.6 GPa. Also, it was shown that the value of the surface modulus of elasticity of materials with nanotubular oxide layer was lower compared to non-anodized materials, while as the depth of indentation during the nanoindentation increased, the influence of the substrate on the modulus of elasticity increased [41]. Zielinski et al. [42] formed nanostructured oxide layer with a diameter in range from 80 to 120 nm on the surface of Ti-13Nb-13Zr alloy and showed that this morphology reduced the value of the surface modulus of elasticity to 25.5 GPa.

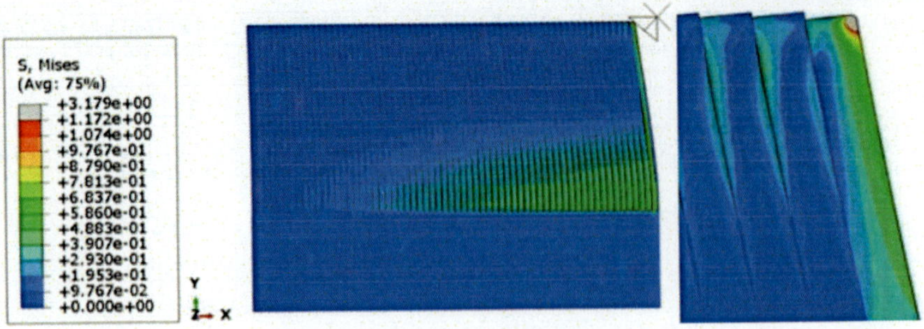

Fig. 10. Equivalent von Mises field for the alloy anodized during 90 min.

After the nanoindentation, nanotubes were damaged, which is shown in Fig. 9. As can be seen, there was no delamination of the nanotubular layer formed by anodization for 90 min. It has already been shown that delamination occurs for thinner oxide layers and for shorter anodization times (15 and 30 min), [34]. As a reason for the impossibility of delamination of thicker oxide layers, in the literature is stated that increased thickness of the nanotubular oxide layer may slow the accumulation of residual deformation. In Fig. 9, an increase in the density of the oxide layer can be observed. This density decreases from the top to the edges of the nanoindenter imprint. The observed behaviour indicates that the nanotubes thicken after nanoindentation, which is probably a consequence of their fracture [41].

3.3 *Results Obtained from Numerical Analysis*

Figure 10 shows the equivalent von Mises stress field for UFG TNZ alloy (anodized during 90 min) exposed to nanoindentation. Stress concentration is observed on contact surfaces and in the roots of the nanotubes.

The influence of the nanostructured surface morphology on the force–displacement curve is shown in the next figures; as mentioned previously, linear elastic behaviour is considered, i.e. damage of the nanotubes is not taken into account. Figure 11 contains the results obtained from the models of layers obtained during 60 min of anodization (nanotube length 1630 nm, diameter 56 nm) and during 90 min (nanotube length 2740 nm, diameter 90 nm). Maximum displacement is pre-defined, as mentioned previously, to be 10% of the nanotube length. Therefore, the final loads are different, but the trend is rather similar.

Unfortunately, the comparison shown in Fig. 11 cannot reveal the influence of the tubes geometry, because both length and diameter are different. Therefore, new models are formed, with only one dimension varied and other kept constant. Three characteristic dimensions which are considered are diameter, wall thickness and distance between the nanotubes, Table 1.

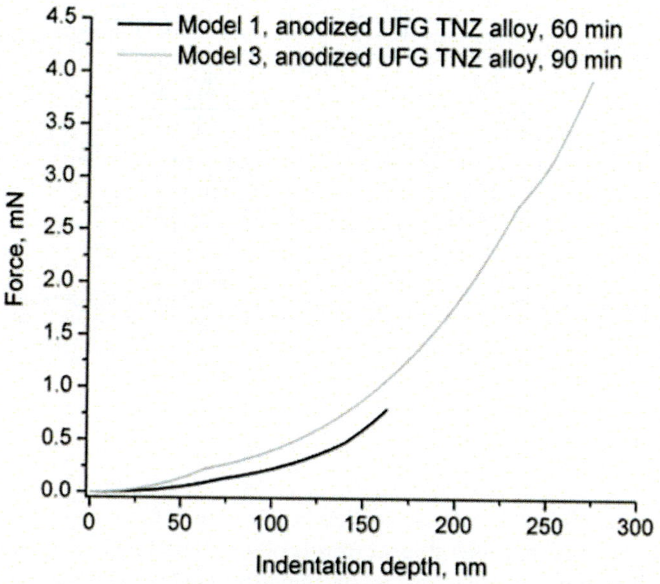

Fig. 11. Force–displacement curve for the models corresponding to the layers after 90 and 60 min of anodization.

The influence of the nanotube diameter and distance between them is shown in Fig. 12. Increase of diameter from 60 to 90 nm increases the force for an order of magnitude, and further increase is observed for diameter 117 mm, Fig. 12a. The influence of the distance between the nanotubes is shown in Fig. 12b; its decrease results in higher force values.

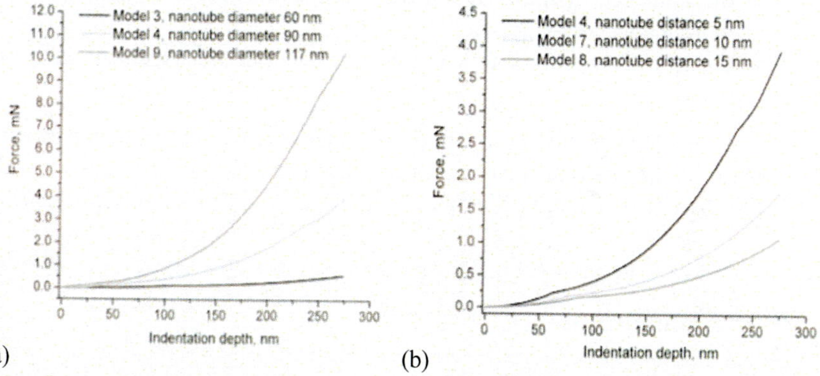

(a) (b)

Fig. 12. Dependence of force–displacement curve on the nanotube diameter (**a**) and distance between the nanotubes (**b**).

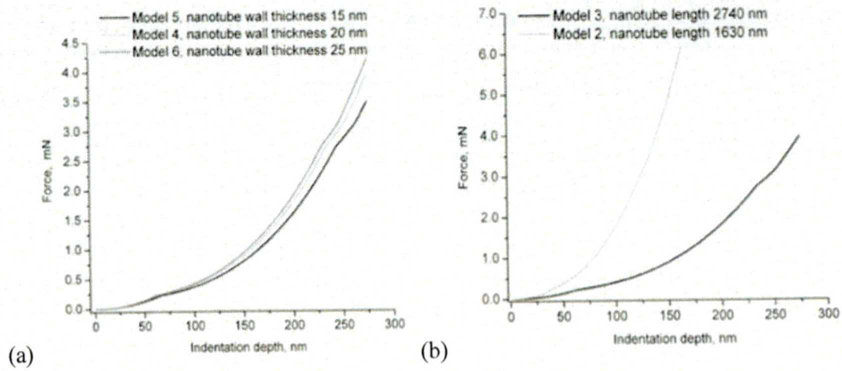

Fig. 13. Dependence of force–displacement curve on the nanotube wall thickness (**a**) and length (**b**).

As mentioned before, during the anodization the tubes gain in length, but their wall thickness becomes thinner (this is observed on SEM micrographs). Therefore, this dimension is also varied in the numerical models. The change of force value is much less significant when compared with the influence of diameter and distance between the tubes, Fig. 13a. Finally, since all the variations of nanotube dimensions are performed with fixed length, the length is also varied. Figure 13b presents the results obtained from the models with lengths of 2740 and 1630 mm, with all other dimensions being the same. Unlike the comparison from Fig. 11, this is isolated influence of the nanotube length on resistance to external loading.

Based on the numerical calculations, it can be concluded that the morphology of the nanotubular layer influences its load carrying capacity in the following manner:

- Increase of nanotube diameter and wall thickness leads to higher force values; the influence of the wall thickness is much lower in comparison with other morphology parameters.
- Increase of nanotube length, as well as distance between them, decreases the force values.

It should be mentioned again that the numerical results presented in this chapter represent linear elastic response of the nanotubular layer. Further work will include the analysis of damage development in this layer during nanoindentation testing.

4 Conclusions

The nanostructured oxide layer formed by anodization exhibited different structure and morphology, depending on the duration of anodization process. The surface modulus of elasticity decreased after anodization, which brings its values closer to those corresponding to the human bones (which lies in the range 10–30 GPa).

This is one of the key factors for successful implantation, i.e. acceptance of the implant by the surrounding tissues, and the alloys with nanostructurally modified surface are therefore more acceptable than their non-modified counterparts. Based on the numerical analysis, it can be said that the following morphology parameters influence the resistance to external loading: length, diameter and distance between the nanotubes, as well as wall thickness. In the numerical models, the effect of each of these dimensions is determined; the latter has the least pronounced influence.

Acknowledgements This work was supported by the Ministry of Education, Science and Technological Development of the Republic of Serbia (Contracts No. 451-03-9/2021-14/200287, 451-03-9/2021-14/200135). The authors of this paper owe great gratitude to Dr Anton Hohenwarter from the Erich Schmid Institute of Material Science, Leoben, Austria, for the preparation of the UFG TNZ alloy.

References

1. Mohammed, M., Khan, Z., Siddiquee, A.: Beta titanium alloys: the lowest elastic modulus for biomedical applications: a review. Int. J. Mater. Metallurgical Eng. **8**(8), 822–827 (2014)
2. Rack, H., Qazi, J.: Titanium alloys for biomedical applications. Mater. Sci. Eng. C **26**(8), 1269–1277 (2006)
3. Nie, L., Zhan, Y., Liu, H., Tang, C.: Novel β-type Zr-Mo-Ti alloys for biological hard tissue replacements. Mater. Des. **53**, 8–12 (2014)
4. Bertrand, E., Gloriant, T., Gordin, D., Vasilescu, S., Drob, P., Vasilescu, C., Drob, I.: Synthesis and characterisation of a new superelastic Ti–25Ta–25Nb biomedical alloy. J. Mech. Behav. Biomed. Mater. **3**(8), 559–564 (2010)
5. Brailovski, V., Prokoshkin, S., Gauthier, M., Inaekyan, K., Dubinskiy, S., Petrzhik, M., Filonov, M.: Bulk and porous metastable beta Ti–Nb–Zr(Ta) alloys for biomedical applications. Mater. Sci. Eng. C **31**(3), 643–657 (2011)
6. Froes, F., Qian, M.: Titanium in Medical and Dental Applications. Woodhead publishing series in biomaterials, Elsevier (2018)
7. Niinomi, M., Nakai, M., Hieda, J.: Development of new metallic alloys for biomedical applications—review. Acta Biomater. **8**(11), 3888–3903 (2012)
8. Barjaktarević, D., Djokić, V., Bajat, J., Dimić, I., Cvijović-Alagić, I., Rakin, M.: The influence of the surface nanostructured modification on the corrosion resistance of the ultrafine-grained Ti–13Nb–13Zr alloy in artificial saliva. Theor. Appl. Fracture Mech. **103**, 102307 (2019)
9. Barjaktarević, D., Cvijović-Alagić, I., Dimić, I., Đokić, V., Rakin, M.: Anodization of Ti-based materials for biomedical applications: a review. Metallurgical Mater. Eng. **22**(3), 129–143 (2016)
10. Damnjanović, I.: Biocompatibility and behavior in corrosive environment of titanium based materials for dental implants application (Ph.D. thesis). University of Belgrade, Faculty of Technology and Metallurgy, in Serbian (2018)
11. Yilmazer, H., Niinomi, M., Nakai, M., Cho, K., Hieda, J., Todaka, Y., Miyazaki, T.: Mechanical properties of a medical β-type titanium alloy with specific microstructural evolution through high-pressure torsion. Mater. Sci. Eng. C **33**(5), 2499–2507 (2013)
12. Wongsa-Ngam, J., Wen, H., Langdon, T.: Microstructural evolution in a Cu–Zr alloy processed by a combination of ECAP and HPT. Mater. Sci. Eng. A **579**, 126–135 (2013)
13. Wongsa-Ngam, J., Kawasaki, M., Langdon, T.: A comparison of microstructures and mechanical properties in a Cu–Zr alloy processed using different SPD technique. J. Mater. Sci. **48**, 4653–4660 (2013)

14. Dimić, I., Cvijović-Alagić, I., Volker, B., Hohenwarter, A., Pippan, R., Veljović, Đ, Rakin, M., Bugarski, B.: Microstructure and metallic ion release of pure titanium and Ti-13Nb-13Zr alloy processed by high pressure torsion. Mater. Des. **5**, 340–347 (2016)
15. Barjaktarević, D., Medjo, B., Štefane, P., Gubeljak, N., Cvijović-Alagić, I, Djokić, V., Rakin, M.: Tensile and corrosion properties of anodized ultrafine-grained Ti-13Nb-13Zr biomedical alloy obtained by high-pressure torsion. Metals Mater. Int. in press (2021)
16. Barjaktarević, D., Dimić, I., Cvijović-Alagić, I., Veljović, Dj., Rakin, M.: Corrosion resistance of high pressure torsion obtained commercially pure titanium in acidic solution. Tech. Gazette **24**(6), 1689–1695 (2017)
17. Lin, Z., Wang, L., Yeung, K., Qin, J.: The ultrafine-grained titanium and biomedical titanium alloys processed by severe plastic deformation (SPD). SOJ Mater. Sci. Eng. **1**(1), 1–6 (2013)
18. Sankar, M., Gopal, V., Alexander, R., Manivasagam, G., Ramalingam, M.: Nanobiomaterials—Classification, Fabrication and Biomedical Applications. Wiley-VCH Verlag GmbH & Co. (2018)
19. Estrin, Y., Vinogradov, A.: Extreme grain refinement by severe plastic deformation: a wealth of challenging science. Acta Mater. **61**(3), 780–786 (2013)
20. Edalati, K., Horita, Z.: A review on high-pressure torsion (HPT) from 1935 to 1988. Mater. Sci. Eng. A **652**, 326–335 (2016)
21. Ashida, M., Chen, P., Doi, H., Tsutsumi, Y., Hanawa, T., Horita, Z.: Microstructures and mechanical properties of Ti-6Al-7Nb processed by high-pressure torsion. Procedia Eng. **81**, 1523–1528 (2014)
22. Sharman, K., Bazarnik, P., Brynk, T., Bulutsuz, A., Lewandowska, M., Huang, Y., Langdon, T.: Enhancement in mechanical properties of a titanium alloy by high-pressure torsion. J. Mater. Res. Technol. **4**(1), 79–83 (2015)
23. Shirooyeh, M., Xu, J., Langdon, T.: Microhardness evolution and mechanical characteristics of commercial purity titanium processed by high-pressure torsion. Mater. Sci. Eng. A **614**, 223–231 (2014)
24. Cho, T., Lee, H., Ahn, B., Kawasaki, M., Langdon, T.: Microstructural evolution and mechanical properties in a Zn–Al eutectoid alloy processed by high-pressure torsion. Acta Mater. **72**, 67–79 (2014)
25. Shahmir, H., Nili-Ahmadabadi, M., Huang, Y., Langdon, T.: Evolution of microstructure and hardness in NiTi shape memory alloys processed by high-pressure torsion. J. Mater. Sci. **49**, 2998–3009 (2014)
26. Minagar, S., Berndt, C., Wang, J., Ivanova, E., Wen, C.: A review of the application of anodization for the fabrication of nanotubes on metal implant surfaces. Acta Biomater. **8**(8), 2875–2888 (2012)
27. Ishizawa, H., Fugino, M., Ogino, M.: Mechanical and histological investigation of hydrothermally treated and untreated anodic titanium oxide films containing Ca and P. J. Biomed. Mater. Res. **29**(11), 1459–1468 (1995)
28. Kulkarni, M., Mazare, A., Schmuki, P., Iglič, A.: Biomaterial Surface Modification of Titanium and Titanium Alloys for Medical Applications. One Central Press, Nanomedicine (2014)
29. Kasuga, T., Hiramatsu, M., Hoson, A., Sekino, T., Niihara, K.: Formation of titanium oxide nanotube. Langmuir **14**(12), 3160–3163 (1998)
30. Zhang, S., Li, W., Jin, Z., Yang, J., Zhang, J., Du, Z., Zhang, Z.: Study on ESR and inter-related properties of vacuum-dehydrated nanotubed titanic acid. J. Solid State Chem. **177**(4–5), 1365–1371 (2004)
31. Tsai, C., Nian, N., Teng, H.: Mesoporous nanotube aggregates obtained from hydrothermally treating TiO_2 with NaOH. Appl. Surf. Sci. **253**(4), 1898–1902 (2006)
32. Ossowska, A., Sobieszczyk, S., Supernak, M., Zielinski, A.: Morphology and properties of nanotubular oxide layer on the Ti–13Zr–13Nb alloy. Surf. Coat. Technol. **258**, 1239–1248 (2014)
33. Feng, X., Macak, J., Schmuki, P.: Flexible self-organization of two size-scales oxide nanotubes on Ti-45Nb alloy. Electrochem. Commun. **9**(9), 2403–2407 (2007)

34. Crawford, G., Chawla, N., Da, K., Bose, S., Bandyopadhya, A.: Microstructure and deformation behavior of biocompatible TiO_2 nanotubes on titanium substrate. Acta Biomater. **3**(3), 359–367 (2007)
35. ISO 14577–4: Metallic materials-Instrumented indentation test for hardness and materials parameters - Part 4: Test method for metallic and non-metallic coatings, ISO (2007)
36. Fischer-Cripps, C.: Nanoindentation. Springer, New York (2004)
37. Raković, D., Uskoković, D.: Biomaterials. Materials Research Society of Serbia, Belgrade, Institute of Technical Sciences of SASA (2010)
38. Merle, B., Maier, V., Göken, M., Durst, K.: Experimental determination of the effective indenter shape and e-factor for nanoindentation by continuously measuring the unloading stiffness. J. Mater. Res. **27**(1), 214–221 (2012)
39. Oliver, W., Pharr, G.: Improved technique for determining hardness and elastic modulus using load and displacement sensing indentation experiments. J. Mater. Res. **7**(6), 1564–1583 (1992)
40. Crawford, G., Chawla, N., Houston, J.: Nanomechanics of biocompatible TiO_2 nanotubes by Interfacial Force Microscopy (IFM). J. Mech. Behav. Biomed. Mater. **2**(6), 580–587 (2009)
41. Xu, Y., Liu, M., Wang, M., Oloyede, A., Bell, J., Yan, C.: Nanoindentation study of the mechanical behavior of TiO_2 nanotube arrays. J. Appl. Phys. **118**(14), 145301 (2015)
42. Zielinski, A., Antoniuk, P., Krzysztofowicz, K.: Nanotubular oxide layers and hydroxyapatite coatings on Ti–13Zr–13Nb alloy. Surf. Eng. **30**(9), 643–649 (2014)
43. Jouanny, I., Labdi, S., Aubert, P., Buscema, C., Maciejak, O., Berger, H., Guipont, V., Jeandin, M.: Structural and mechanical properties of titanium oxide thin films for biomedical application. Thin Solid Films **518**(12), 3212–3217 (2010)

Occurrence and Ecotoxicological Risk Assessment of Emerging Contaminants in Urban Wastewater Treatment Plant

Ivana Matić Bujagić, Eleonora Gvozdić, Tatjana Đurkić, and Svetlana Grujić

Abstract This is the first study of a broad range of chemical classes of emerging contaminants conducted by analyzing influent and effluent samples from the wastewater treatment plant of the city Topola, in Serbia. The list of compounds is extensive and this paper provides a better understanding of the environmental burden from different classes of emerging contaminants. The samples were prepared using an optimized solid-phase extraction method and analyzed by liquid chromatography-tandem mass spectrometry. Removal patterns of selected compounds are discussed based on their physico-chemical properties and detected concentrations. Significant removal efficiencies, exceeding 70%, were found for the majority of investigated pharmaceuticals, pesticides, steroids, and sweeteners. Ecotoxicological risk assessment was performed by using two complementary methods: (1) an individual substance approach, based on the calculation of risk quotients (RQs) for each substance as the ratio of Predicted Environmental Concentration (PEC) and Predicted No Effect Concentration (PNEC), and (2) mixture risk assessment ("the cocktail effect") based on the summation of individual RQs. The classical approach (ERA method with individual substances) identified amlodipine as the riskiest substance in WWTP effluent. The mixture ERA approach revealed new risks, which were not recognized by the classical ERA method, indicating that individually "safe" emerging compounds can contribute to a significant risk of the whole effluents.

Keywords Emerging contaminants · Wastewater treatment plant · Liquid chromatography-tandem mass spectrometry · Ecotoxicological risk assessment

I. Matić Bujagić (✉)
Academy of Applied Technical Studies Belgrade, Belgrade Polytechnic College, Katarine Ambrozić 3, 11000 Belgrade, Serbia
e-mail: imatic@politehnika.edu.rs

E. Gvozdić
Innovation Center of the Faculty of Technology and Metallurgy, Karnegijeva 4, 11000 Belgrade, Serbia

T. Đurkić · S. Grujić
Faculty of Technology and Metallurgy, University of Belgrade, Karnegijeva 4, 11000 Belgrade, Serbia

N. Mitrovic et al. (eds.), *Current Problems in Experimental and Computational Engineering*, Lecture Notes in Networks and Systems 323,
https://doi.org/10.1007/978-3-030-86009-7_8

1 Introduction

The list of chemical compounds that can be frequently found in the literature as "emerging substances" is constantly growing. The presence of these compounds in the aquatic environment and wastewater has been well documented [1–6] and is known to pose an adverse long-term risk to human health and aquatic ecosystems even at low concentrations [7–10]. Urban wastewater treatment plants (WWTPs) have been designed to remove high levels of conventional pollutants such as oil and grease, coliform fecal bacteria, nitrogen, phosphorus, and organic matter. However, many studies have shown that the removal of emerging contaminants remains very poor, and therefore the effluents from wastewater treatment plants still represent one of the main sources of emerging substances discharges into the aquatic environment [11–13]. As a consequence, these pollutants are continuously released in trace amounts (typically ranging from ng/L to μg/L) into receiving watercourses. It should be also noted that these substances are usually not detected individually, but rather as a complex mixture, so their "cocktail effects" should be taken into account when assessing the risks to humans and the environment.

In the WWTPs, emerging substances undergo various processes, such as adsorption onto suspended matter, biodegradation, or chemical degradation. The removal efficiency of these substances varies significantly depending on their physico-chemical properties, operational parameters of the plant, and the type of treatment process applied. The main aim of this work is to determine the occurrence and fate of a broad range of chemical classes of emerging compounds in WWTP of the city Topola, in Serbia. The list of compounds analyzed in this paper is extensive and comprises 23 different pharmaceuticals, 16 pesticides, 20 steroids (combination of steroid hormones and sterols), and 8 sweeteners (seven artificial and one natural). Finally, the present study is aimed to develop and apply a new ecotoxicological risk assessment (ERA) method for emerging substances released from WWTP effluents into freshwater watercourses. So far, the majority of published papers [14–16] were based on the analysis of substance measured in environmental waters and did not consider the risk posed by the release of the pollutants by WWTP discharges. The ERA method applied in this work will assess the risk of each pollutant alone based on the comparison of Predicted Environmental Concentrations (PEC) and Predicted No Effect Concentration (PNEC) values, following European guidelines [17], but also it will take into account the "cocktail effect" due to the mixture of emerging substance in WWTP effluents in the territory, using a recently developed procedure [18–20].

2 Materials and Methods

2.1 Chemicals and Reagents

Analytical standards of selected pharmaceuticals were supplied from Hemofarm (STADA Group, Vršac, Serbia), whereas pesticide standards were obtained from Riedel-de Haën (Seelze, Germany). Steroid analytical standards were purchased from Steraloids Inc. (Newport, US) and analytical standards of artificial sweeteners were obtained from Sigma-Aldrich (Buchs, Switzerland), except sucralose which was purchased from TCI Europe (Zwijndrecht, Belgium). All standards of the investigated compounds were of high purity grade (>95%). The list of 67 initially selected analytes, their chemical classes and physico-chemical properties are presented in Table 1.

The stock standard solutions were prepared in methanol at a concentration of 100 μg/mL. The working standard solutions in the concentration range 10–1,000 ng/mL were prepared by mixing the appropriate amounts of the stock standard solutions and diluting with methanol. All solutions were preserved at −4 °C. All solvents used were HPLC grade from J. T. Baker (Center Valley, US) or Sigma-Aldrich (St. Louis, US). Ammonium acetate and concentrated acetic acid were of the analytical grade. Deionized water was obtained by passing the distilled water through a GenPure ultrapure water system (TKA, Niederelbert, Germany).

2.2 Studied Area

The samples were collected from a small wastewater treatment plant in town Topola, in the Republic of Serbia. The incoming load of the facility in population equivalent is 8000, and the annual mean incoming flow rate of the WWTP influent is 1089 m^3/day. The annual mean flow rate of the WWTP effluent is 978 m^3/day. Only households and catering facilities are connected to the city sewerage network, and therefore mechanical and biological treatments are applied in the plant. The secondary treatment process is based on an activated sludge system. Wastewater effluents are discharged to the river Kamenica. The average flow of the recipient i.e. receiving watercourse Kamenica at the time of the sample collection was 30 m^3/s.

2.3 Sample Preparation Procedure

Wastewater samples were prepared for analysis using solid-phase extraction (SPE). The used SPE protocol has been previously developed for the isolation and preconcentration of selected pharmaceuticals and pesticides from the water matrix [21]. The optimized method showed high recoveries for all investigated analytes

Table 1. Analytes selected for investigation: chemical class, molecular weight (M_w) and water solubility (WS).

Analyte	Chemical class	M_w, g/mol	WS[a], mg/L
Pharmaceuticals			
Trimethoprim	Used in combination with sulfonamide antibiotics	290	400
4-Acetylaminoantipyrine (4-AAA)	The final metabolite of analgesic/antipyretic metamizole	245	–
4-Formylaminoantipyrine (4-FAA)	The final metabolite of analgesic/antipyretic metamizole	231	–
Sulfamethoxazole	Sulfonamide antibiotic	253	610
Azithromycin	Macrolide antibiotic	748	7.1
Erythromycin	Macrolide antibiotic	733	1.4
Midecamycin	Macrolide antibiotic	814	Soluble in acidic water
Clarithromycin	Macrolide antibiotic	748	1.693[b]
Roxithromycin	Macrolide antibiotic	837	181
Doxycycline	Tetracycline antibiotic	444	630[b]
Metoprolol	Antihypertensive, β-blocker	267	4,780
Bisoprolol	Antihypertensive, β-blocker	325	2,240
Enalapril	Antihypertensive, ACE inhibitor	376	16,400
Cilazapril	Antihypertensive, ACE inhibitor	417	27.5[b]
Amlodipine	Antihypertensive, calcium channel blocker	408	75.3
Atorvastatin	Antihyperlipemic, statin	558	$1.1 \cdot 10^{-3}$
Simvastatin	Antihyperlipemic, statin	418	0.03
Clopidogrel	Anticoagulant	321	50.8
Bromazepam	Sedative, benzodiazepin	315	175
Lorazepam	Sedative, benzodiazepin	320	80.0
Diazepam	Sedative, benzodiazepin	284	50.0
Carbamazepine	Antiepileptic	236	17.7[b]
Diclofenac	Analgesic/antipyretic	296	2.4
Pesticides			
Acephate	Organophosphate insecticide	183	$8.2 \cdot 10^{5}$

(continued)

Table 1. (continued)

Analyte	Chemical class	M_w, g/mol	WS[a], mg/L
Monocrotophos	Organophosphate insecticide	233	$1.0 \cdot 10^6$
Dimethoate	Organophosphate insecticide	229	23,300
Malathion	Organophosphate insecticide	330	143
Carbendazim	Benzimidazole fungicide	191	29.0
Imidacloprid	Neonicotinoid insecticide	255	610
Acetamiprid	Neonicotinoid insecticide	222	4,200
Monuron	Phenylurea herbicide	198	230
Diuron	Phenylurea herbicide	232	42.0
Linuron	Phenylurea herbicide	249	75.0
Carbaryl	Carbamate insecticide	201	110
Carbofuran	Carbamate insecticide	221	320
Simazine	Triazine herbicide	201	6.2
Atrazine	Triazine herbicide	215	34.7
Propazine	Triazine herbicide	229	8.6
Tebufenozide	Diacylhydrazine insecticide	352	0.83
Steroids			
Estriol	Steroid hormone	288	500
Estrone (E1)	Steroid hormone	270	30.0
Equilin	Steroid hormone	268	1.4
Norethindrone	Steroid hormone	298	7.0
17α-Ethinylestradiol (EE2)	Steroid hormone	296	11.3
17β-Estradiol (E2)	Steroid hormone	272	3.6
17α-Estradiol	Steroid hormone	272	3.9
Levonorgestrel	Steroid hormone	312	2.1
Mestranol	Steroid hormone	310	0.30
Epicoprostanol	Human/animal sterol	388	$3.4 \cdot 10^{-4}$ [b]
Epicholestanol	Human/animal sterol	388	$3.5 \cdot 10^{-5}$
Coprostanol	Human/animal sterol	388	$2.0 \cdot 10^{-2}$
Cholestanol	Human/animal sterol	388	$8.8 \cdot 10^{-5}$
Cholesterol	Human/animal sterol	386	$9.5 \cdot 10^{-2}$
Cholestanone	Human/animal sterol	386	$2.9 \cdot 10^{-4}$ [b]
Desmosterol	Plant sterol	384	$2.0 \cdot 10^{-4}$

(continued)

Table 1. (continued)

Analyte	Chemical class	M_w, g/mol	WS[a], mg/L
Stigmasterol	Plant sterol	412	$1.1 \cdot 10^{-5}$
Campesterol	Plant sterol	400	$2.8 \cdot 10^{-5}$
β-Sitosterol	Plant sterol	414	$1.3 \cdot 10^{-5}$
Sitostanol	Plant sterol	416	$9.8 \cdot 10^{-6}$
Artificial sweeteners			
Acesulfame	Sulfamate ester	163	$9.1 \cdot 10^{5}$
Saccharin	Benzisothiazole	183	789.2
Cyclamate	Salt of cyclamic acid	179	$1.0 \cdot 10^{6}$
Sucralose	Disaccharide	397	$2.3 \cdot 10^{4}$
Aspartame	Dipeptide	294	564.7
Neohesperidin dihydrochalcone; NHDC	Dihydrochalcone	612	$2.0 \cdot 10^{3}$
Neotame	Dipeptide	378	14.4
Stevioside	Diterpene glycoside	805	$4.5 \cdot 10^{3}$

[a] *Source* https://www.srcinc.com
[b] *Source* US EPA. [2012]. Estimation Programs Interface Suite™ for Microsoft® Windows, v 4.11. US EPA, Washington, DC, US

(ranging from 63.9% to 141.7%). Briefly, the volume of 100 mL of the wastewater was adjusted at pH = 6. The OASIS HLB cartridges, used for extraction and preconcentration of the target analytes, were preconditioned with 5 mL of methanol/dichloromethane mixture (1:1, *v/v*) followed by 5 mL of deionized water. Wastewater samples were loaded onto cartridges, and afterwards the cartridges were dried under vacuum for 10 min. The elution of analytes was performed with 15 mL of methanol/dichloromethane mixture (1:1, *v/v*). Extracts were evaporated and reconstituted to 1 mL with methanol. The final extracts were filtered through 0.45 μm polyvinylidene difluoride (PVDF) filters, acquired from Roth (Karlsruhe, Germany), into the autosampler vials and analyzed.

2.4 Calibration

The standard addition method was used for calibration. This calibration approach is often used when it is necessary to take into account the matrix effect (i.e. ion suppression or enhancement) and the incomplete analyte extraction. Each water sample was split into six aliquots. Four aliquots were used for the preparation of the calibration solutions by spiking the wastewater samples with working standard solution at the concentrations of 1–250 μg/L (for pharmaceuticals, pesticides, steroid hormones, and sweeteners) and 10–5000 μg/L (for sterols).

2.5 LC-MS²Analysis of Emerging Contaminants

After extraction, samples were analyzed by liquid chromatography-tandem mass spectrometry using DionexUltiMate® 3000 LC system (Thermo Fisher Scientific, Waltham, US) coupled to linear ion trap LTQ XL (Thermo Fisher Scientific). For efficient separation of all analytes, three chromatographic columns were used. Separation of pharmaceuticals, pesticides, and steroid hormones was performed using reverse-phase Zorbax Eclipse® XDB-C18 column (75 mm × 4.6 mm, 3.5 μm, Agilent Technologies, Santa Clara, US), whereas LiChrospher RP-18 EC column (250 mm × 4.6 mm, 5 μm, Cronus, SMI-LabHut Ltd., UK) was employed for the chromatographic separation of sterols. Separation of artificial sweeteners was achieved on Luna® C8 column (3.0 mm × 150 mm, 3 μm) from Phenomenex, Torrance, US. In front of the separation columns, a precolumn was installed (12.5 mm × 4.6 mm, 5 μm, Agilent Technologies).

For chromatographic separation of pharmaceuticals and pesticides the mobile phase consisted of water (A), methanol (B), and 10% acetic acid (C). When analyzing pharmaceuticals, mobile phase gradient (with the flow rate of 0.6 mL/min) changed as follows: 0 min, A 65%, B 33%, C2%; 12 min, B 98%, C2%; 18 min, B 100%. In the case of pesticides, mobile phase gradient changed in the following manner: 0 min, A 66%, B 33%, C 1%; 7.5 min, A 41%, B 58%, C 1%; 25 min, B 100%. The flow rate of the mobile phase was 0.5 mL/min. The mobile phase for chromatographic separation of steroids consisted of water (A) and methanol (B). Steroid hormones were separated using the following mobile phase gradient (flow rate 0.8 mL/min): 0 min, A 45%, B 55%; 13 min, B 100%. For sterols mobile phase was changed as follows: 0 min, B 100%; 12 min, A 10%, B 90%; 15 min, B 100%. The flow rate of the mobile phase was held at 1.5 mL/min. The mobile phase for the separation of sweeteners consisted of water (A), methanol (B), and 0.1 mol/L aqueous solution of ammonium acetate (D). The mobile phase gradient was changed in the following manner: 0 min, A 84%, B 15%, D 1%; 8 min, A 84%, B 15%, D 1%; 13 min, A 34%, B 65%, D 1%; 15 min, B 100%; 20 min, B 100%. The flow rate was 0.33 mL/min. In all chromatographic methods the initial conditions were re-established and held for 10 min. An aliquot of 10 μL of the final extract was injected into the LC system.

Two ionization interfaces of mass spectrometer were used for obtaining stabile ions of the selected analytes. All pharmaceuticals and pesticides were successfully ionized by the electrospray ionization (ESI) technique in the positive mode. The same ionization technique in the negative mode was used for the selected sweeteners. The optimal ESI source working parameters for monitoring all ions were source voltage of 4.5 kV and capillary temperature of 290 °C. Atmospheric pressure chemical ionization (APCI) in the positive mode was applied in the steroid analysis. The optimized APCI parameters were capillary temperature of 200 °C and vaporizer temperature

Table 2. MS operating parameters for selected pharmaceuticals: analytes' fragmentation reactions for quantification and confirmation purposes and optimal collision energies (CE).

Pharmaceuticals	Precursor ion (m/z)	Quantification reaction	CE (%)	Conformation reaction	CE (%)
Trimethoprim	$291[M+H]^+$	291 → 230	44	291 → 123	44
4-AAA	$246[M+H]^+$	246 → 228	28	246 → 204	28
4-FAA	$232[M+H]^+$	232 → 204	30	232 → 214	30
Metoprolol	$268[M+H]^+$	268 → 191	37	268 → 218	37
Sulfamethoxazole	$254[M+H]^+$	254 → 188	34	254 → 156	34
Azithromycin	$749[M+H]^+$	749 → 591	30	591 → 434	28
Bisoprolol	$326[M+H]^+$	326 → 116	31	326 → 222	31
Doxycycline	$445[M+H]^+$	445 → 428	25	445 → 460	25
Enalapril	$377[M+H]^+$	377 → 234	30	377 → 303	30
Erythromycin	$734[M+H]^+$	734 → 576	26	734 → 716	26
Bromazepam	$316[M+H]^+$	316 → 288	36	288 → 261	35
Amlodipine	$409[M+H]^+$	409 → 238	25	409 → 294	25
Midecamycin	$814[M+H]^+$	814 → 614	25	814 → 596	25
Carbamazepine	$237[M+H]^+$	237 → 194	34	237 → 219	34
Clarithromycin	$748[M+H]^+$	748 → 590	24	748 → 558	24
Roxithromycin	$837[M+H]^+$	837 → 679	23	837 → 558	23
Lorazepam	$321[M+H]^+$	321 → 303	32	303 → 275	26
Diazepam	$285[M+H]^+$	285 → 257	40	257 → 228	39
Atorvastatin	$559[M+H]^+$	559 → 466	25	559 → 440	25
Diclofenac	$296[M+H]^+$	296 → 278	28	278 → 250	22
Clopidogrel	$322[M+H]^+$	322 → 212	28	212 → 184	23
Simvastatin	$419[M+H]^+$	419 → 285	21	419 → 199	21
Cilazapril	$418[M+H]^+$	418 → 211	25	211 → 183	32

of 400 °C. Fragmentation reactions of the precursor ion to the most intense fragment ion were used for identification and quantification of each analyte. Additional transitions were used for the confirmation of positive results. Detailed information on five separate analytical methods, including mass spectrometric parameters for the data acquisition, as well as fragmentation reactions for quantification and conformation of all selected analytes can be seen in Tables 2, 3, 4, 5 and 6.

2.6 *Removal Efficiency of Selected Emerging Contaminants*

To determine the removal efficiency of selected emerging contaminants during the wastewater treatment process, the following equation Eq. (1) was applied:

Table 3. MS operating parameters for selected pesticides: analytes' fragmentation reactions for quantification and confirmation purposes and optimal collision energies (CE).

Pesticides	Precursor ion (m/z)	Quantification reaction	CE (%)	Conformation reaction	CE (%)
Acephate	$184[M+H]^+$	184 → 143	40	184 → 113	40
Monocrotophos	$224[M+H]^+$	224 → 193	38	224 → 167	38
Carbendazim	$192[M+H]^+$	192 → 160	34	160 → 132	35
Imidacloprid	$256[M+H]^+$	256 → 210	25	256 → 175	25
Acetamiprid	$223[M+H]^+$	223 → 126	36	223 → 187	36
Dimethoate	$230[M+H]^+$	230 → 199	26	199 → 171	22
Monuron	$199[M+H]^+$	199 → 72	30	–	–
Carbaryl	$202[M+H]^+$	202 → 145	25	145 → 117	31
Simazine	$202[M+H]^+$	202 → 124	36	202 → 132	36
Carbofuran	$222[M+H]^+$	222 → 165	32	165 → 123	27
Atrazine	$216[M+H]^+$	216 → 174	38	174 → 146	35
Diuron	$233[M+H]^+$	233 → 72	34	–	–
Propazine	$230[M+H]^+$	230 → 188	35	230 → 146	35
Linuron	$249[M+H]^+$	249 → 182	35	249 → 160	35
Malathion	$331[M+H]^+$	331 → 285	24	285 → 127	20
Tebufenozide	$375[M+Na]^+$	375 → 225	34	375 → 319	34

$$RE(\%) = \frac{C_i - C_e}{C_i} \times 100 \quad (1)$$

With: C_i—the quantified concentration of the pollutant in the WWTP influent, in μg/L; C_e—the quantified concentration of the pollutant in the WWTP effluent, in μg/L.

2.7 *Ecotoxicological Risk Assessment for the Receiving Watercourse*

For the collected WWTP effluent sample, an ecotoxicological risk assessment was performed by linking it to each pollutant, but also with the mixture of pollutants ("the cocktail effect") that is being released into the receiving watercourse. ERA methods are usually carried out by comparison of the Predicted Environmental Concentration of a substance in watercourses and Predicted No Effect Concentration at which no pharmacological effect is expected to occur for a specific organism. A PNEC value is generally derived using ecotoxicity testing data.

Table 4. MS operating parameters for selected steroids: analytes' fragmentation reactions for quantification and confirmation purposes and optimal collision energies (CE).

Steroids	Precursor ion (m/z)	Quantification reaction	CE (%)	Conformation reaction	CE (%)
Steroid hormones					
Estriol	$271[M–H_2O + H]^+$	271 → 253	20	271 → 197	20
Estrone	$271[M + H]^+$	271 → 253	20	271 → 197	20
Equilin	$269[M + H]^+$	269 → 251	23	269 → 211	23
Norethindrone	$299[M + H]^+$	299 → 281	23	299 → 263	23
17α-Ethinylestradiol	$279[M–H_2O + H]^+$	279 → 133	25	279 → 205	25
17β-Estradiol	$255[M–H_2O + H]^+$	255 → 159	22	255 → 133	22
17α-Estradiol	$255[M–H_2O + H]^+$	255 → 159	22	255 → 133	22
Levonorgesterel	$313[M + H]^+$	313 → 295	22	313 → 277	22
Mestranol	$293[M–H_2O + H]^+$	293 → 147	26	293 → 173	26
Sterols					
Epicoprostanol	$371[M–H_2O + H]^+$	371 → 149	24	371 → 261	24
Epicholestanol	$371[M–H_2O + H]^+$	371 → 149	24	371 → 261	24
Coprostanol	$371[M–H_2O + H]^+$	371 → 149	24	371 → 261	24
Cholestanol	$371[M–H_2O + H]^+$	371 → 149	24	371 → 261	24
Cholesterol	$369[M–H_2O + H]^+$	369 → 243	24	369 → 287	24
Cholestanone	$387[M + H]^+$	387 → 369	19	387 → 243	19
Desmosterol	$367[M–H_2O + H]^+$	367 → 257	26	367 → 161	26
Stigmasterol	$395[M–H_2O + H]^+$	395 → 297	24	395 → 311	24
Campesterol	$383[M–H_2O + H]^+$	383 → 243	25	383 → 257	25
β-Sitosterol	$397[M–H_2O + H]^+$	397 → 243	25	397 → 257	25
Sitostanol	$399[M–H_2O + H]^+$	399 → 149	24	399 → 163	24

Table 5. MS operating parameters for selected sweeteners: analytes' fragmentation reactions for quantification and conformation purposes and optimal collision energies (CE).

Sweeteners	Precursor ion (m/z)	Quantification reaction	CE (%)	Confirmation reaction	CE (%)
Acesulfame	$162[M–H]^-$	162 → 82	27	162 → 102	27
Saccharin	$182[M–H]^-$	182 → 106	37	182 → 62	37
Cyclamate	$178[M–H]^-$	178 → 80	36	178 → 96	36
Sucralose	$433[M + Cl]^-$	433 → 397	14	433 → 395	14
Aspartame	$293[M–H]^-$	293 → 261	29	293 → 200	29
NHDC	$611[M–H]^-$	611 → 491	20	611 → 387	20
Neotame	$377[M–H]^-$	377 → 345	18	377 → 200	18
Stevioside	$641[M–C_6H_{11}O_5–H]^-$	641 → 479	22	641 → 317	22

Table 6. Median pollutant concentrations in effluent (C_e), Predicted Environmental Concentrations (PEC) of each pollutant, Predicted No Effect Concentrations (PNEC) and the associated Risk Quotients (RQ).

Analyte	C_e, μg/L	PEC, μg/L	PNEC, μg/L	PNEC reference	RQ
Trimethoprim	15.01	$5.66 \cdot 10^{-3}$	0.0058	[19]	$9.76 \cdot 10^{-1}$
4-AAA	10.78	$4.07 \cdot 10^{-3}$	– [a]		–
4-FAA	21.19	$8.00 \cdot 10^{-3}$	–		–
Metoprolol	6.17	$2.33 \cdot 10^{-3}$	0.1	[19]	$2.33 \cdot 10^{-2}$
Sulfamethoxazole	0.84	$3.18 \cdot 10^{-4}$	0.59	[19]	$5.39 \cdot 10^{-4}$
Azithromycin	2.41	$9.10 \cdot 10^{-4}$	0.0094	[26]	$9.68 \cdot 10^{-2}$
Bisoprolol	8.18	$3.09 \cdot 10^{-3}$	72	[27]	$4.29 \cdot 10^{-5}$
Bromazepam	4.93	$1.86 \cdot 10^{-3}$	17.4	[28]	$1.07 \cdot 10^{-4}$
Amlodipine	6.19	$2.34 \cdot 10^{-3}$	0.00028	[29]	8.35
Roxithromycin	3.19	$1.20 \cdot 10^{-3}$	0.01	[19]	$1.20 \cdot 10^{-1}$
Diazepam	34.08	$1.29 \cdot 10^{-2}$	0.1	[26]	$1.29 \cdot 10^{-1}$
Clopidogrel	36.38	$1.37 \cdot 10^{-2}$	1.6	[30]	$8.58 \cdot 10^{-3}$
Simvastatin	0.24	$8.98 \cdot 10^{-5}$	22.8	[31]	$3.94 \cdot 10^{-6}$
Acephate	71.83	$2.71 \cdot 10^{-2}$	110	[32]	$2.46 \cdot 10^{-4}$
Carbendazim	10.41	$3.93 \cdot 10^{-3}$	1.5	[33]	$2.62 \cdot 10^{-3}$
Imidacloprid	4.90	$1.85 \cdot 10^{-3}$	121	[34]	$1.53 \cdot 10^{-5}$
Propazine	9.30	$3.51 \cdot 10^{-3}$	0.04	[35]	$8.77 \cdot 10^{-2}$
Mestranol	21.64	$8.17 \cdot 10^{-3}$	130	[36]	$6.28 \cdot 10^{-5}$
Acesulfame	0.21	$7.92 \cdot 10^{-5}$	2200	[37]	$3.60 \cdot 10^{-8}$
Sucralose	4.20	$1.58 \cdot 10^{-3}$	930	[38]	$1.70 \cdot 10^{-6}$
RQ_{mix}					9.79

[a]The PNEC values were not available in the literature

Predicted Environmental Concentration (PEC) calculation

The PEC value of each detected emerging contaminant in the receiving watercourse was calculated by taking into account the WWTP and watercourse flow rates and the concentration of each analyte quantified by LC-MS^2 effluent analysis using the following Eq. (2):

$$PEC = \frac{\text{WWTP flow rate}}{\text{Watercourse flow rate}} \times C_e \tag{2}$$

With PEC—Predicted Environmental Concentration of the pollutant in μg/L; WWTP flow rate—the flow rate of the WWTP in m^3/s; Watercourse flow rate—the flow rate of the WWTP's receiving watercourse in m^3/s and C_e—the detected concentration of the pollutant in the WWTP effluent.

Ecotoxicological risk assessment of the individual micropollutant

Ecotoxicological risk assessment is usually performed by the calculation of the risk quotient (RQ) (EC, 2003) using Eq. (3):

$$RQ = \frac{PEC}{PNEC} \quad (3)$$

With RQ—Risk Quotient of the pollutant detected WWTP effluent; PEC—Predicted Environmental Concentration of the pollutant in μg/L and PNEC—Predicted No Effect Concentration of the pollutant in μg/L. PNECs are usually calculated based on critical concentrations, e.g. EC50 (median effective concentration), LC50 (median lethal concentration), and NOEC (no-observed-effect concentration) [22].

An RQ value below 1 is associated with insignificant ecotoxicological risk, and an RQ value above 1 to a potential ecotoxicological risk for watercourses. The environmental risk ranking categories are as follows: $RQ < 0.01$ is insignificant, < 0.1 low risk, $0.1 \leq RQ \leq 1$ medium risk and $RQ > 1$ high risk [16].

Ecotoxicological risk assessment of the micropollutant mixture

The risk quotient of the micropollutant mixture was calculated according to the procedure suggested by Gosset et al. [19] following Eq. (4):

$$RQ_{mix} = \sum_{1}^{n} \frac{PEC}{PNEC} = \sum_{1}^{n} RQ \quad (4)$$

With RQ_{mix}—Risk quotient of the mixture of pollutants; RQ—Risk quotient of the individual pollutant detected in WWTP. Similar to comparing RQ of individual substances to 1, the same is done for RQ_{mix}.

3 Results and Discussion

All four classes of the investigated organic contaminants were detected in the influent and effluent samples of the investigated WWTP in various concentration ranges summarized in Figs. 1, 2 and 3. The widespread occurrence of sterols (Fig. 1) was detected in WWTP influent, with the maximum concentration of cholesterol (4128 μg/L), followed by coprostanol (3705 μg/L). Since cholesterol is the most abundant sterol in the human organism, it is expected to have high concentrations in sewage-contaminated samples. A high concentration of coprostanol can be explained by the fact that coprostanol is a sterol produced in the digestive tract of humans and higher vertebrates by hydrogenation of cholesterol and it comprises 40–60% of the total sterols excreted in human feces [23]. However, none of the monitored sterols were detected in effluent samples.

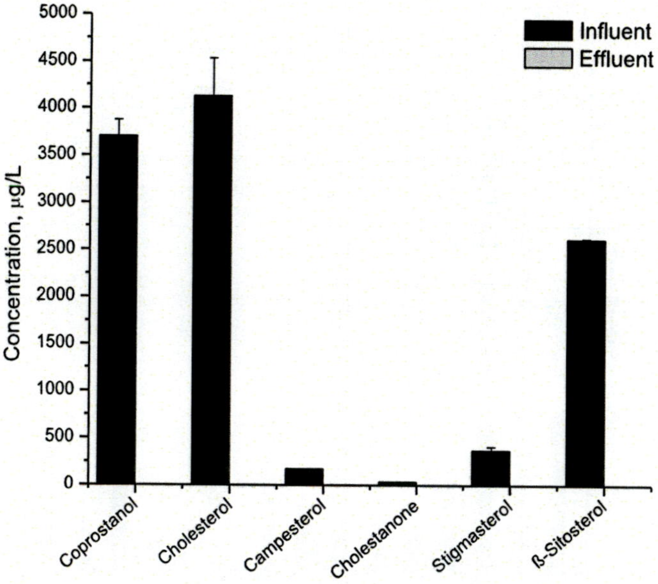

Fig. 1. Representation of the mean influent and effluent concentrations ± standard deviation of detected sterols.

Notable concentrations in the WWTP influent were also recorded in the case of metamizole metabolites, 4-FAA and 4-AAA (up to 107 µg/L), bisoprolol (83 µg/L), and clopidogrel (61 µg/L) (Fig. 2).

Pesticides acephate and carbendazim were found at the concentrations of 254 µg/L and 83 µg/L, respectively, whereas out of all monitored sweeteners, saccharin was detected at the highest concentration of 39 µg/L. It can be also concluded that for the majority of compounds the concentrations in the effluent sample are significantly lower, indicating partial removal of these compounds in WWTP.

3.1 *Removal Efficiency of the Detected Micropollutants*

Overall, removal efficiency (RE) of the target compounds was over 70% for 30 detected analytes (Fig. 4). Differences in RE of pharmaceuticals (ranging from 18 to 100%) can be attributed to different physical and chemical properties of the compounds, to distinct mechanisms of degradation, sorption/sedimentation processes and uptake by active sludge [24]. Complete removal of sterols in WWTP (100%) can be associated with their physico-chemical properties, i.e. low water solubility and polarity, indicating that their primary mechanism of removal is adsorption onto active sludge particles. High RE was also observed in the case of pesticides and sweeteners, except for sucralose (RE = 42%). For sucralose, sorption cannot

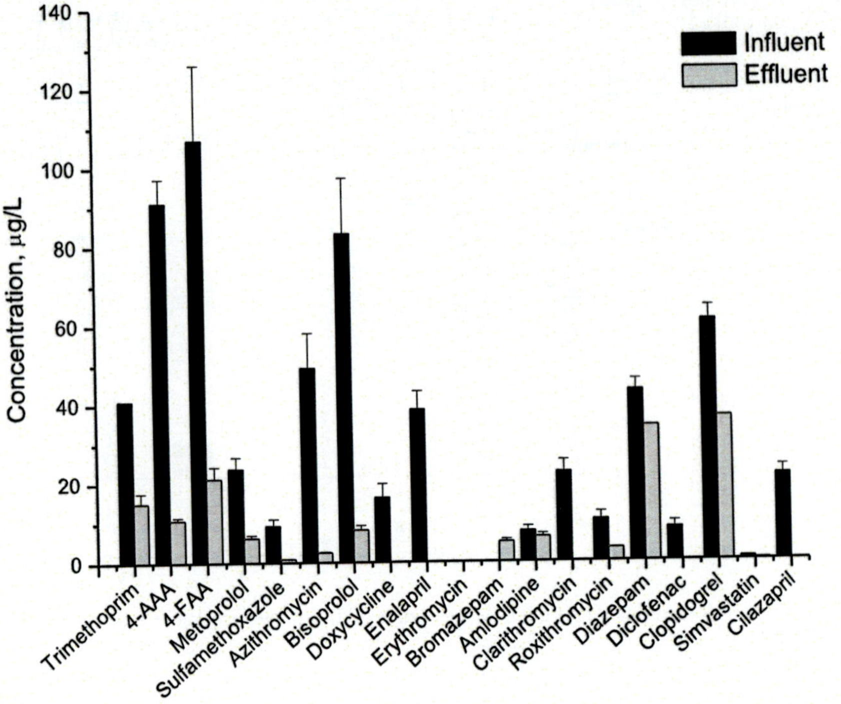

Fig. 2. Representation of the mean influent and effluent concentrations ± standard deviation of detected pharmaceuticals.

be considered as its dominant removal mechanism due to the high value of water solubility and low molecular weight (Table 1).

However, some organic compounds detected in WWTP showed a negative removal rate (Fig. 4). For example, propazine and mestranol were enriched in the process of wastewater treatment. By analyzing the effects of various processes on the removal of target micropollutants, it can be concluded that the biological treatment stage could not only decompose some organic compounds but also increase the concentration of some precursor compounds [25].

3.2 *Ecotoxicological Risk Assessment of the Detected Emerging Contaminants*

The ecotoxicity of each of the studied pollutants detected in the present paper is presented by the PNEC values. The wide variability of PNEC values can be noted in Table 2. The three most toxic compounds recorded were pharmaceuticals amlodipine, azithromycin, and trimethoprim, with PNEC values between 0.28 and 9.4 ng/L. Their

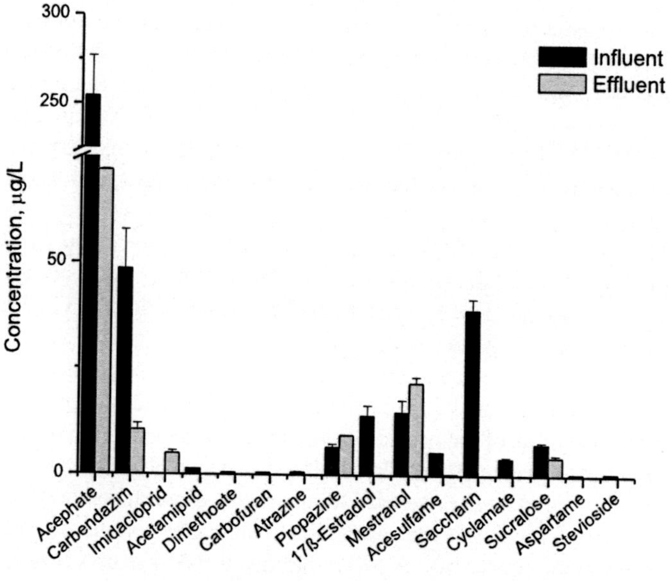

Fig. 3. Representation of the mean influent and effluent concentrations ± standard deviation of detected pesticides, steroid hormones and sweeteners.

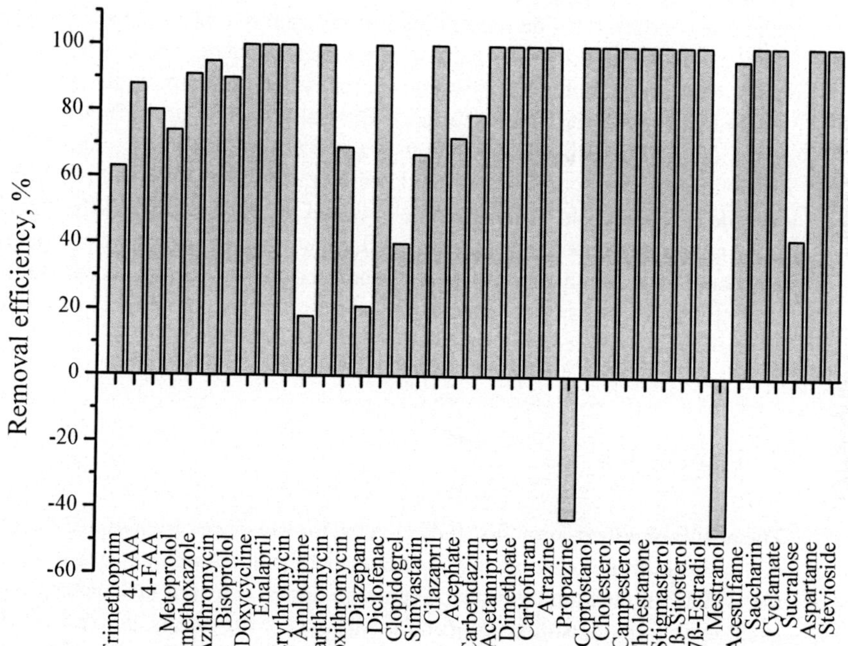

Fig. 4. Removal efficiencies of each investigated analyte detected in WWTP.

very high toxicity is directly linked to the significant toxicity of these molecules for aquatic organisms, such as fish, green algae, and daphnid. Other ecotoxic micropollutants were the pesticide propazine (PNEC = 40 ng/L), the pharmaceuticals (e.g. roxithromycin PNEC = 10 ng/L; metoprolol PNEC = 100 ng/L; and sulfamethoxazole PNEC = 590 ng/L). The least ecotoxic pollutants detected were artificial sweeteners acesulfame and sucralose, with PNEC values of 2200 and 930 μg/L, respectively.

The risk quotients (RQ) calculated for each analyte detected in WWTP effluent and risk quotient of the mixture of pollutants (RQ_{mix}) are reported in Table 6. RQ values were obtained by comparing the PEC values calculated by Eq. (2) and PNEC values from the literature survey.

Among all detected compounds in the WWTP effluent, amlodipine was identified as the riskiest pollutant with an RQ value of 8.35. In the case of this compound, even with the low value of the predicted/measured concentration in the receiving streams, the high toxicity resulted in a low PNEC value, which led to a high RQ value.

The medium risk was recorded for pharmaceuticals trimethoprim, roxithromycin, and diazepam, indicating that, even though these compounds are partially removed in the treatment process, there is still a significant risk to aquatic organisms. RQ values lower than 0.1 were obtained for pesticide propazine and pharmaceuticals metoprolol and azithromycin, showing that these compounds are of low environmental risk. For the remaining emerging pollutants detected in WWTP effluent, the calculated RQ values were below 0.01, demonstrating insignificant risk to the aquatic environment.

In this paper ecotoxicological risk associated with the whole mixture of previously identified pollutants is also considered, since previous studies [18, 19] have shown that the classic single substance ERA approach is not sufficient to reliably assess the risk associated with a complex mixture of pollutants that are independently affecting aquatic biota. Calculated RQ_{mix} (9.79, Table 6) has significantly exceeded the threshold value of 1. The obtained result demonstrates the significance of the mixture approach, revealing that individually "safe" emerging compounds can contribute to a significant risk of the whole effluents. Several recent studies [19, 20, 39] have drawn similar conclusions about "the cocktail effect" of mixtures of micropollutants in receiving waters.

4 Conclusion

The results obtained from this study showed the presence of widespread contamination by emerging contaminants, including both influent and effluent samples from the WWTP in the vicinity of city Topola, in Serbia. Most of the substances investigated were found at similar concentrations, except for sterols, which were detected at significantly high levels (>10–100 fold). The removal efficiency of the analyzed WWTP varied, depending on the compound, in the range of 18–100%.

The ecotoxicological risk assessment of the individual micropollutant recognized pharmaceutical amlodipine as the compound of environmental concern, while for the majority of the selected emerging compounds the risk was determined as insignificant when considered individually. However, in order to assess environmental risk properly, the "the cocktail effect" of the entire mixture should be taken into account.

The overall results showed that an ecotoxicological risk cannot be excluded in the investigated area. Even though only one substance individually exceeded the ERA threshold, the combination of the detected compounds in the WWTP effluent poses an environmental risk. There is still a need for obtaining data on the acute and chronic ecotoxicity of different trophic levels to achieve a more robust and reliable ERA.

Furthermore, this work highlights the need to include certain emerging pollutants (e.g. pharmaceuticals) in regular monitoring programs at the national and international levels. Additionally, the scale of the study should be increased and expanded to WWTP with higher treatment capacity to confirm the current result obtained for this territory at a national scale. Each territory has specific pollution characteristics, which is why further research is required.

Acknowledgements This work was supported by the Ministry of Education, Science and Technological Development of the Republic of Serbia (Contract No. 451-03-9/2021-14/200135).

References

1. Antić, N., Radišić, M., Radović, T., Vasiljević, T., Grujić, S., Petković, A., Dimkić, M., Laušević, M.: Pesticide residues in the Danube River Basin in Serbia—a Survey during 2009–2011. Clean: Soil, Air, Water **43**(2), 197–204 (2015)
2. Matić, I., Grujić, S., Jauković, Z., Laušević, M.: Trace analysis of selected hormones and sterols in river sediments by liquid chromatography-atmospheric pressure chemical ionization–tandem mass spectrometry. J. Chromatogr. A **1364**, 117–127 (2014)
3. Jauković, Z.D., Grujić, S.D., Matić Bujagić, I.V., Laušević, M.D.: Determination of sterols and steroid hormones in surface water and wastewater using liquid chromatography-atmospheric pressure chemical ionization-mass spectrometry. Microchem. J. **135**, 39–47 (2017)
4. Jauković, Z.D., Grujić, S.D., Vasiljević, T.M., Petrović, S.D., Laušević, M.D.: Cardiovascular drugs in environmental waters and wastewaters: Method optimization and real sample analysis. J. AOAC Int. **97**(4), 1167–1174 (2014)
5. Matić Bujagić, I., Grujić, S., Jauković, Z., Laušević, M.: Sterol ratios as a tool for sewage pollution assessment of river sediments in Serbia. Environ. Pollut. **213**, 76–83 (2016)
6. Gvozdić, E., Matić Bujagić, I., Đurkić, T., Grujić, S.: Trace analysis of artificial sweeteners in environmental waters, wastewater and river sediments by liquid chromatography–tandem mass spectrometry. Microchem. J. **157**, 105071 (2020)
7. Hoeger, B., Köllner, B., Dietrich, D.R., Hitzfeld, B.: Water-borne diclofenac affects kidney and gill integrity and selected immune parameters in brown trout (Salmo trutta f. fario). Aquatic Toxicol. **75**(1), 53–64 (2005)
8. Kim, K.-H., Kabir, E., Jahan, S.A.: Exposure to pesticides and the associated human health effects. Science of Total Environment **575**, 525–535 (2017)
9. Grenni, P., Ancona, V., Caracciolo, A.B.: Ecological effects of antibiotics on natural ecosystems: A review. Microchem. J. **136**, 25–39 (2018)
10. Gagné, F., Blaise, C., Lachance, B., Sunahara, G.I., Sabik, H.: Evidence of coprostanol estrogenicity to the freshwater mussel Elliptio complanata. Environ. Pollut. **115**, 97–106 (2001)

11. Kay, P., Hughes, S.R., Ault, J.R., Ashcroft, A.E., Brown, L.E.: Widespread, routine occurrence of pharmaceuticals in sewage effluent, combined sewer overflows and receiving waters. Environ. Pollut. **220**(Pt B), 1447–1455 (2017)
12. Paíga, P., Santos, L.H.M.L.M., Delerue-Matos, C.: Development of a multi-residue method for the determination of human and veterinary pharmaceuticals and some of their metabolites in aqueous environmental matrices by SPE-UHPLC–MS/MS. J. Pharmaceutical Biomed. Anal. 135, 75–86 (2017)
13. Martins, C.C., Cabral, A.C., Barbosa-Cintra, S.C.T., Dauner, A.L.L., Souza, F.M.: An integrated evaluation of molecular marker indices and linear alkylbenzenes (LABs) to measure sewage input in a subtropical estuary (Babitonga Bay, Brazil). Environ. Pollut. **188**, 71–80 (2014)
14. Kuzmanović, M., López-Doval, J.C., De Castro-Català, N., Guasch, H., Petrović, M., Muñoz, I., Ginebreda, A., Barceló, D.: Ecotoxicological risk assessment of chemical pollution in four Iberian river basins and its relationship with the aquatic macroinvertebrate community status. Sci. Total Environ. **540**, 324–333 (2016)
15. Matić, B.I., Grujić, S., Laušević, M., Hofmann, T., Micić, V.: Emerging contaminants in sediment core from the Iron Gate I Reservoir on the Danube River. Sci. Total Environ. **662**, 77–87 (2019)
16. Styszko, K., Proctor, K., Castrignanò, E., Kasprzyk-Hordern, B.: Occurrence of pharmaceutical residues, personal care products, lifestyle chemicals, illicit drugs and metabolites in wastewater and receiving surface waters of Krakow agglomeration in South Poland. Sci. Total Environ. 768, 144360 (2021)
17. EC, European Commission.: Technical guidance document on risk assessment, part II (2003). https://echa.europa.eu/documents/10162/16960216/tgdpart2_2ed_en.pdf, Last accessed 15 Apr 2021
18. Backhaus, T., Faust, M.: Predictive Environmental Risk Assessment of Chemical Mixtures: A Conceptual Framework. Environ. Sci. Technol. **46**(5), 2564–2573 (2012)
19. Gosset, A., Polomé, P., Perrodin, Y.: Ecotoxicological risk assessment of micropollutants from treated urban wastewater effluents for watercourses at a territorial scale: Application and comparison of two approaches. Int. J. Hygiene Environ. Health 224, 113437 (2020)
20. Gosset, A., Wiest, L., Fildier, A., Libert, C., Giroud, B., Hammada, M., Hervé, M., Sibeud, E., Vulliet, E., Polomé, P., Perrodin, Y.: Ecotoxicological risk assessment of contaminants of emerging concern identified by "suspect screening" from urban wastewater treatment plant effluents at a territorial scale. Sci. Total Environ. 778, 146275 (2021)
21. Matić Bujagić, I., Grujić, S., Đurkić, T., Laušević, M.: Methods for trace analysis of pharmaceuticals and pesticides in water and sediment samples (Chap. 9). In: Dimkić M (eds.) Alluvial Aquifer Processes, pp. 495–530. IWA Publishing (2021)
22. Papadakis, E.N., Vryzas, Z., Kotopoulou, A., Kintzikoglou, K., Makris, K.C., Papadopoulou-Mourkidou, E.: A pesticide monitoring survey in rivers and lakes of northern Greece and its human and ecotoxicological risk assessment. Ecotoxicol. Environ. Saf. **116**, 1–9 (2015)
23. Daughton, C.G.: Real-time estimation of small-area populations with human biomarkers in sewage. Sci. Total Environ. **414**, 6–21 (2012)
24. Vystavna, Y., Frkova, Z., Marchand, L., Vergeles, Y., Stolberg, F.: Removal efficiency of pharmaceuticals in a full scale constructed wetland in East Ukraine. Ecol. Eng. 108 (Part A), 50–58 (2017)
25. Li, Y., Niu, X., Yao, C., Yang, W, Lu, G.: Distribution, removal, and risk assessment of pharmaceuticals and their metabolites in five sewage plants. Int. J. Environ. Res. Public Health 16(23), 4729 (2019)
26. Grill, G., Khan, U., Lehner, B., Nicell, J., Ariwi, J.: Risk assessment of down-the-drain chemicals at large spatial scales: Model development and application to contaminants originating from urban areas in the Saint Lawrence River Basin. Sci. Total Environ. **541**, 825–838 (2016)
27. Andrade Godoy, A., Domingues, I., Bastos de Carvalho, L., Caloto Oliveira, A., de Jesus Azevedo, C.C., Marins Taparo, J., Kushim Assano, P., Mori, V., de Almeida Vergara Hidalgo, V., Arsénia Nogueira, A.H., Kummrow, F.: Assessment of the ecotoxicity of the pharmaceuticals bisoprolol, sotalol, and ranitidine using standard and behavioral endpoints. Environ. Sci. Pollut. Res. 27(5), 5469–5481 (2020)

28. Cunha, D.L., Mendes, M.P., Marques, M.: Environmental risk assessment of psychoactive drugs in the aquatic environment. Environ. Sci. Pollut. Res. **26**(1), 78–90 (2019)
29. Food and Drug Administration, https://www.accessdata.fda.gov/drugsatfda_docs/nda/2015/205003Orig1s000EA.pdf, Last accessed 15 Apr 2021
30. Escher, B.I., Baumgartner, R., Koller, M., Treyer, K., Lienert, J., McArdellet, C.S.: Environmental toxicology and risk assessment of pharmaceuticals from hospital wastewater. Water Res. **45**(1), 75–92 (2011)
31. Pereira, A.M.P.T., Silva, L.J.G., Meisel, L.M., Lino, C.M., Pena, A.: Environmental impact of pharmaceuticals from Portuguese wastewaters: geographical and seasonal occurrence, removal and risk assessment. Environ. Res. **136**, 108–119 (2015)
32. European Food Safety Authority Journal. Guidance on tiered risk assessment for plant protection products for aquatic organisms in edge-of-field surface waters. EFSA J. 11(7), 3290 (2013), https://efsa.onlinelibrary.wiley.com/doi/pdf/https://doi.org/10.2903/j.efsa.2013.3290, Last accessed 15 Apr 2021
33. European Food Safety Authority Journal. Conclusion on the peer review of the pesticide risk assessment of the active substance carbendazim. EFSA J. 8(5), 1598 (2010). https://efsa.onlinelibrary.wiley.com/doi/pdf/https://doi.org/10.2903/j.efsa.2010.1598, Last accessed 15 Apr 2021
34. Souza, J.C.G., Ribeiro, A.R., Barbosa, M.O., Ribeiro, C., Tiritan, M.E., Pereira, M.F.R., Silva, A.M.T.: Monitoring of the 17 EU Watch List contaminants of emerging concern in the Ave and the Sousa Rivers. Sci. Total Environ. **649**, 1083–1095 (2019)
35. Masiá, A., Campo, J., Navarro-Ortega, A., Barceló, D., Picó, Y.: Pesticide monitoring in the basin of Llobregat River (Catalonia, Spain) and comparison with historical data. Sci. Total Environ. **503–504**, 58–68 (2015)
36. Deo, R.P.: Pharmaceuticals in the surface water of the USA: A review. Current Environ. Health Rep. **1**, 113–122 (2014)
37. Belton, K., Schaefer, E., Guiney, P.D.: A review of the environmental fate and effects of acesulfame-potassium. Integrated Environ. Assessment Manag. **16**(4), 421–437 (2020)
38. Tollefsen, K.E., Nizzetto, L., Huggett, D.B.: Presence, fate and effects of the intense sweetener sucralose in the aquatic environment. Sci. Total Environ. **438**, 510–516 (2012)
39. Riva, F., Zuccato, E., Davoli, E., Fattore, E., Castiglioni, S.: Risk assessment of a mixture of emerging contaminants in surface water in a highly urbanized area in Italy. J. Hazard. Mater. **361**, 103–110 (2019)

A Study on the Performance of LDPE/PCL Double Layered Packaging Films Containing Coper Oxide/Nanocellulose Composite

Danijela Kovačević, Steva Lević, and Nenad Đorđević

Abstract Double-layered films, consisting of an LDPE (low density polyethylene) layer coated with PCL (polycaprolactone) layer filled with inorganic/organic composites have been manufactured. The intention was to obtain the packaging material with multiple potential uses. As an organic compound in the composites, nanocellulose (NC) and maleic anhydride modified nanocelulose (NCMA) has been used. Inorganic/organic composites have been prepared by deposition of copper oxides on the surface of cellulose nanocrystals. The composites have been drayed in two ways: under vacuum and by freeze-drying (lyophilisation). The prepared composites have been added as filler into PCL. Thereafter, PCL/ copper oxide/nanocellulose has been applied as a coating on the surface of the LDPE foil in order to obtain double layered hybrid film as a potential packaging material. The influence of filler type and content, NC modification and copper oxide/nanocellulose composite drying procedure on structural and mechanical properties of hybrid double layered films has been investigated.

Keywords Packaging · Low density polyethylene · Nanocellulose

1 Introduction

Improving the performance of existing food packaging materials through use of biodegradable components and nanomaterials led to development of new advanced materials for packaging purposes [1–5]. The development of this type of materials is a step towards production of packaging materials having specific tailored end-use properties. Requirements for environmentally acceptable materials, which will provide health security, extend shelf life, facilitate product distribution and provide

D. Kovačević (✉) · N. Đorđević
The Academy of Applied Technical Studies Belgrade, 11000 Belgrade, Serbia
e-mail: dkovacevic@politehnika.edu.rs

S. Lević
Faculty of Agriculture, University of Belgrade, 11000 Belgrade, Serbia

N. Mitrovic et al. (eds.), *Current Problems in Experimental and Computational Engineering*, Lecture Notes in Networks and Systems 323,
https://doi.org/10.1007/978-3-030-86009-7_9

more valuable information for the customers are some of the main reasons for a great number of studies in this area [6–9].

Low density polyethylene (LDPE) is well known and widely used packaging material. The main advantages of this material are strength, stability, light weight and its resistance to water [10]. However, some disadvantages such as poor barrier properties to gases (oxygen, carbon dioxide) organic vapor and water vapor limit the use of LDPE. Improvement of LDPE performance can go in two directions: functionalization of PE surface or layer application (*i.e.* bio-coating). Adsorption of chitosan-colloidal systems onto to the O_2 plasma treated PE foil resulted in a reduction of both oxygen permeability and wetting contact angle. Also, the study indicated that adhesion of chitosan significantly improved antibacterial properties of functionalized PE foil [11]. Layer-by-layer (LbL) self-assembly of PE and nano-sized graphene have been produced in order to obtain material with *enhanced oxygen barrier properties* [12].

The extensive use of nanoparticles and nanotechnology in different research areas has become more frequently over the last decade [13–15]. Cellulose nanocrystals (CNCs) obtained from natural cellulose sources have shown potential for widespread use. Inherent features of CNC i.e. high surface area, rigidity, thermal stability, anisotropic mechanical features and particularly nontoxicity opens the possibility of applying this material in food packaging. Addition of CNC resulted in improvement of antimicrobial properties [16] and it has been demonstrated that chemical grafting of cellulose whiskers improve their compatibilization with polycaprolactone matrix [17]. Polycaprolactone (PCL) is synthetic biodegradable polymer widely used in biomedical areas (tissue engineering, drug delivery) [18] and can also be used in packaging, particularly in food packaging [19]. The use of biodegradable polymer in packaging as a substitute for petroleum-based *polymer* is increasing due to demands for application of environmentally friendly materials [20]. A complete replacement of a common plastic packaging with biodegradable polymers and biofilms cannot be implemented mainly due to some poor properties of biopolymers (mechanical strength, structural integrity…) [21]. Polycaprolactone has been usually used as one layer of multilayer packaging materials or matrix with addition of fillers (antimicrobial components and components controlling biodegradation rate) [22].

There is a growing demand for addition of active compounds in order to improve mechanical, antibacterial, thermal and barrier properties of the packaging materials [23, 24]. Addition of inorganic/organic compounds as an active component in packaging materials can be found in a great number of studies [1, 2, 4, 5]. Incorporation of inorganic/organic compounds in biodegradable polymers, among other things, can contribute to the improvement of mechanical properties, considering that biodegradable layer may impair the inherent mechanical properties of PE foil.

In the present work, organic–inorganic hybrid coatings consisting from PCL, nanocellulose, modified nanocellulose and copper (II) oxide have been prepared in different compositions and applied to LDPE foil. Our previous study examined structural and thermal properties of nanocellulose/coper oxide composites and detailed examined composite morphology [4]. Moreover, barrier and antimicrobial properties of PE hybrid double layered films have been investigated. The study showed an improvement in antimicrobial and as well in oxygen barrier properties [4]. In addition to these two features, mechanical properties have proven to be extremely important factor for bio-based *packaging materials.* The aim of this publication has been to investigate structural and mechanical properties of the PE hybrid double layered films. The continuation of this research in terms of structural and mechanical properties of the hybrid films has been shown in order to make our previous research more comprehensive and complete.

2 Experimental

2.1 Materials

All chemicals used in preparation of hybrid films, sulfuric acid, acetic acid, maleic acid anhydride, perchloric acid, glacial acetic acid, copper(II) acetate, sodium hydroxide and organic solvents toluene, chlorophorm, methanol, ethanol and dichloromethane were purchased from commercial suppliers Sigma Aldrich and Fluka, p.a. grade. Reagents and solvents were used as received. Millipore, deionized (DI) water (18 MS cm resistivity) was used for sample washing and solution preparation. Polycaprolactone, average molecular weight of ~45,000, was purchased from Sigma-Aldrich. Polyethylene foil LG SP 311 (LLDPE) was supplied by Macchi (three layer folio 28/44/28%).

2.2 Preparation Procedures

Nanocellulose nanocrystals (NC) were prepared by acid hydrolysis of commercially available cellulose according to established procedure [25]. Afterwards, NC was modified by esterification with maleic anhydride (MA) according to slightly modified literature method and obtained functionalized nanocellulose was marked as NCMA [26].

A detailed preparation procedure of NC, NCMA and copper oxide nanocellulose composite have already been shown in our previous publication [4, 5] and schematically presented on Fig. 1.

2.3 Characterization of Hybrid Films

Raman spectra were collected with a XploRA Raman spectrometer from Horiba Jobin Yvon. The system employed laser at 532 nm (maximum output power 20–25 mW).

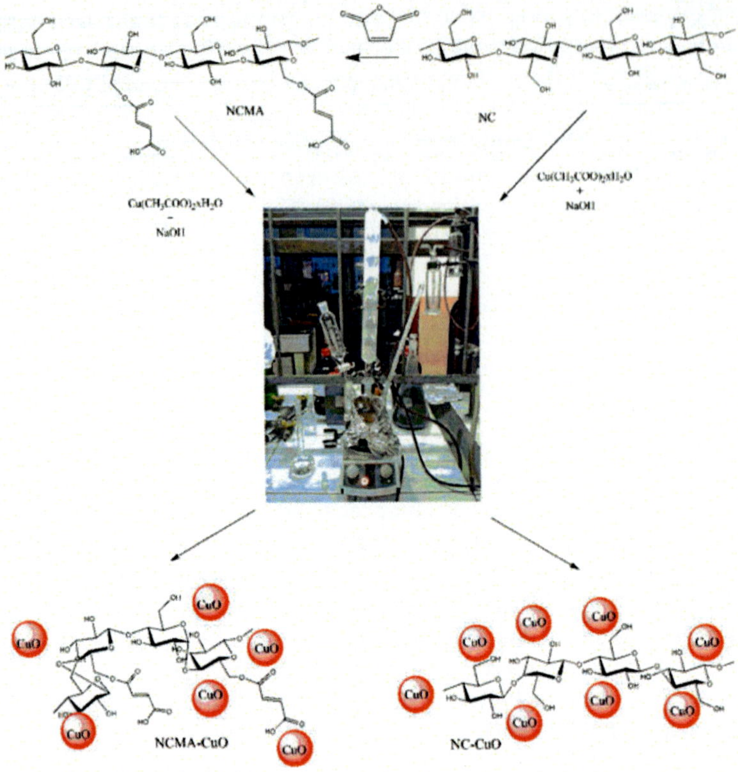

Fig. 1. Preparation procedure of NC, NCMA and deposition of CuO on the surface of NC and NCMA in order to obtain NC-CuO and NCMA-CuO.

Attenuated Total Reflectance-Fourier transform infrared spectroscopy (ATR-FTIR) analysis was performed using the IRAffinity-1 FTIR spectrophotometer (Schimadzu, Japan). The wavelength range was from 3500 to 600 cm^{-1}, while the resolution was 4 cm^{-1}.

Tensile properties of the prepared hybrid layered films were examined on Shimadzu Universal Testing Machine AG-Xplus (Shimadzu, Kyoto, Japan) with 1 kN load cell. The films were cut into 15 mm × 150 mm strips and conditioned at 23 °C and 50% relative humidity (50% RH) for 48 h before testing, and all the tests were also performed under these conditions. The specific tensile strength, elongation at break, and Young's modulus were determined using ISO Standard 527-3 and a Lorentzen & Wettre strength tester (Sweden).

Preparation of composite film on polyethylene surface (PE/nanocellulose composite). Polycaprolactone (PCL) solution of 10 wt. % was prepared by dissolving PCL pellets in chloroform. After the PCL was dissolved by using combination of both mixing and ultrasound treatment, the solutions were used for preparation of PCL based composite dispersion. CPE/nanocellulose composite hybrid structures were

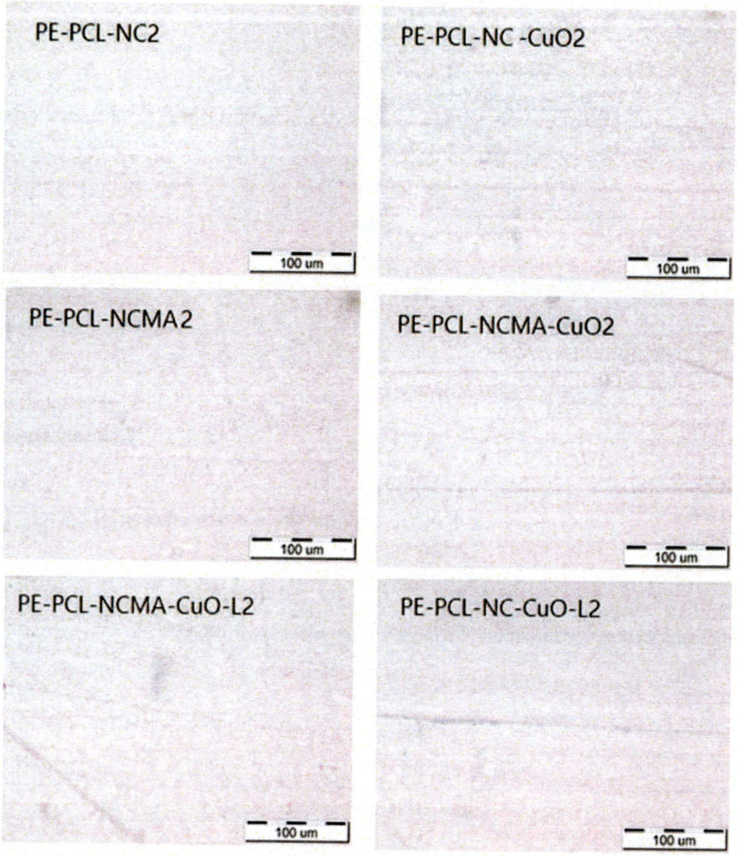

Fig. 2. Doubly-layered films under the optical microscope.

prepared by extruding PE, and then coating it with a thin film of differently modified nanocellulose. Polycaprolactone (PCL) was used as a binder in order to provide uniform film formation on polyethylene surface (Fig. 3) as well as the appearance of the films under the optical microscope (Fig. 2). Composition of PE hybrid films is presented in Table 1 [4].

3 Results and Discussion

The aim of this paper was more detailed investigation on structural and mechanical properties of PE based hybrid double layered films. Our previous study [4] reported complete characterization of ingornic/organic fillers inside PCL layer and partial characterization of hybrid films. Structural, thermal and morphological properties of all prepared nanocellulose/coper oxide composites have also been reported in the

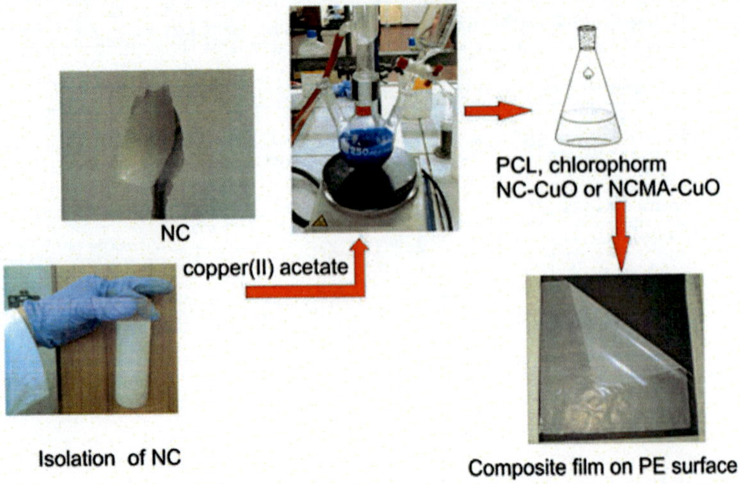

Fig. 3. Preparation procedure of PCL composite film on PE surface [4].containing CuO have been confirmed

Table 1. Prepared PE hybrid samples coated with PCL/nanocellulose/coper oxide films having different content of modified NC in PCL composite layer [4].

	Samples	Filler in PCL composite layer	PErcentage of filler in PCL composite layer
1	PE	/	0
2	PE-PCL	/	0
3	PE-PCL-NC0.5	NC	0.5
4	PE-PCL-NC2	NC	2
5	PE-PCL-NCMA0.5	NCMA	0.5
6	PE-PCL-NCMA2	NCMA	2
7	PE-PCL-NC-CuO0.5	NC-CuO[a]	0.5
8	PE-PCL-NC-CuO2	NC-CuO[a]	2
9	PE-PCL-NCMA-CuO0.5	NCMA-CuO[a]	0.5
10	PE-PCL-NCMA-CuO2	NCMA-CuO[a]	2
11	PE-PCL-NC-CuO-L0.5	NC-CuO-L[b]	0.5
12	PE-PCL-NC-CuO-L2	NC-CuO-L[b]	2
13	PE-PCL-NCMA-CuO-L0.5	NCMA-CuO-L[b]	0.5
14	PE-PCL-NCMA-CuO-L2	NCMA-CuO-L[b]	2

[a]Samples were drying under vacuum at 80°C for 6 h to obtain black powder (NC-CuO and NCMA-CuO)

[b]Samples were exposed by freeze-drying (lyophilisation) in order to obtain samples NC-CuO-L and NCMA-CuO-L

above mentioned paper. Considering that antimicrobial and barrier features of all packaging materials are of great importance, they have also been investigated and reported for all prepared films [4]. The continuation of this research in terms of structural and mechanical properties of the hybrid films is shown in order to make research more comprehensive and complete.

3.1 Raman Analysis

Raman analysis was employed in order to get inside into structural composition of the selected samples. Raman spectra of PE and PE/PCL were presented on Fig. 4.

The dominant signals in both samples, pristine PE and PE with PCL, are signals corresponding to PE. The peaks in the area of 1290–1450 cm^{-1} are attributed to CH_2 bending and twisting regions, while the Raman shifts in the region 2800–3000 cm^{-1} corresponds to CH stretching. The symmetric CH_2 stretching mode is located at 2846 cm^{-1} and asymmetric CH_2 stretching mode is band at 2880 cm^{-1}. All signals in the spectrum are in accordance with reported Raman bands in the literature for LLDPE [27].

The only difference between PE and PE-PCL spectrum is reflected in intensity of the PE signals. The signals attributed to PCL are not observed. These signals appear to be too weak to be detected among dominant PE signals. The fingerprint region for PCL in the Raman spectrum ranges from 850–1750 cm^{-1}. Expected signals (skeletal

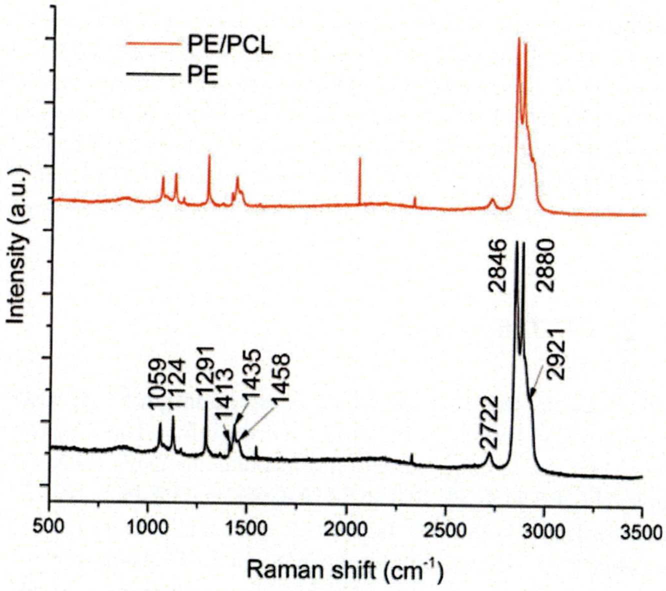

Fig. 4. Raman spectra of PE and PE-PCL samples.

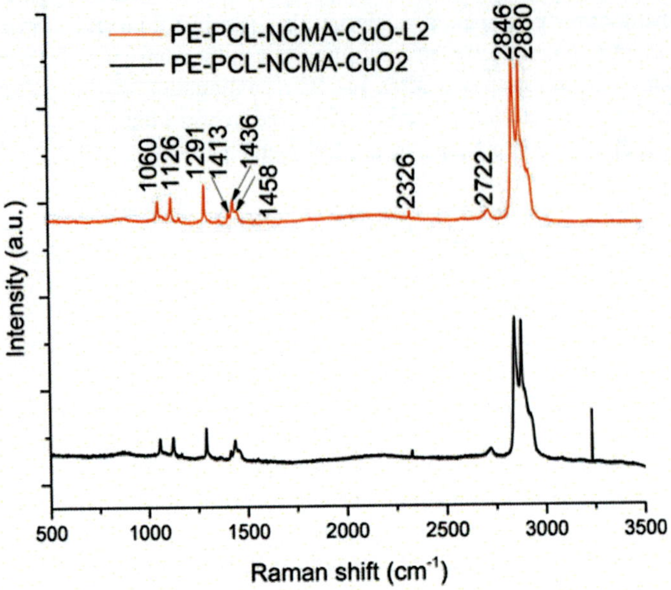

Fig. 5. Raman spectra of PE-PCL-NCMA-CuO2 and PE-PCL-NCMA-CuO2-L2.

modes C–C stretch, C–O stretch, and C–COO stretch in the range of 850–1100 cm^{-1} and C=O stretching in the region of 1710–1750 cm^{-1}) have not been detected [28].

The similar trend is observed in Raman spectrum of selected PE-PCL-NCMA-CuO2 and PE-PCL-NCMA-CuO2-L2 films (Fig. 5). It is noticed that the intensity of these signals even decreased, when fillers were added in PCL layer, comparing with pristine PE, and even with PE-PCL. Raman spectra of samples NC, NCMA and all prepared nanocellulose/coper oxide composite samples have been presented in already published study [4]. On these spectra, signals attributed to NC groups have been observed and three active peaks at 297 (Ag), 339 (Bg) and 631 (Bg) cm^{-1} in Raman spectra of all composite sample containing CuO have been confirmed [4].

3.2 *ATR-FTIR Analysis*

Attenuated Total Reflectance-Fourier transform infrared spectroscopy (ATR-FTIR) was employed to investigate the variation in the infrared optical properties of selected film samples and to compare with results obtain by using Raman analysis. Selected samples (PE, PE-PCL, PE-PCL-NCMA-CuO2 and PE-PCL-NCMA-CuO2-L2) have been recorded on both sides. Figure 6 shows samples recorded on the of PCL layer side, while Fig. 7 shows samples recorded on the PE layer side.

Contrary to the Raman analysis, which did not show the presence of signals corresponding to PCL, ATR-FTIR clearly demonstrated difference in spectra of pristine

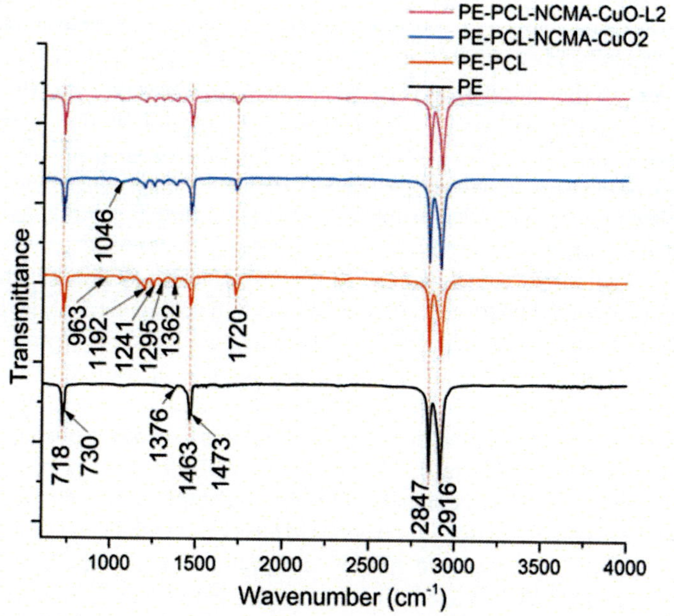

Fig. 6. ATR-FTIR spectra of samples PE, PE-PCL, PE-PCL-NCMA-CuO2 and PE-PCL-NCMA-CuO2-L2, recorded on the PCL layer.

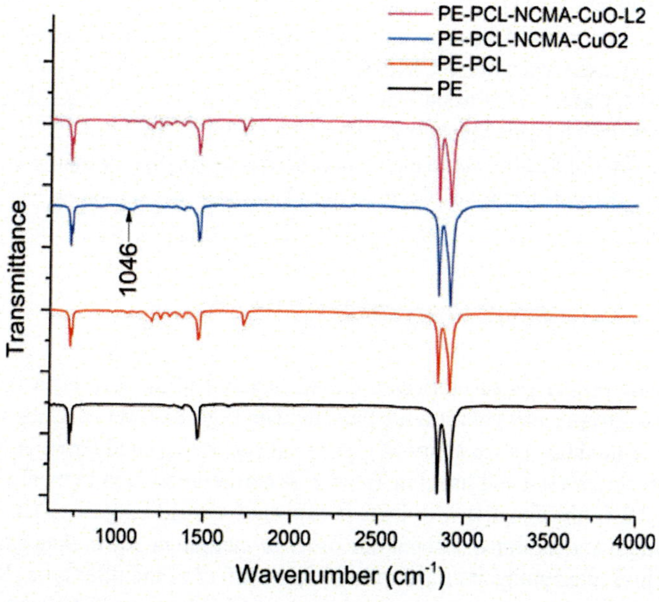

Fig. 7. ATR-FTIR spectra of samples PE, PE-PCL, PE-PCL-NCMA-CuO2 and PE-PCL-NCMA-CuO2-L2, recorded on the PE layer.

PE and PE having PLC coatings. In ATR-FTIR spectra of double-layered films, presented on Fig. 3 both signals from PE and PCL have been present.

The most evident change is observation of characteristic peaks attributed to carbonyl groups at 1720 cm^{-1} in all samples with PCL coatings. Intensity of the peak at 1720 cm^{-1} decreases with the addition of composites in PCL layer. The signal showed decrease in intensity when NCMA-CuO sample conventionally drayed was added and then the value of intensity decreases further, upon lyophilized NCMA-CuO sample was incorporated in PCL layer. The aliphatic groups at 2916 cm^{-1} (C–H asymmetric stretching) and 2847 cm^{-1} (CH_2 symmetric stretching) have been observed for the pristine PE film and in the film containing PCL layer. Considering that both constituent (PE and PCL) have these groups, these signals overlap [29].

The band at 1463 cm^{-1} is due to the vibration of angular deformation of CH. Absorption band at 1241 cm^{-1} is attributed to an asymmetric stretching of C–O–C bonds, while their symmetric stretching is (in this case due to overlapping) displayed as a shoulder at 1160 cm^{-1}. Absorption band at 1191 cm^{-1} is attributed to O–C–O stretching vibrations, the signal at 730 cm^{-1} is contributed to vibration of -CH_2 bonds while a peak at 718 cm^{-1} corresponds to CH_2 bond in the vinyl group $(–CH_2–)_4$ [1].

On Fig. 7 are presented ATR-FTIR spectra of PE, PE-PCL, PE-PCL-NCMA-CuO2 and PE-PCL-NCMA-CuO2-L2, recorded on the PCL layer. The group of author also recorder both side, PE and PCL layer and reported slightly different results. Their study on PCL layer showed peaks originating only from PCL, while ATR-FTIR spectra recorded on PE layer demonstrated only absorption band attributed to PE [1].

The intensity of absorption band attributed to C-O stretching at 1046 cm^{-1} even the increased when filer NCMA-CuO has been added to PCL layer (sample PE-PCL-NCMA-CuO2). This trend was observed regardless of which side the film sample was recorded indicating some interactions between PCL matrix layer and NCMA-CuO filler indicating some interaction between biopolymer matrix and filler (copper oxide/nanocellulose).

3.3 Mechanical Properties

Tensile properties of fabricated double layered films have been investigated. The stress–strain curves of films are presented on Figs. 8, 9 and 10. Corresponding values of the modulus of elasticity (E), stress at break (σ_B) and strain at break (ε_B) have been determined and listed in Table 2. A certain reduction in mechanical properties of LDPE films coated with PCL is expected, considering that PCL has low tensile strength and modulus is about one order of magnitude lower than LDPE [2]. Hence, the presented study attempts to investigate effect of the filler type (NC, NCMA and nanocellulose/copper oxide composite) and loadings in PCL layer on mechanical properties of final films. Modulus of elasticity is the highest in the case of PE foil, indicating that neat PE is stiffer than other investigated samples. Ultimate elongation

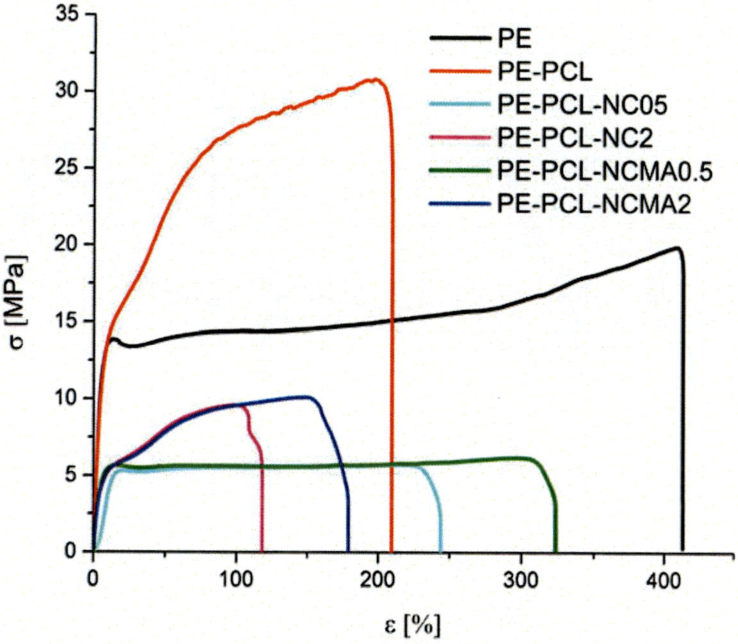

Fig. 8. The dependence stress–strain of prepared double-layered films—samples containing different concentration of functionalized and non-functionalized NC.

of about 414% can be observed for neat PE. PCL layer did not significantly decreased modulus of elasticity (185 MPa).

Stress at break of the sample PE-PCL increased dramatically, comparing with PE foil and reached value of 27.2 MPa, while the value of strain at break is only half of value obtain for PE foil indicating that PE-PCL acts more like a brittle material, reaching higher values of tensile strength but a lower maximum strain. Addition of NC and NCMA in amount of 0.5 wt% (see Fig. 6) to PCL layer resulted in films with the lowest value of stress at break, whereby with increasing organic fraction in hybrid coating (2 wt.%) values are slightly higher. Simultaneously, increased loadings of NC and NCMA decreased values of strain at break in cases of both samples (PE-PCL-NC2 and PE-PCL-NCMA2).

It has been reported in literature that brittle behavior of the lignocellulose fibers can reduce value of elongation at break. Brittle behavior of the cellulosic materials causes decreased deformation capability of composite filled cellulosic material [30]. The common to all four samples with functionalized and non-functionalized NC is gradually fracture of film (contrary to both PE and PCL).

Figure 9 shows stress-strain curves of samples with different loadings of nanocellulose/copper oxide composites in PCL layer. Films having lower concentration of NC-CuO and NCMA-CuO in PCL layer exhibited very similar behavior and stress-strain curves for those two samples almost overlap. Lower loadings of nanocellulose/copper oxide composites caused decrease in both values, stress at break and

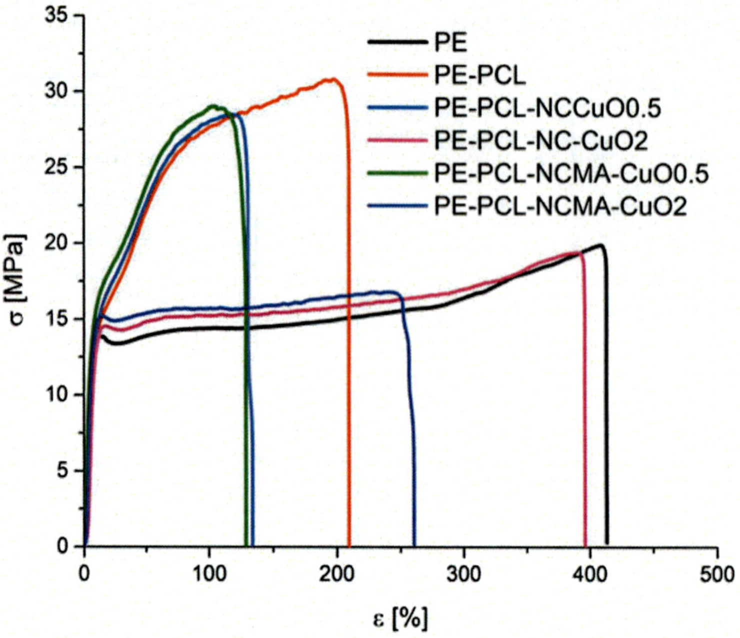

Fig. 9. The dependence stress-strain of prepared double-layered films—samples containing different concentration of NC-CuO in PCL layer (conventional drying).

strain at break, comparing to PE-PCL sample. Addition of higher amount of NC-CuO and NCMA-CuO decreased stress at break, but substantially increase strain at break especially for sample PE-PCL-NC-CuO2 in which reached 395%, close to value of neat PE (414%).

The influence of drying conditions of nanocellulose/copper oxide composite on films mechanical characteristics has also been evaluated. Figure 10c presents stress-strain curves of the films in which PLC is loaded with nanocellulose/copper oxide composite drayed by lyophilization. Samples with lower amounts of filler inside PCL layer demonstrated similar trend in stress-strain curves to samples drayed on conventional way. PE-PCL-NC-CuO-L0.5 sample showed a bit higher values of tensile strength (18.4 MPa) and strain at break (141%), comparing to PE-PCL-NCMA-CuO-L0.5. Increasing reinforcement from 0.5 wt.% to 2 wt.% in PCL leads to lower tensile strength while the maximum strain is reduced by 38% for PE-PCL-NCMA-CuO-L2 and by 27% for PE-PCL-NC-CuO-L0.5.

In summary, all samples having higher concentration of nanocellulose/copper oxide composites exhibited higher value of elongation at break, comparing to PE-PCL sample, regardless of drying condition, particularly ultimate elongation at break, close to a value for neat LDPE, reached sample PE-PCL-NC-CuO2. Higher amount of reinforcement in PCL enhanced interfacial adhesion between filler and matrix material, contributed to interlayer PE-PCL compatibility and enables better stress

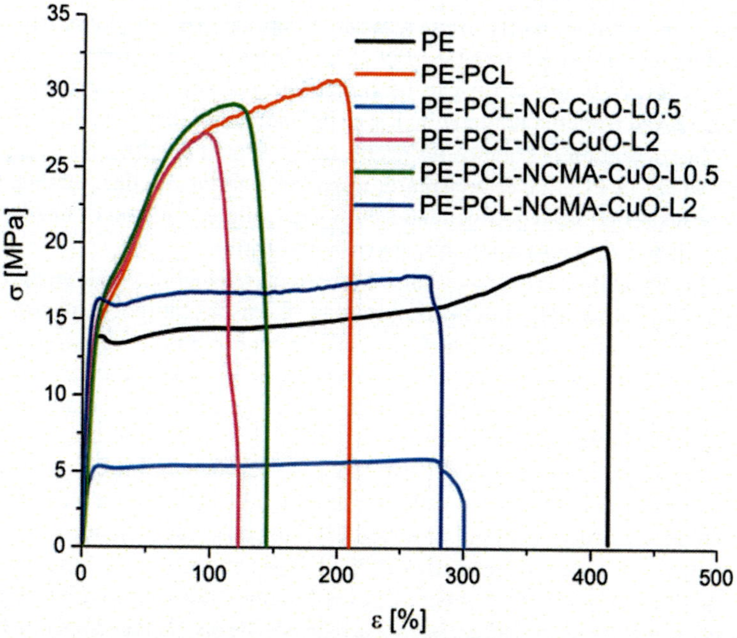

Fig. 10. The dependence stress-strain of prepared double-layered films—samples containing different concentration of NC-CuO in PCL layer (liophilization).

Table 2. Values of the modulus of elasticity (E), stress at break (σB) and strain at break (εB).

	Samples	E [MPa]	εB (%)	σB (Mpa)
1	PE	210.25	414	19.8
2	PE-PCL	185.12	208	27.2
3	PE-PCL-NC0.5	53.13	244	2.51
4	PE-PCL-NC2	58.10	118	6.023
5	PE-PCL-NCMA0.5	76.98	323	3.47
6	PE-PCL-NCMA2	76.61	180	3.86
7	PE-PCL-NC-CuO0.5	208.13	130	24
8	PE-PCL-NC-CuO2	149.15	395	18.8
9	PE-PCL-NCMA-CuO0.5	222.66	124	25
10	PE-PCL-NCMA-CuO2	172.89	256	13.5
11	PE-PCL-NC-CuO-L0.5	195.15	113	10
12	PE-PCL-NC-CuO-L2	205.00	300	13.26
13	PE-PCL-NCMA-CuO-L0.5	162.84	141	18.35
14	PE-PCL-NCMA-CuO-L2	199.65	282	14.7

transfer from the matrix to filler. When comparing tensile properties of PCL reinforced with nanocellulose (see Fig. 8) and nanocellulose/copper oxide composites (see Figs. 8, 9 and 10) it can be concluded that stress is much more effectively transferred from the polymer matrix to the inorganic filler.

Incorporation of composites nanocellulose/CuO nanoparticles used in this study (2 wt.%) in PCL layer apparently improves interfacial adhesiveness at the interface matrix polyester—filler and LDPE layer. Higher amount of filler nanoparticles contributed to a restriction of matrix chain mobility.

Lower amount (concentration, 0.5 wt.%) caused higher tensile strength, regarding neat PE, but without significant variations of tensile strength comparing to PE-PCL film. The presented study shows that drying conditions (vacuum oven *vs*, liophilization), i.e. morphology of CuO [4] nanoparticles influence mechanical properties by changing level of filler-matrix interaction and consequently interaction between two constitute layer.

Although sample PE-PCL-NCMA-CuO-L2 demonstrated some decrease in mechanical properties, the excellent antifungal (inhibition of 97%) and moderate antibacterial properties are observed [4]. However, the formation of such a multi-component system to perfectly fulfill all expected features is currently *challenging especially in terms of* compatibility, both among the constituent layers, as well as between polymer and fillers inside each layer. The optimal composition of composite layer and mutual compatibility crucially determines resulting material performances.

4 Conclusion

A novel double layered hybrid films, consisting of LDPE layer coated with biodegradable polymer PCL with inorganic/organic filler has been produced. Inorganic/organic filler inside PCL layer was copper oxides/nancellulose composites. The nanocellulose (NC) has been used as unmodified (obtain from acid hydrolysis of cotton fibers) and modified with maleic anhydride (NCMA). Upon precipitation of copper oxides on modified and unmodified NC and draying of resulting composites, incorporation of copper oxides/nancellulose composites inside PCL matrix and coating of LDPE foil has been conducted. The structural and mechanical characterization of double layered hybrid films has been performed. The results of Raman analysis reviled that dominant signals correspond to LDPE and signals of all others components appear to be too weak to be detected. Contrary to the Raman analysis, ATR-FTIR clearly demonstrated difference in spectra of pristine PE and PE having PLC coating. Modulus of elasticity is the highest in the case of PE foil, and ultimate elongation of about 414% can be observed for neat PE. Stress at break of the sample PE-PCL increased dramatically. Addition of NC and NCMA in amount of 0.5 wt% to PCL layer resulted in films with the lowest value of stress at break, whereby with increasing organic fraction in hybrid coating (2 wt.%) values are slightly higher. Lower loadings of nanocellulose/copper oxide composites caused decrease in both values, stress at break and strain at break, comparing to PE-PCL sample. Addition of higher amount of

NC-CuO and NCMA-CuO decreased stress at break, but substantially increase strain at break especially for sample PE-PCL-NC-CuO2 in which reached 395%, close to value of neat PE (414%). The results in this study demonstrated that changes in each filler concentration and modification in nanocellulose/copper (II) oxide composites drying conditions can enable tailoring material properties towards specifically set goals.

References

1. Reščček, A., Ščetar, M., Hrnjak-Murgić, Z., Dimitrov, N., Galić, K.: Polyethylene/polycaprolactone nanocomposite films for food packaging modified with magnetite and casein: oxygen barrier, mechanical, and thermal properties. Polym.-Plast. Technol. Eng. **55**(14), 1450–1459 (2016)
2. Reščček, A., Katančić, Z., Krehula, L. K., Ščetar, M., Hrnjak-Murgić, Z., Galić K.: Development of double layered PE/PCL films for food packaging modified with zeolite and magnetite nanoparticles. Adv. Polym. Technol. **37**(3), 21727(1–6) (2018)
3. Reščček, A., Krehula, L. K Katančić, Z., Hrnjak-Murgić, Z.: Active bilayer PE/PCL films for food packaging modified with zinc oxide and casein. Croat. Chem. Acta **88**(4), 461–473 (2015)
4. Đorđević, N., Marinković, A. D., Živković, P. Kovačević D. V., Dimitrijević, S., Kokol, V., Uskoković, P.S.: Improving the packaging performance of low-density polyethylene with PCL/nanocellulose/copper(II)oxide barrier layer. Sci. Sinter. **50**(2), 149–161 (2018)
5. Đorđević, N., Marinković, A. D., Nikolić, J. B., Drmanić, S. Ž. Rančić M., Brković D. V., Uskoković, P.S.: A study of the barrier properties of polyethylene coated with ananocellulose/magnetite composite film. **81**(5), 589–605 (2016)
6. Arora, A., Padua, G.W.: Review: nanocomposites in food packaging. J. Food Sci. **75**(1), R43–R49 (2010)
7. Huang, J.-Y., Li, X., Zhou, W.: Safety assessment of nanocomposite for food packaging application. Trends Food Sci. Technol. **45**(2), 187–199 (2015)
8. Silva, N., Blumberga, D.: Why biopolymer packaging materials are better. Environ. Clim. Technol. **23**(2), 366–384 (2019)
9. Fabra, M.J., López-Rubio, A., Lagaron, J.M.: Biopolymers for food packaging applications. Smart Polym. Appl. 476–509 (2014)
10. Siročić, A.P., Reščček, A., Ščetar, M., Krehula, L.K., Hrnjak-Murgić, Z.: Development of low density polyethylene nanocomposites films for packaging. Polym. Bull. **71**, 705–715 (2014)
11. Kraševac Glaser, T., Plohl, O., Vesel, A., Ajdnik, U., Poklar Ulrih, N., Knez Hrnčič, M., Bren, U., Fras Zemljič, L.: Functionalization of polyethylene (PE) and polypropylene (PP) material using chitosan nanoparticles with incorporated resveratrolas potential active packaging. Materials **12**, 2118(1–20) (2019)
12. Heo, J., Choi, M., Hong, J.: Facile surface modification of polyethylene film via spray-assisted layer-by-layer self-assembly of graphene oxide for oxygen barrier properties. Sci. Rep. **9**, 2754 (2019)
13. Brković, D.V., Pavlović, V.B., Pavlović, V.P., Obradović, N., Mitrić, M., Stevanović, S., Vlahović, B., Uskoković, P.S., Marinković, A.D.: Structural properties of the multiwall carbon nanotubes/poly(methyl methacrylate) nanocomposites: effect of the multiwall carbon nanotubes covalent functionalization. Polym. Compos. **38**(S1), E472–E489 (2017)
14. Brković, D.V., Kovačević, V.V., Sretenović, G.B., Kuraica, M.M., Trišović, N.P., Valentini, L., Marinković, A.D., Kenny, J.M., Uskoković, P.S.: Effects of dielectric barrier discharge in air on morphological and electrical properties of graphene nanoplatelets and multi-walled carbon nanotubes. J. Phys. Chem. Solids **75**(7), 858–868 (2014)

15. Duncan, T.V.: Applications of nanotechnology in food packaging and food safety: barrier materials, antimicrobials and sensors. J. Colloid Interf. Sci. **363**(1), 1–24 (2011)
16. Jebali, A., Hekmatimoghaddam, S., Behzadi, A., Rezapor, I., Mohammadi, B.H., Jasemizad, T., Yasini, S.A., Javadzadeh, M., Amiri, A., Soltani, M., Rezaei, Z., Sedighi, N., Seyfi, M., Rezaei, M., Sayadi, M.: Antimicrobial activity of nanocellulose conjugated with allicin and lysozyme. Cellulose **20**(6), 2897–2907 (2013)
17. Siqueira, G., Bras, J., Dufresne, A.: Cellulose whiskers versus microfibrils: influence of the nature of the nanoparticle and its surface functionalization on the thermal and mechanical properties of nanocomposites. Biomacromol **10**(2), 425–432 (2009)
18. Milosevic, M., Stojanovic, D.B., Simic, V. Grkovic, M., Bjelovic, M., Uskokovic, P.S., Kojic, M,: Preparation and modeling of three-layered PCL/PLGA/PCL fibrous scaffolds for prolonged drug release. Sci. Rep. **10**, 11126 (2020)
19. Sou Lyu, J., Lee, J.-S., Han, J.: Development of a biodegradable polycaprolactone film incorporated with an antimicrobial agent via an extrusion process. Sci. Rep. **9**, 20236 (2019)
20. Tang, X.Z., Kumar, P., Alavi, S., Sandeep, K.P.: Recent advances in biopolymers and biopolymer-based nanocomposites for food packaging materials. Crit. Rev. Food Sci. Nutr. **52**(5), 426–442 (2012)
21. Baker, S.R., Banerjee, S., Bonin, K., Guthold, M.: Determining the mechanical properties of electrospunpoly-ε-caprolactone (PCL) nanofibers using AFM and a novelfiberanchoring technique. Mater. Sci. Eng. C **59**, 203–212 (2016)
22. Cesur, S., Koroglu, C., Tansel Yalcin, H.: Antimicrobial and biodegradable food packaging applications of polycaprolactone/organo nanoclay/chitosanpolymeric composite films. J. Vinyl Add. Tech. **24**(4), 376–387 (2017)
23. Nedović, V., Kalusević, A., Manojlović, V., Lević, S., Bugarski, B.: An overview of encapsulation technologies for food applications. Proc. Food Sci. **1**, 1806–1815 (2011)
24. Milinčić, D. D., Popović, D. A., Lević, S.M., Kostić, A. Ž., Tešić, Ž. Lj., Nedović, V. A., Pešić, M. B.: Application of polyphenol-loaded nanoparticles in food industry. Nanomaterials **9**(11), 1629(1–21) (2019)
25. Bondeson, D., Mathew, A., Oksman, K.: Optimization of the isolation of nanocrystals from microcrystalline cellulose by acid hydrolysis. Cellulose **13**(2), 171–180 (2006)
26. Kim, D.Y., Nishiyama, Y., Kuga, S.: Surface acetylation of bacterial cellulose. Cellulose **9**(3–4), 361–367 (2002)
27. Furukawa, T., Sato, T., Kita, Y., Matsukawa, K., Yamaguchi, H., Ochiai, S., Siesler, H.W., Ozaki, Y.: Molecular structure, crystallinity and morphology of polyethylene/polypropylene blends studied by raman mapping, scanning electron microscopy, wide angle X-ray diffraction, and differential scanning calorimetry. Polym. J. **38**(11), 1127–1136 (2006)
28. Kotula, A.P., Snyder, C.R., Migler, K.B.: Determining conformational order and crystallinity in polycaprolactone via Raman spectroscopy. Polymer **117**, 1–10 (2017)
29. Yuan, S., Xiong, G., Wang, X., Zhang, S., Choong, C.: Surface modification of polycaprolactone substrates using collagen conjugated poly(methacrylic acid) brushes for the regulation of cell proliferation and endothelialisation. J. Mater. Chem. **22**, 13039–13049 (2012)
30. Habibi, Y.W., El-Zawawy, K., Ibrahim, M.M., Dufresne, A.: Processing and characterization of reinforced polyethylene composites made with lignocellulosic fibers from Egyptian agro-industrial residues. Compos. Sci. Technol. **68**(7–8), 1877–1885 (2008)

Configuring a Class of Machines Based on Reconfigurable 2DOF Planar Parallel Mechanism

Goran Vasilic, Sasa Zivanovic, Branko Kokotovic, Zoran Dimic, and Milan Milutinovic

Abstract The parallel 2DOF (Degrees of Freedom) mechanism presented in this paper has been the basis of much research by many authors. There are many significant results for the presented mechanism, and some of them are reported in this paper. The main goal of the research regarding the parallel mechanism is to create a hardware and software system that will be used to configure machine tools with three or more DOFs. The software system consists of two parts. One part is a set of applications intended for machine analysis and defining optimal configuration, and the other part is a control system of the machine adapted to the hardware of the machine, its configuration and purpose. For the presented mechanism, the kinematic model of the mechanism is described first. Based on the kinematic model, equations representing solutions of kinematics problems are derived. The derived equations are in a generalized form, with some variable parameters of the machine, and in such a form they correspond to every possible configuration of the reconfigurable mechanism. The equations are initially used to analyze some basic configurations, and then to analyze some configurations that have not been analyzed and presented so far. Also, equations in this form that are applicable for all possible configurations of the mechanism, are part of both parts of the software system. The final result of the presented procedures is one machine that has optimized parameters in accordance with the appropriate production process and with a configured control system that corresponds to the configuration of the machine.

Keywords Parallel mechanism · Hybrid mechanism · Complex machine tool · Reconfigurable machine tool · Inverse and direct kinematics problem

G. Vasilic (✉) · M. Milutinovic
Department of Traffic, Mechanical and Protection Engineering, Academy of technical vocational studies, Nade Dimic 4, 11080 Belgrade, Serbia
e-mail: gvasilic@tehnikum.edu.rs

G. Vasilic · S. Zivanovic · B. Kokotovic
Department for Production Engineering, University of Belgrade, 11000 Belgrade, Serbia

Z. Dimic
LOLA Institute, Kneza Viseslava 70A, 11030 Belgrade, Serbia

N. Mitrovic et al. (eds.), *Current Problems in Experimental and Computational Engineering*, Lecture Notes in Networks and Systems 323,
https://doi.org/10.1007/978-3-030-86009-7_10

1 Introduction

The concept of new generation machine tools differs from that of the previous generation. First of all, machine tools of a new generation are based on parallel mechanisms unlike the traditional machine tools, which are based on serial mechanisms. The reason for using parallel mechanisms to build machine tools is that the parallel mechanisms have many advantages compared to serial mechanisms like: higher accuracy, higher velocity, smaller mass of moveable parts, better flexibility, good dynamics characteristics [1–3], etc. In addition to their advantages, the mechanisms have some disadvantages: irregular shape of workspace, complex geometry of the mechanism, complex solutions of the kinematics problems, complex control algorithm, the existence of singularities, variable resolution in workspace [4–7], etc. Cited characteristics of the parallel mechanism largely depend on the concept and geometry of the mechanism. As parallel mechanisms provide the possibility to create many configurations that differ in some mechanism parameters (number of DOFs, type of activated joint, geometric parameters, etc.), many researches are focused on the analysis and configuring of parallel mechanisms to define the optimal mechanism configuration for the intended application. The most analyzed parallel mechanisms are 2DOF planar mechanisms [8–10], 3DOF planar mechanisms [11–13] and 3DOF spatial parallel mechanisms [14–16]. As can be seen, even if the geometry of mechanisms is relatively simple, the analysis and configuration of the mechanism is complex. Analyzing and configuring of parallel mechanisms with more than 3DOFs is more difficult, and for this reason a new group of mechanisms has been defined that includes mechanisms consisting of 2(or more) DOF parallel mechanisms with one or more serial added axes. Using the advantages of hybrid mechanisms, many authors provide solutions for the development of machine tools that are based on a combination of serial and parallel mechanisms. Papers [17–19] show hybrid machine tools based on a parallel mechanism with translatory actuated joints, while in papers [20–23] authors propose hybrid mechanisms that include the parallel mechanism based on rotary axes.

The mechanism considered in this paper is based on the existing similar mechanisms [2, 9]. Unlike the existing mechanisms, the geometry of the mechanism shown in this paper can be changed in a rapid and easy way [24, 25], and for this reason the presented mechanism is classified as a reconfigurable mechanism. The presented mechanism was the subject of some earlier research by the authors [26, 27] but, in this paper, the authors present new research results of the considered mechanism. The new research results show the possibilities and advantages of using the configuration of mechanisms that have not been considered so far. This fact enables the expansion of the building program so far defined in [26]. For the proposed new configurations, solutions of inverse and direct kinematics problems will be derived. The derived solutions are the basis of the analyses presented in this paper, and also the solutions of the kinematics problem will be used to configure the control systems of the proposed machine tools based on the analyzed reconfigurable parallel mechanism.

2 Description of a Reconfigurable 2DOF Parallel Mechanism

The planar parallel mechanism considered in this paper is built according to the modular principle, and for that reason it is referred to as MOMA (MOdular MAchine tool). Each module represents a set of elements that makes up a separate unit. The planar parallel mechanism modules are as follows:

- Actuated axes modules shown in Fig. 1a. Each of the actuated axes consists of guides, stepper motors, screws, and sliders. Characteristic point of the actuated axis is reference point R_i at which the actuated axis slider is when in its initial position.
- Legs module shown in Fig. 1b. This module is composed of two legs that can be of equal or different length. Legs are mutually connected by a joint, each with its one end. This articulated connection makes the parallel mechanism platform. With another end, each of the legs is articulated to the slider of an actuated axis each.
- Mechanism bases shown in Fig. 1c and d on which the actuated axes of the parallel mechanism are positioned and fixed together with the mechanism legs. Mechanism Base1 enables orientation of each of the actuated axes in three different ways by rotation about point R_i by angle g_i.

Depending on the mechanism base used, various configurations of the planar parallel mechanism can be accomplished. Thus, by positioning and fixing the actuated axes on Base 1, the mechanism configurations presented in Fig. 1e. can be obtained, named MOMA-M1, MOMA-M4 and MOMA-M5. By positioning and then fixing the actuated axes on Base 2, configurations of the mechanisms MOMA-M2 and MOMA-M3 are obtained (Fig. 1f).

Based on above description of the parallel mechanism MOMA, it can be concluded that the configuration can be rapidly and easily changed, therefore the mechanism can be characterized as a reconfigurable mechanism. The mechanism configuration can be changed in one of the following manners:

- by changing the legs length,
- by changing the actuated axes orientation: (i) by changing the orientation angle on Base 1; (ii) by using the Base 2 mechanism.

Based on possibilities provided by described hardware, 33 different mechanism configurations can be obtained, classified according to the mechanism type. Accordingly, a building program was generated and it was the basis for classifying all configurations according to the mechanism type [27].

To date research of presented configurations has produced significant results in terms of analysis, configuring, programming and control of 2DOF reconfigurable parallel mechanism MOMA [27]. Previous results obtained will be used for the analysis of new, non-analyzed configurations of 2DOF reconfigurable parallel mechanism MOMA in the framework of this paper. The goal of new configurations analyses is

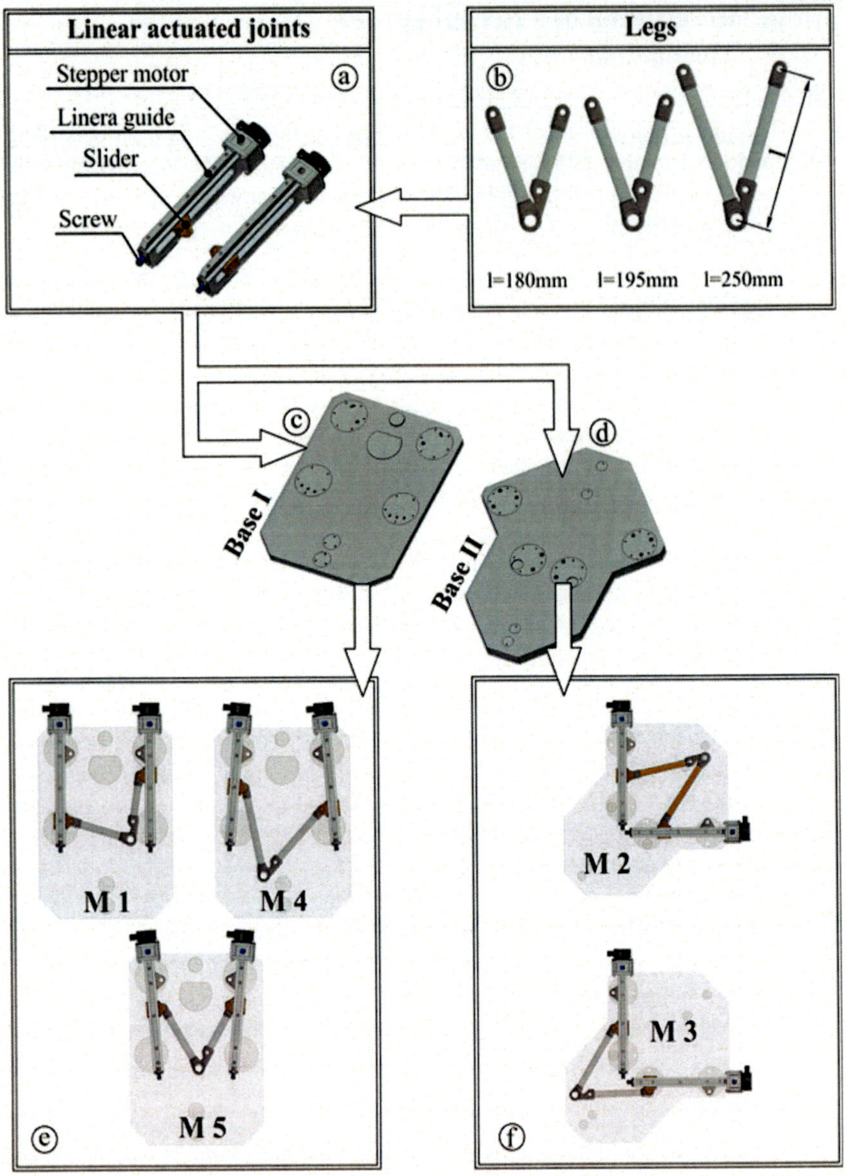

Fig. 1. Modules and hardware configuring of parallel mechanism MOMA

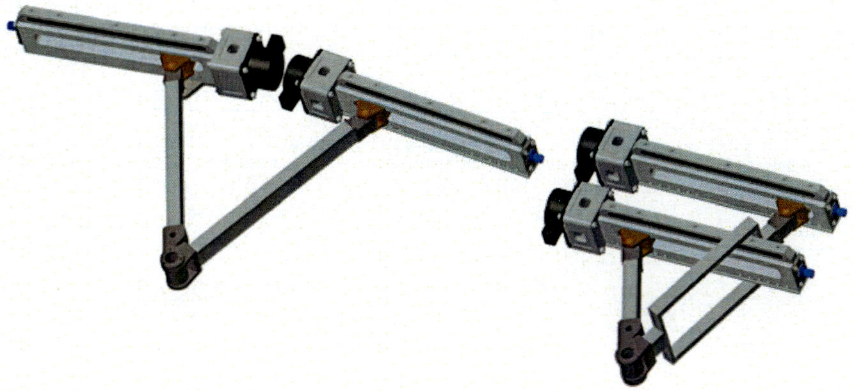

Fig. 2. A new configuration of 2DOF reconfigurable parallel mechanism MOMA

to compare and improve some characteristics compared to already defined configurations. The configurations analyzed in this paper are presented in Fig. 2 and will be included in the extended building program.

3 Geometric Model and Kinematics of Reconfigurable 2DOF Parallel Mechanism MOMA

For the upcoming analyses of new configurations of the reconfigurable parallel mechanism, it is necessary to define the geometric parameters of the mechanism. Geometric parameters essential and vectors necessary for further analyses are presented in Fig. 3.

Labels in Fig. 3 mean as follows:

- *{B}*—Fixed coordinate system related to the parallel mechanism base
- *P*—Mechanism platform
- ${}^{B}\mathbf{p}_{P}$—Position vector of the mechanism platform in the coordinate system *{B}*. Vector is defined by the platform coordinates x_P and y_P in the coordinate system *{B}*. Coordinates x_P and y_P represent external coordinates of the mechanism.
- R_i—Reference points of the actuated axis. Positions of the reference points are defined by their coordinates in the fixed system {B}, i.e. x_{Ri} and y_{Ri}
- ${}^{B}\mathbf{p}_{Ri}$—Vector defining the reference point position in the coordinate system *{B}*
- Z_i—Articulated connection between the leg end and the actuated axis slider
- *pi*—Distance of the actuated axis slider Z_i from the reference point R_i
- α_i—Orientation angle of the actuated axis $\alpha_i = {}^{3\pi}/_{2} + \gamma_i$
- γ_i—Auxiliary angle of the actuated axis orientation
- l_i—Mechanism legs lengths
- ${}^{B}\boldsymbol{a}_i$—Unit vectors of the actuated axis orientation defined by the orientation angle of the actuated axis a_i

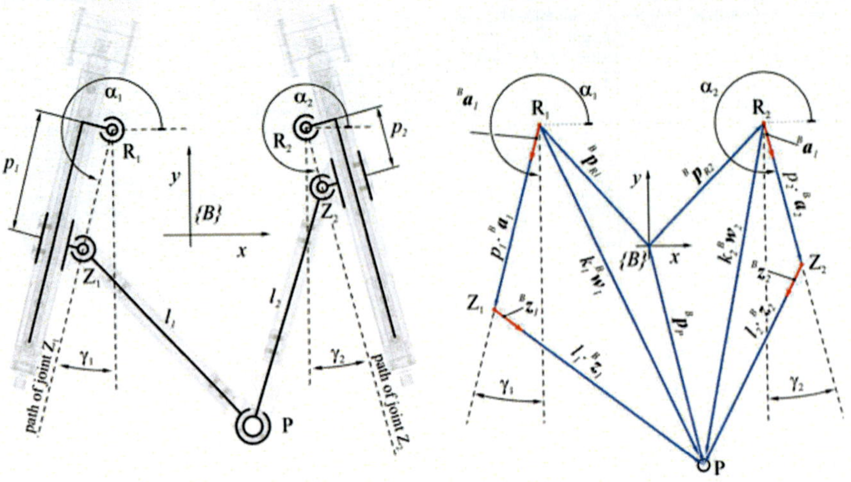

Fig. 3. Kinematic model of 2DOF reconfigurable parallel mechanism MOMA

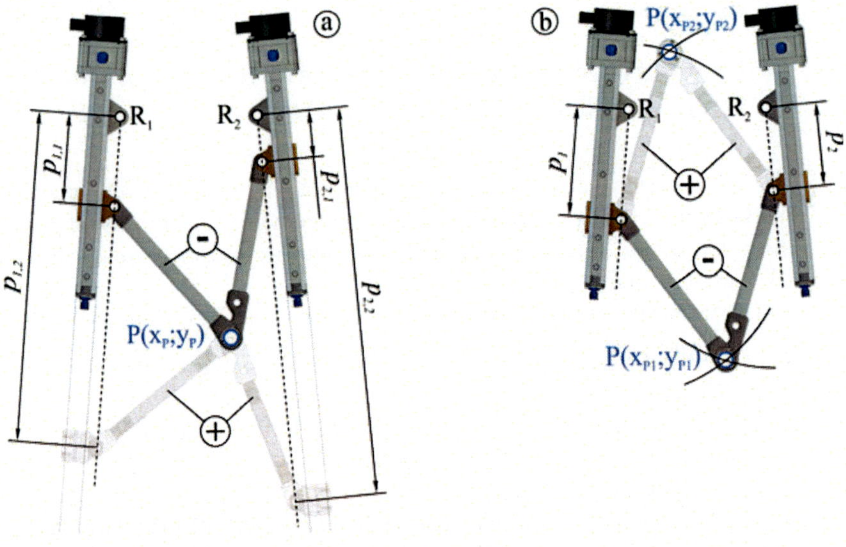

Fig. 4. Multiple solutions of: **a** Inverse kinematics problem; **b** direct kinematics problem

- ${}^{B}z_i$—Unit vectors defining the legs orientations
- $l_i \cdot {}^{B}z_i$—Legs vectors defined by legs length and orientation of legs unit vectors
- $p_i \cdot {}^{B}a_i$—Vector of the mechanism internal coordinates defined by the scalar value of internal coordinates p_i and unit vector orientation ${}^{B}\boldsymbol{a}_i$.

Kinematic model from Fig. 3 is a generalized kinematic model and by assigning concrete values to the geometric parameters a_i, x_{Ri} and y_{Ri}, and l_i the kinematic model describes a precisely defined mechanism configuration. In this way, based on a generalized kinematic model, generalized equations are generated to be further used for configuring and analyzing any configuration of the mechanism. Using the mechanism kinematic model, the following initial vector equations can be written:

$$\begin{aligned} {}^{B}\boldsymbol{p}_{R1} + {}^{B}\boldsymbol{a}_1 \cdot p_1 + {}^{B}z_1 \cdot l_1 = {}^{B}\boldsymbol{p} \\ {}^{B}\boldsymbol{p}_{R2} + {}^{B}\boldsymbol{a} \cdot_2 p_2 + {}^{B}z_2 \cdot l_2 = {}^{B}\boldsymbol{p}_P \end{aligned} \tag{1}$$

and by their rearrangement the following implicit generalized equations are obtained:

$$\begin{aligned} & f_1 : p_1^2 - 2p_1\left[a_{x1} \cdot (x_P - x_{R1}) + a_{y1}(y_P - y_{R1})\right] \\ & +(x_P - x_{R1})^2 + (y_P - y_{R1})^2 + l_1^2 = 0 \\ & f_2 : p_2^2 - 2p_2\left[a_{x2} \cdot (x_P - x_{R2}) + a_{y2}(y_P - y_{R2})\right] \\ & +(x_P - x_{R2})^2 + (y_P - y_{R2})^2 + l_2^2 = 0 \end{aligned} \tag{2}$$

3.1 Solution of Inverse Kinematics Problem

The solution of inverse kinematic mechanism yields the dependence of the mechanism internal coordinates p_i on the mechanism external coordinates x_{Pi} and y_{Pi}. This actually means that for specified coordinates x_{Pi} and y_{Pi} the mechanism, via the actuator of the driven axis, brings the actuated axis slider to the precisely defined distance from the reference point defined by quantity p_i. Solving Eq. (2) with respect to internal coordinates, the required dependence, i.e. $p_i = p_i\ (x_{Pi};y_{Pi})$, is obtained, given by the following equations:

$$\begin{aligned} & p_1 = a_{x1} \cdot (x_P - x_{R1}) + a_{y1} \cdot (y_P - y_{R1}) \\ & \pm\sqrt{\left[a_{x1}(x_P - x_{R1}) + a_{y1}(y_P - y_{R1})\right]^2 - (x_P - x_{R1})^2 - (y_P - y_{R1})^2 + l_1^2} \\ & p_2 = a_{x2} \cdot (x_P - x_{R2}) + a_{y2} \cdot (y_P - y_{R2}) \\ & \pm\sqrt{\left[a_{x2}(x_P - x_{R2}) + a_{y2}(y_P - y_{R2})\right]^2 - (x_P - x_{R2})^2 - (y_P - y_{R2})^2 + l_2^2} \end{aligned} \tag{3}$$

Equations (3) represent the solution of inverse kinematics problem of 2DOF reconfigurable parallel mechanism MOMA. In the equations a double sign "±" figures, which actually means that for each platform position there are two solutions each of the inverse kinematics problem for each actuated axis of the mechanism. Double solutions of the inverse kinematics problem are presented in Fig. 5a, and the sign "−" will be used in further analysis of the mechanism instead of the double sign "±".

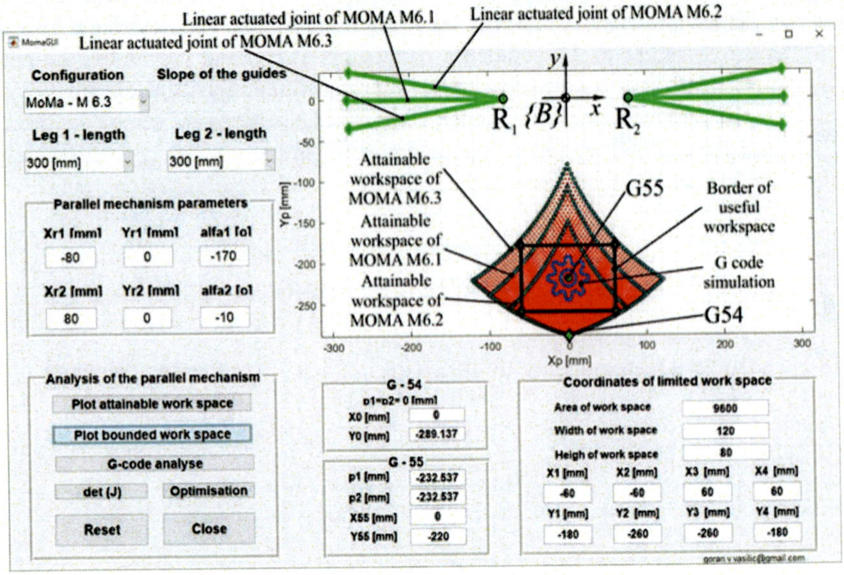

Fig. 5. Workspace and G code analysis—MOMA M6.3 configuration

As it is seen from Fig. 5a. the sign "–" in Eq. (3) for specified platform coordinates requires guides of a shorter length.

3.2 *Solutions of Direct Kinematics Problem*

The solutions of direct kinematics problem are obtained by solving Eq. (2) with respect to external coordinates x_{Pi} and y_{Pi} thereby obtaining their dependence on internal coordinates p_i, i.e. $x_{Pi} = x_{Pi}(p_{1};p_2)$ and $y_{Pi} = y_{Pi}(p_{1};p_2)$. To solve direct kinematics problem more easily, shifts given by Eq. (4) are introduced.

$$\begin{aligned}
&t_1 = 2(x_{R1} + p_1 \cdot a_{x1}) \quad t_2 = 2(y_{R1} + p_1 \cdot a_{y1})\\
&t_3 = -\left[p_1^2 + 2p_1\left(x_{R1} \cdot a_{x1} + y_{R1} \cdot a_{y1}\right) + x_{R1}^2 + y_{R1}^2 - l_1^2\right]\\
&t_4 = 2(x_{R2} + p_2 \cdot a_{x2}) \quad t_5 = 2(y_{R2} + p_2 \cdot a_{y2})\\
&t_6 = -\left[p_2^2 + 2p_2\left(x_{R2} \cdot a_{x2} + y_{R2} \cdot a_{y2}\right) + x_{R2}^2 + y_{R2}^2 - l_2^2\right]\\
&t_7 = t_4 + t_1 \quad t_8 = \frac{t_6 - t_3}{t_7} \quad t_9 = \frac{t_2 - t_5}{t_7}\\
&t_{10} = \frac{t_2 + t_4 + t_5 - t_1}{t_7} \quad t_{11} = t_9(2t_8 - t_1) + t_2 \quad t_{12} = t_8^2 - t_8 t_1 + t_3
\end{aligned} \tag{4}$$

Considering mentioned shifts, Eq. (5) are obtained, representing the solution of direct kinematics problem of the reconfigurable 2DOF parallel mechanism MOMA. In Eq. (5) a double sign "±" also figures, which means that for given values of internal coordinates p_i, the mechanism platform can have two possible positions, as shown in Fig. 5b. In order that the solutions of direct kinematics problem agree with the solutions of inverse kinematics problem, further mechanism analyses employ the sign "−" instead of the double sign "±".

$$\begin{aligned} y_P &= \frac{-v_{11} \pm \sqrt{v_{11}^2 - 4 \cdot v_{10} \cdot v_{12}}}{2v_{10}} \\ x_P &= v_8 + y_P \cdot v_9 \end{aligned} \tag{5}$$

4 Analyses of Reconfigurable 2DOF Parallel Mechanism

The upcoming analyses of a 2-axis reconfigurable parallel kinematic mechanism are based on previously derived equations of inverse and direct kinematics problem. Since the previously derived equations are generalized and in a general form, they hold for each mechanism configuration with beforehand defined values of the mechanism geometric quantities.

4.1 Workspace Analysis

Workspace analysis involves defining a series of points at which the mechanism can position the mechanism platform. The set of points that define the workspace shape and size is obtained based on equations that represent the solutions of direct kinematics problem (5). That is, by substituting the values of internal coordinates from $p_i = 0$ to $p_i = p_{max}$, introduced shifts (4) and then incorporated in Eq. (5), coordinates of the points are obtained, making the mechanism workspace attainable in the coordinate system *{B}*. For the values of internal coordinates $p_i = 0$, the actuated axes sliders Z_i are at reference points R_i, while for the values of internal coordinates $p_i = p_{max}$ the sliders are at the actuated axis pitch end.

4.2 Jacobian Matrix and Singularity of Reconfigurable 2DOF Parallel Mechanism MOMA

As it is known, one of the most essential characteristics of parallel mechanisms is the existence of singularity, as well as mechanism non-uniform characteristics in terms of kinematics, stiffness, accuracy, etc. at different points of the workspace. The basis for characteristics assessment of each parallel mechanism is Jacobian matrix of the mechanism (6)

$$J = J_p^{-1} \cdot J_x = \begin{bmatrix} \frac{\partial f_1}{\partial p_1} & \frac{\partial f_1}{\partial p_2} \\ \frac{\partial f_2}{\partial p_1} & \frac{\partial f_2}{\partial p_2} \end{bmatrix}^{-1} \cdot \begin{bmatrix} \frac{\partial f_1}{\partial x_p} & \frac{\partial f_1}{\partial y_p} \\ \frac{\partial f_2}{\partial x_p} & \frac{\partial f_2}{\partial y_p} \end{bmatrix} \tag{6}$$

where *Jp* and *Jx* are Jacobian matrices of inverse and direct kinematics. Considering implicit Eq. (2) and Jacobian matrix (6), the generalized Jacobian matrices *Jp* and *Jx* of the reconfigurable 2DOF parallel mechanism MOMA are of the following form

$$J_P = 2\begin{bmatrix} p_1 - a_{x1}(x_P - x_{R1}) - a_{y1}(y_P - y_{R1}) & 0 \\ 0 & p_2 - a_{x2}(x_P - x_{R2}) - a_{y2}(y_P - y_{R2}) \end{bmatrix}$$

$$J_x = 2\begin{bmatrix} (x_P - x_{R1}) - p_1 \cdot a_{x1} & (y_P - y_{R1}) - p_1 \cdot a_{y1} \\ (x_P - x_{R2}) - p_2 \cdot a_{x2} & (y_P - y_{R2}) - p_2 \cdot a_{y2} \end{bmatrix} \tag{7}$$

and as such will be used for the analysis of proposed configurations of 2DOF parallel mechanism MOMA.

4.3 MOMA—Gui, Software for Analysis of Reconfigurable 2 DOF Parallel Mechanism

Geometric parameters of the proposed configurations of a reconfigurable 2DOF mechanism from Fig. 3 are given in Table 1. Like configurations of the MOMA-M1, MOMA-M4 and MOMA-M5 mechanisms [27], configuration of the MOMA-M6 mechanism has its 3 subtypes. According to previous research results, the mechanism maximum operation is in symmetric configurations. So, for the basic configuration of type M6, the configuration of M6.1. is taken, where the actuated axes guides are horizontal. In configurations M6.2 and M6.3, the actuated axes guides are inclined to one side or the other to keep the mechanism symmetry.

Table 1. Geometric parameters of proposed configurations of reconfigurable 2DOF parallel mechanism MOMA

Configuration		x_{R1} (mm)	y_{R1} (mm)	x_{R2} (mm)	y_{R2} (mm)	l_1 (mm)	l_2 (mm)	α_1 (°)	α_2(°)
M6	M6.1	−80	0	80	0	300	300	0	180
	M6.2	−80	0	80	0	300	300	−190	10
	M6.3	−80	0	80	0	300	300	−170	−10
M7		0	50	0	−50	195	150	0	0

Configurations of the mechanisms M6 and M7 are temporary configurations being considered. For that reason, some disregarded facts, such as length of the actuated axes guides, legs length and shape, etc., are conditioning the configuring of the mechanism. The goal of this procedure is to determine the justification for building new constituent elements, whereby the building program would be considerably expanded.

For the analysis of proposed configurations of MOMA-M6 and MOMA-M7, an already developed application MOMA-Gui was used [27]. Generalized equations representing solutions of kinematics problems (3), (4) and (5), as well as derived Jacobian matrices (6) and (7) were implemented in the application. The application was customized to proposed new configurations of the mechanism so that within the application drop-down menus it is simple and easy to choose an option for desired configuration. The choice of desired configuration enables automatic loading of geometric parameters of the mechanism, which are used for mechanism analyses.

Figures 5 and 6 display the results of analyses of the mechanism configurations, whose parameters are given in Table 1. Mentioned figures show theoretically attainable workspaces defined in a previously described manner. In Fig. 7 two units are noted, representing work-attainable workspace of the mechanism MOMA-M7. Each unit represents theoretically attainable workspace for each of the two solutions of direct kinematics problem. By extending the actuated axes pitch, two partial workspaces would merge into one workspace, whereby the entire workspace would be extended, and avoiding singularity and changing the sign of kinematics problems can be realized in one of the manners as described in [23]. Figures 5 and 6 also show useful workspaces of the regular shape for configurations of the mechanisms

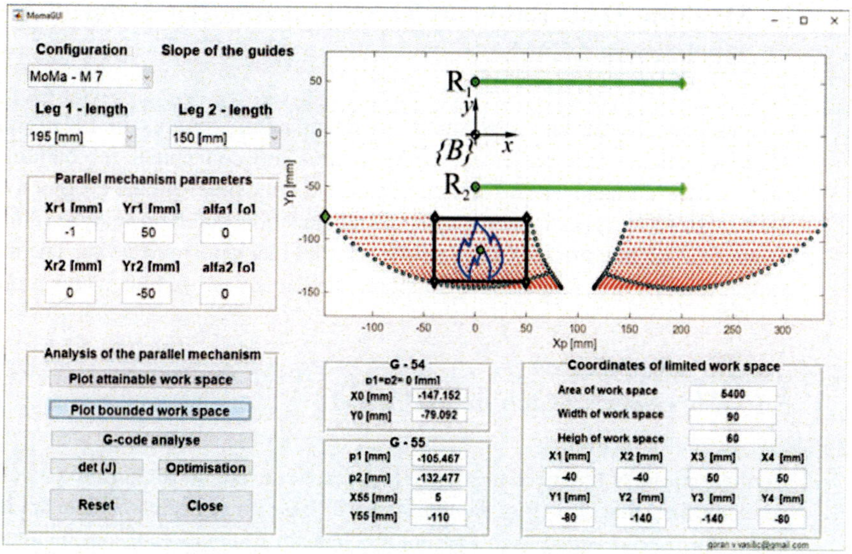

Fig. 6. Workspace and G code analysis—MOMA M7 configuration

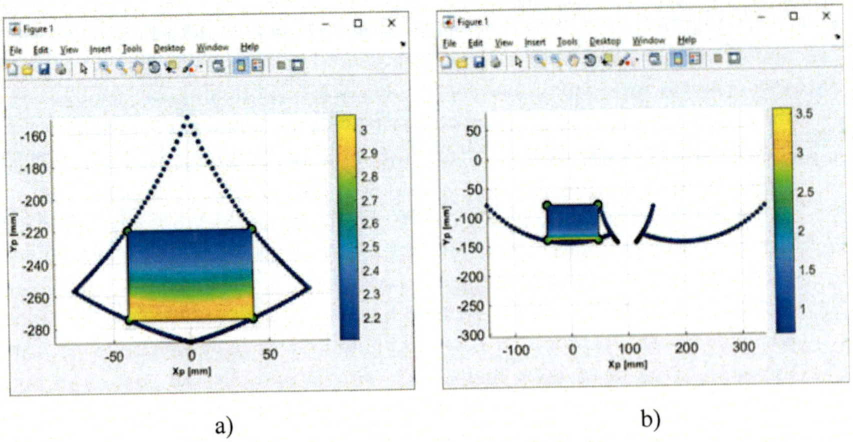

Fig. 7. Distribution of det(J) in useful workspace of: **a** MOMA M6.3; **b** MOMA M7 configuration

MOMA-M6.2 and MOMA-M7 and in indicated figures they are framed in black color.

For previously defined useful workspaces, simulation of the loaded G-code is presented first, and thereafter the analysis of distribution of the Jacobian matrix determinant $det(J)$ of analyzed mechanism configuration MOMA. Distributions of the Jacobian matrix of mentioned configurations are given in diagram forms in Fig. 7.

5 Machine Tools Based on Reconfigurable 2DOF Parallel Mechanism MOMA

Obtained results for the analyses of different configurations of the mechanism MOMA indicate that presented configurations are convenient for building both simpler hybrid machines and complex multi-axis machines. In the chapter to follow the authors give suggestions for building machine tools based on analyzed configurations of the reconfigurable 2DOF planar parallel mechanism MOMA. The machine tools presented are completed by supplemented equations of kinematics problems.

5.1 *Three-Axis Machine MOMA-3*

Configurations of the mechanism MOMA-6 are convenient for building 3DOF horizontal milling machine (Fig. 8a) that has in its base one parallel mechanism MOMA and a third serial added axis. Parallel mechanism provides tool motion along axes "x" and "y" while serial added actuated axis provides motion along horizontal "z" axis.

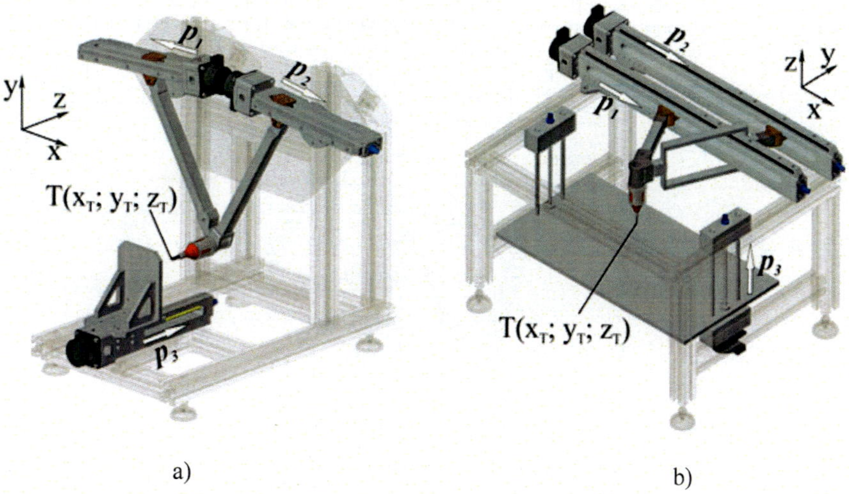

Fig. 8. CAD model of proposed 3DOF hybrid machines based on MOMA mechanism: **a** Horizontal milling machine tool; **b** vertical milling machine tool with long x-axis

Configuration of the mecha nism MOMA-7 is convenient for building 3DOF vertical milling machine (Fig. 8b). As in the previous case, the parallel mechanism provides tool motion along axes "x" and "y" while serial added actuated axis ensures workpiece motion along vertical "z" axis. Configuration of the mechanism MOMA-7 reminds of 3DOF parallel mechanism [28, 29] therefore, based on [30], it is convenient for building the machine tool with a long "x" axis travel. Configuration of the mechanism MOMA-M7, except for building a vertical milling machine as shown in Fig. 8b, can be also used for building 3D printer such as RepRap printer [31].

For presented machine tools the solutions of kinematics problems for the third actuated axis are trivial, which means that $p_3 = z$ (IKP solution) and $z = p_3$ (DKP solution) hold, respectively. Previous equalities together with Eqs. (3–5) are a constituent part of kinematics problems of presented 3-axis hybrid machines based on a reconfigurable 2-axis mechanism MOMA.

5.2 *Four-Axis Complex Machine MOMA-W*

The complex 4-axis machine MOMA-W (W-wire) is intended for machining process by wire cutting. Machine tool MOMA-W was built based on hybrid machines that have two parallel mechanisms [1, 32, 33]. Machine tool MOMA-W has been the subject of earlier research [34]. It consists of two parallel mechanisms whose platforms are mutually connected by wire (Fig. 9).

In the complex machine tool MOMA-W wire represents a tool for machining process (cutting). Each of the platforms enables translatory movement of one end

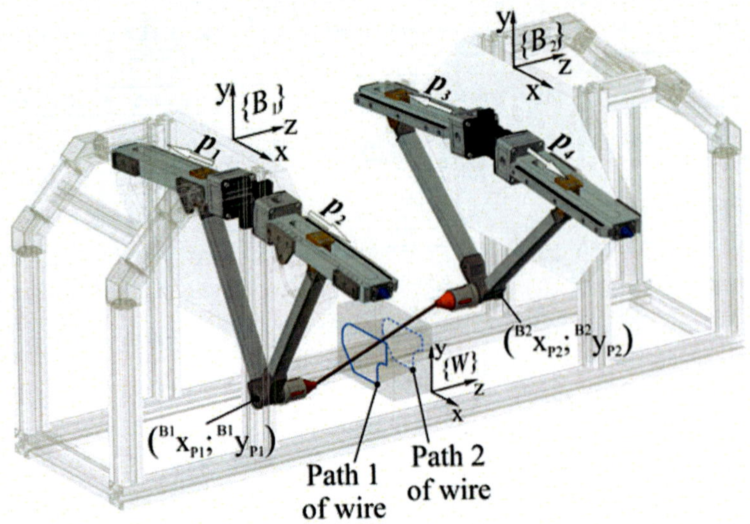

Fig. 9. CAD model of 4DOF complex machine tool for wire cutting based on MOMA mechanism

of the wire along two axes *x* and *y*. Synchronized two wire ends movement allows the accomplishment of the wire translatory movement in the directions of axes *x* and *y*, as well as rotation of the wire about the same axes. In this way, the tool (wire) of the machine is provided with 4DOFs. During machining, parallel mechanisms work independently but their movements ensure tool motion according to the programmed paths (Fig. 9—Path 1 of wire and Path 2 of wire). Also, during work, the distance between the two platforms changes but this fact has no impact on the mechanism kinematics. For this reason, the connection between the two mechanisms cannot be characterized as a serial connection, therefore the whole mechanism cannot be characterized as a hybrid mechanism. It is defined as a complex multi-axis mechanism.

As above mentioned, by programming machine tool MOMA-W the tool (wire) is assigned motion along two contours consisting of two series of points. Coordinates of the points of both contours are defined in the coordinate system *{W}*. In Fig. 10, points of the first contour are denoted by C_1 and of the second contour by C_2. Coordinates of the points C_1 and C_2 in the coordinate system *{W}* are defined by vectors ${}^{W}\boldsymbol{p}_{c1}$ and ${}^{W}\boldsymbol{p}_{c2}$. However, coordinates of the same points in the coordinate system *{B₁}* are defined by vectors ${}^{B1}\boldsymbol{p}_{c1}$ and ${}^{B2}\boldsymbol{p}_{c2}$.

Since wire motion is provided by parallel mechanisms, it is necessary to connect the coordinates of the contour points with the coordinates of the mechanisms' platforms, which are realized by the mechanism internal coordinates. As the mechanism possesses four actuated axes, the mechanism internal coordinates are denoted by p_1, p_2, p_3 and p_4 and they are shown in Fig. 9. Considering the connection between the coordinate systems *{W}*, *{B1}* and *{B2}* defined by position vectors ${}^{B1}\boldsymbol{p}_W$ and ${}^{B2}\boldsymbol{p}_W$, coordinates of the points C_1 and C_2 in the coordinate system $\{B_i\}$ $i = 1,2$ are defined

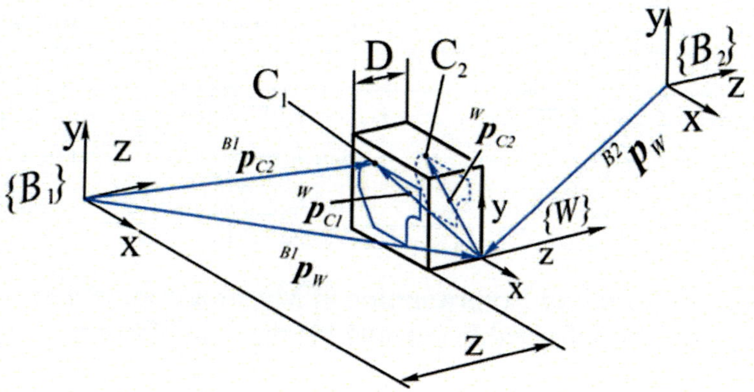

Fig. 10. Coordinates of points C_1 and C_2 in c.s *{W}*, *{B1}* and *{B2}*; Connection between c.s *{W}*, *{B1}* and *{B2}*

by the following expressions

$$\begin{aligned} {}^{Bi}p_{C1} &= {}^{Bi}p_W + {}^{W}p_{C1} \\ {}^{Bi}p_{C2} &= {}^{Bi}p_W + {}^{W}p_{C2} \end{aligned} \tag{8}$$

According to Figs. 9 and 10, based on programmed points and their coordinates in the coordinate system *{W}*, and considering the transformations of coordinates given by Eq. (8), applying Eq. (9) the coordinates of the platforms x_{Pi} and y_{pi} are defined in the coordinate systems *{Bi}* (i = 1, 2), which provide the wire desired position and orientation.

$$\begin{aligned} x_{Pi} &= \frac{-{}^{Bi}z_{C1}}{{}^{Bi}z_{C2} - {}^{Bi}z_{C2}}\left({}^{Bi}x_{C2} - {}^{Bi}x_{C1}\right) + {}^{Bi}x_{C1} \\ y_{Pi} &= \frac{-{}^{Bi}z_{C1}}{{}^{Bi}z_{C2} - {}^{Bi}z_{C2}}\left({}^{Bi}y_{C2} - {}^{Bi}y_{C1}\right) + {}^{Bi}y_{C1} \end{aligned} \tag{9}$$

Incorporating Eq. (9) into Eq. (2), four equations are obtained. Solving them with respect to p_i (i = 1, 2, 3, 4), the values of internal coordinates p_1, p_2, p_3 and p_4 are obtained, depending on the coordinates of the programmed points C_1 and C_2, which represents the solution of inverse kinematics problem of the complex multi-axis machine tool MOMA-W. The solutions of inverse kinematics problem of the complex machine tool MOMA-W are given by Eq. (10) where Eq. (9) are incorporated. Also, by solving the system of Eq. (2) with respect to x_{ci} and y_{ci}, dependence of stated coordinates on internal coordinates p_1, p_2, p_3 and p_4 is obtained, representing the solution of direct kinematics problem of the complex machine tool MOMA-W.

$$A_i = a_{xi} \cdot (x_P - x_{Ri})$$
$$B_i = a_{yi} \cdot (y_P - y_{Ri})$$
$$p_i = A_i + B_i \pm \sqrt{(A_i + B_i)^2 - (x_P - x_{R1})^2 - (y_P - y_{R1})^2 + l_1^2} \;;\quad (i = 1 \div 4) \tag{10}$$

6 Control and Programming of Machine Tools Based on Reconfigurable Parallel Mechanism MOMA

Using the results reported by authors in [27, 35, 36], an open architecture system based on PC platform, Linux operating system and EMC2 software is imposed as a convenient system for control and programming of presented machine tools based on reconfigurable planar parallel mechanism MOMA. EMC2 software works in real time and is used for control and programming of industrial robots and machine tools. Since the system is of open architecture, it is possible to configure and customize the control system to every machine tool. By implementing the derived equations in EMC2 software, which represent the solutions of kinematics problems (3), (4) and (5), the control system is enabled to create control of the actuated axes according to the kinematics of machine tool itself.

The structure of the control system Linux EMC2 is presented in Fig. 11, and the basic modules of the control system, presented in the figure are as follows:

- EMCTASK (motion controller)—interprets and distributes program commands on the machine (G-code);
- EMCIO (discrete I/O controller)—module used to control movements not related to the actuated axes movements (tools change, switching the cooling and lubrication system on and off, etc.);
- EMCMOT—module working in real time. It uses implemented equations of kinematics problems and realizes the actuated axes movement;
- HAL (hardware abstraction layer)—module providing data transfer in real time from EMC2 to virtual machine but also to the hardware control part of the real machine.

7 Conclusion

The framework of this paper involves new configurations of the planar reconfigurable mechanism MOMA. For the presented configurations the solutions of kinematics problems were derived and, based on the solutions, the analyses of proposed

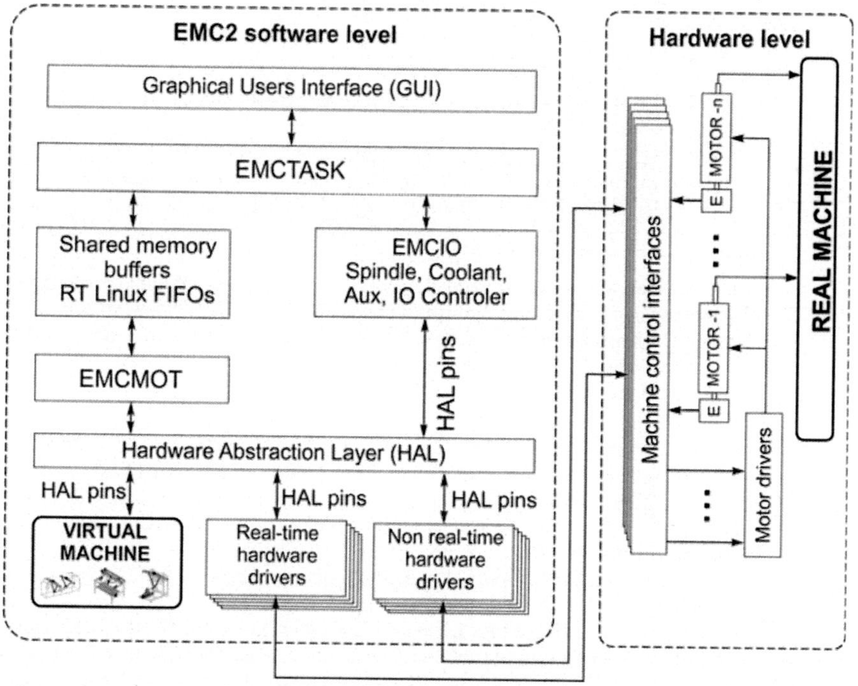

Fig. 11. Simplified structure of control system Linux EMC2

configurations were conducted. The configurations were analyzed using the developed software MOMA-GUI that was expanded and customized for the needs of this and future research.

The results of analyses indicate that the proposed configurations can be used to build machine tools of lesser or greater complexity that will produce acceptable results during exploitation. As shown in the paper, machine tools based on presented configurations of the reconfigurable mechanism MOMA can be used for different machining processes. Thus, it has been achieved that, except for hardware reconfigurability, the mechanism MOMA can be configured according to the machining process that it is intended for.

For the presented 3DOF and 4DOF machine tools based on the reconfigurable mechanism MOMA, the reconfigurable control and programming system is also developed, based on derived equations of kinematics problems. The goal of future research is to create unique control for all machine tools that have the reconfigurable mechanism MOMA in their base, where unique control is founded on generalized equations of kinematics problems.

Acknowledgements The presented research was supported by the Ministry of Education, Science and Technological Development of the Republic of Serbia by contract no. 451-03-9/2021-14/200105 dated 5 February 2021.

References

1. Tanev, T.K.: Kinematics of a hybrid (parallel-serial) robot manipulator. Mech. Mach. Theory **35**(9), 1183–1196 (2000)
2. Wu, J., Wang, J., Li, T., Wang, L.: Dynamic analysis of the 2DOF parallel manipulatorpf a heavy duty machine tool. Int. J. Adv. Manuf. Technol. **34**(3–4), 413–420 (2007)
3. Stan, S.D., Maties, V., Balan, R.: Optimal design of parallel kinematics machines with 2 degrees of freedom, Parallel Manipulators, towards New Applications, Huapeng Wu, IntechOpen (2008)
4. Jin, Y., Chanel, H., Paccot, F.: Parallel Robots. In: Nee, A. (Eds.) Handbook of Manufacturing Engineering and Technology. Springer, London (2015)
5. Simaan, N.: Analysis and Synthesis of Parallel Robots for Medical Applications, Research thesis, The Technion—Israel Institute of Technology (2009)
6. Miller, K.: Optimal design and modeling of spatial parallel manipulators. Int. J. Robot. Res. **23**(2), 127–140 (2004)
7. Zhenqi, Z., Jinsong, L., Zhongxue, G., Hui, Z.: Kinematic and dynamic modelling for real-time control of Tau parallel robot. Mech. Mach. Theory **40**(9), 1051–1067 (2005)
8. Majou, F., Wenger, P., Chablat, D.: Design of 2-Dof Parallel Mechanisms for Machining Applications. In: Lenarčič, J., Thomas, F. (eds.) Advances in Robot Kinematics, pp. 319–328. Springer, Dordrecht (2002)
9. Wu, J., Wang, J., Li, T., Wang, L.: Analysis and application of a 2-DOF planar parallel mechanism. ASME J. Mech. Des. **129**(4), 434–437 (2007)
10. Lee, J.H., Nam, Y.J., Park, M.K.: Kinematics and optimization of a 2-DOF parallel manipulator with a passive constraining leg and linear actuators. J. Mech. Sci. Technol. **24**, 19–23 (2010)
11. Lee, J.J., Dong, Y., Jeon, Y.H., et al.: Design of a planar 3-DOF parallel micromanipulator. J. Mech. Sci. Technol. **27**, 2985–2990 (2013)
12. Wu, J., Wang, J., Wang, L.: Optimal kinematic design and application of a redundantly actuated 3DOF planar parallel manipulator. ASME. J. Mech. Des. **130**(5): 054503 (2008)
13. Wu., Jun, Wang, J., Wang, L., Li, T.: Dynamics and control of a planar 3-DOF parallel manipulator with actuation redundancy. Mech. Mach. Theory **44**(4), 835–849 (2009)
14. Ruggiu, M., Kong, X.: Reconfiguration analysis of a 3-DOF parallel mechanism. Robotics **8**(3) (2019)
15. Wang, L., Xie, F., Liu, X., Wang, J.: Kinematic calibration of the 3-DOF parallel module of a 5-axis hybrid milling machine. Robotica **29**(4), 535–546 (2011)
16. Milutinovic, D., Slavkovic, N., Kokotovic, B., Milutinovic, M., Zivanovic, S., Dimic, Z.: Kinematic modeling of reconfigurable parallel robots based on DELTA concept. J. Prod. Eng. **15**(2), 71–74 (2012). ISSN 1821-4932. http://www.jpe.ftn.uns.ac.rs/papers/2012/no2/15-Milutinovic1-JPE.pdf
17. Yuyao, S., Jun, W., Guang, Y., Tian, H.: Dynamic characteristic prediction of a 5-DOF hybrid machine tool by using scale model considering the geometric distortion of bearings. Mech. Mach. Theory **145**, 103679 (2020)
18. Xie, F., Liu, X., Zhang, H., et al.: Design and experimental study of the SPKM165, a five-axis serial-parallel kinematic milling machine. Sci. China Technol. Sci. **54**, 1193–1205 (2011)
19. Weiyao, B., Fugui, X., Xin-Jun, L., Xuan, L.: Optimal design of a novel 4-degree-of-freedom parallel mechanism with flexible orientation capability. Proc. Inst. Mech. Eng. Part B J. Eng. Manuf. **233**(2), 632–642 (2019)
20. You, W., Zhen, Y., Zhuang, F., Jian, F. and Hui, Z.: Kinematics and dynamics analysis of a novel five-degrees-of-freedom hybrid robot. Int. J. Adv. Robot. Syst. **14**(3), (2017)
21. Ning, L., Junjun, W.: Kinematics and application of a hybrid industrial robot—delta-RST. Sens. Transd. **169**(4), 186–192 (2014)
22. Yeshmukhametov, A., Kalimoldayev, M., Mamyrbayev, O. and Amirgaliev, Y.: Design and kinematics of serial/parallel hybrid robot. In: 3rd International Conference on Control, Automation and Robotics (ICCAR), Nagoya, Japan, pp. 162–165 (2017)

23. Hesselbach, J., Hoffmeister, H.W., Loohß, T., et al.: Parallel kinematic concept for stationary high performance cutting in wood machining centers. Prod. Eng. Res. Devel. **1**, 205–212 (2007)
24. Koren, Y., Shpitalni, M.: Design of reconfigurable manufacturing system. J. Manuf. Syst. **29**(4), 130–141 (2010)
25. Landers, R.G., Min, B.K., Koren, Y.: Reconfigurable machine tools. CIRP Ann. **50**(1), 269–274 (2001)
26. Vasilic, G., Zivanovic, S., Kokotovic, B.: Modelling and analysis of 3-axis reconfigurable hybrid kinematics mechanism with translatory actuated joints. In: Proceedings of 5th International Conference on Advanced Manufacturing Engineering and Technologies. NEWTECH. Lecture Notes in Mechanical Engineering. Springer, Cham (2017)
27. Vasilic, G., Zivanovic, S., Kokotovic, Dimic, Z.: Configuring and analysis of a class of generalized reconfigurable 2-axis parallel kinematic machine. J. Mech. Sci. Technol. **33**, 3407–3421 (2019)
28. Glavonjic, M., Milutinovic, D., Zivanovic, S., Dimic, Z., Kvrgic, V.: Desktop 3-axis parallel kinematic milling machine. Int. J. Adv. Manuf. Technol. **46**, 51–60 (2010)
29. Zivanovic, S., Glavonjic, M., Dimic, Z.: Methodology for configuring desktop 3-axis parallel kinematic machine. FME Trans. **37**, 107–115 (2009)
30. Milutinovic, D., Glavonjic, M., Kvrgic, V., Zivanovic, S.: A new 3-DOF spatial parallel mechanism for milling machines with long X travel. CIRP Ann. **54**(1), 345–34 (2005)
31. http://microfabricator.com/articles/view/id/52d324e59aad9d2f3f000015/reprap-tuga-noncartesian-3d-printer-from-portugal (2021)
32. Zheng, X.Z., Bin, H.Z., Luo, Y.G.: Kinematic analysis of a hybrid serial-parallel manipulator. Int. J. Adv. Manuf. Technol. **23**(11–12), 925–930 (2004)
33. Romdhane, L.: Design and analysis of a hybrid serial manipulator. Mech. Mach. Theory **34**(7), 1037–1055 (1999)
34. Vasilic, G., Zivanovic, S.: Configuring and analysis of complex multi-axis reconfigurable machine for wire cutting process. Mech. Mach. Theory **149**, 103833 (2020)
35. Slavkovic, N., Zivanovic, S., Kokotovic, B., Dimic, Z., Milutinovic, M.: Simulation of compensated tool path through virtual robot machining model. J. Braz. Soc. Mech. Sci. Eng. **42**(374), 1–17
36. Rakic, A., Zivanovic, S., Dimic, Z., Knezevic, M.: Digital twin control of multi-axis wood CNC machining center based on LinuxCNC. BioResources **16**(1), 1115–1130

Strategic Importance and Sustainable Governance of High-Tech Business Incubators: Evidence from Serbia

Andjelija Djordjevic and Marko Mihic

Abstract This paper analyzes symbiotic relationship between high-tech start-ups, governance of business incubators dedicated to developing them and overall impact on Serbian economy and tech landscape. It was written based on previously conducted research, relevant industry reports, a review of academic literature and many articles on the business incubation. As high-tech business incubators are a part of economic development, this paper aims to examine the role of business incubator on SMEs and high-tech start-ups and explore which types of support drive this impact, but also to investigate impact on wider economic and social landscape in Republic of Serbia. Furthermore, the overall benefits of government investing in incubators are elaborated with review of private vs public ownership and differences in opportunities that they provide. The need for strategic governance and systematic support is rationalized by inspecting efficiency and effectiveness through large set of KPIs. Results should provide proper model for development of innovation ecosystem and entrepreneurship support by using business incubator as a tool. If managed properly, incubators can become a key component of national development strategy since they can help high-tech companies to survive and develop in MOST critical initial phase, by providing wide range of services such as management support, financial resources, technical services or simply offering adequate office space.

Keywords High-Tech start-ups · Business incubator · Entrepreneurship ecosystem · Performance management

1 Introduction

Business incubators give significant contribution in developing high-growth innovative businesses which is why many governments are taking necessary steps to include them in their economic agenda. They create employment opportunities,

A. Djordjevic (✉) · M. Mihic
Department of Management and Project Management, Faculty of Organizational Sciences, University of Belgrade, 11000 Belgrade, Serbia
e-mail: andjelija.djordjevic@fon.bg.ac.rs

N. Mitrovic et al. (eds.), *Current Problems in Experimental and Computational Engineering*, Lecture Notes in Networks and Systems 323,
https://doi.org/10.1007/978-3-030-86009-7_11

promote different technologies and nurture entrepreneurial drive [1]. Value that is being created this way is perceived worldwide as a mainstay of economic development programs. In order for government to ensure sustainable value creation and building of wealth from investing in these business accelerators, there's a need for understanding of how business incubators function and how they can impact national development and growth [2].

Future flourishing of Serbian economy regarding business incubators, involves managing macro financial aspects by initiating different programs, drafting policies or funding and operating multi-year economic development plans and projects that will generate employment or kick-start industry. This can be supported by microfinance management through an individual-focused, community-based approach to provide money and financial services to individuals or small businesses that lack access to mainstream or conventional resources [3].

Looking at business incubators from Serbian perspective, we can state that they are important part of an overall-national and even global-entrepreneurship and business support eco-system. They should be perceived as a tool that can be used in providing economic prosperity through efficiency of their operations. This can be achieved by synchronization of different political, legal, institutional and organizational elements [4].

It is notable that according to Statistical office of the Republic of Serbia, estimates on the overall economic activity in the Republic of Serbia in 2019, measured by the real trends of Gross domestic product (GDP), indicate a growth of 4.0% when related to the year 2018. In addition, Ministry of Finance has also stated that IT industry contributes with 10% to GDP of Serbia, and that this is a sector that has higher growth rates each year.

Furthermore, in recent years government has introduced public with digitalization process as one of the priorities, since it has contributed with almost EUR 1 billion to export of the Republic of Serbia and it is prominent that a lot of young people are building their future around this line of work [5].

A report published by Vojvodina ICT Cluster, shows that the investment rate in IT sector goes over EUR 80 per capita, with a growth tendency of up to EUR 150 by the year 2025. While this is well below the EU average of EUR 800, it still shows that there is huge interest in digitization and that there is rationale behind investing in accelerators that would further encourage development of this sector. Authors draw attention to the fact that this industry has been recognized as the area with highest potential for development of SMEs and that this trend needs to be put to positive use as soon as possible.

However, research also shows that average Serbian tech startup needs a whole decade to transform into large company. IT sector is one of the most promising in Serbia and in order for government to support it in a right way it is significant to have a good overview of IT landscape [6]. This could be achieved by creating business incubators set up as public-private partnerships, which will enable a good outline for entrepreneurs, investors and anyone who can benefit from relevant information regarding IT sector [7].

The same study claims, that although there is a record number of 337 software companies founded in 2019 in Serbia, there is also data that shows that 8 out of 10 leading Serbian software companies have foreign owners. On a global tech map, Republic of Serbia has been somewhere in between for a while, and there is a obvious need to comprehend bigger picture and for government to get more involved in setting the firm ground for stronger development of tech startups and bridging the gap between Serbia and developed countries [8].

Serbian startups have, in the last 10 years, raised more than EUR 143 millions of investment for their businesses, but more than 80% of this amount was raised by just two companies. The challenge of raising funds for startups to fund their growth becomes even more apparent when considering the fact that more than 50% of startups in the country haven't received any investments and are entirely self-funded. When it comes to personal and third-party financing, it seems that most startup founders have personal financial support at formation, while the majority of them are not aware of the possibility of third-party financial support. This further confirms the fact that early stage financing is quite limited and rather absent within the startup ecosystem in the Republic of Serbia.

There is also a lag behind the global average when it comes to local connectedness. The analysis reveals that the way in which founders and investors help each other and the number of quality relationships between founders, as well as investors and experts is below global levels.

The government has recently introduced several tax incentives which are oriented towards the knowledge and digital economy, such as IP box and R&D deductions, as well as deductions on taxes and social contributions for founders and foreign employees. The most important incentive for the startup ecosystem is a 30% tax credit for investments in startups by companies.

Although several regulatory improvements have been made, some of the large impediments to startup growth still remain. Practically the only source of funding available in the Serbian economy is landing money from the banks, with equity crowd funding not being regulated and VC having been introduced only recently. Another big obstacle for local startups lies in the fact that the possibilities for the digitalization of financial transactions in Serbia are quite limited, primarily due to the Foreign Exchange Act [9].

It is difficult to compare the startup ecosystem in Serbia to those in other countries. The problems startups and investors are facing are different, the opportunities and challenges vary and depend not only on the maturity of the ecosystem, but on its specialization [10].

According to the Study on the current situation of BIs in the Republic of Serbia, development of business incubators in any country usually consists of several steps. Most of the authors agree on three main phases: (1) development of basic infrastructure and physical space and basis services; (2) service improvement and interaction with and between clients; (3) enhancing strategic linkages and partnerships within innovation ecosystem.

In this study, authors identify position of Serbian ecosystem to be somewhere between first and second generation of business incubators [4]. Available office spaces and premises for tenants should provide economies of scale and job creation in first phase, and then in second phase by offering coaching and mentorship, incubators should deliver value adding services and training. By including networking and linkages, tenants will get enhanced access to external resources in phase three [11].

Efficient functioning of accelerators depends on cooperation among coordinated national business regulative, involvement of municipal and regional governing entities and competencies of business incubator managers. In order to ensure future prosperity, transparency of business incubator operations must be achieved on the level of individual incubators and the level of business incubator ecosystem as a whole. Tracking of expenditures and revenues on each of those levels could help shed a light on this topic.

2 Private Versus Public Operation of BI

Business incubators are certainly not a new concept, but many experts agree that their importance has never been greater. Since accelerators and business incubators nowadays have undeniable impact on economy, they have been recognized by governments as well as private entities as a driving force of technological innovation and progress as a whole. This growth has been additionally inflated by worldwide acknowledgment of business incubators as a valuable asset for developing and nurturing young business ventures [12].

Putting aside different legal definitions of business incubators, we can identify two main types of ownership—private and public. Private technological incubators began operating in developed countries thanks to the rapidly growing private capital sector, which traditionally had not funded such projects. Upon further research it can be concluded that there are differences but also a lot of similarities between those two types of incubators [10].

The main difference is the one that also emphasize the importance of existence of public incubator programs. Based on different empirical analysis and findings, it can be concluded that private incubators should not fully replace public ones. The main reason behind this statement can be found in the fact that private sector has tendency to focus on certain selected areas or fields, that are the most lucrative, whereas public incubators must sponsor a large variety of activities in order to facilitate fulfillment of national objectives (developing rural areas, incentivize specific population groups, advancing of peripheral regions, developing particular kind of technology, etc.). Without government support such activities would be out of reach.

The other notable difference lies within main sources of income. Public funding can come from national and regional public bodies which fund the core activities of the business incubators, different programs and projects implemented by public authorities or different types of development funds. Private funding is usually associated with income from clients such as SMEs and entrepreneurs, income from housing and incubator services and different types of private sponsorships [13].

On the other hand, there seems to be a lot of similarities when it comes to operation of incubators. Unfortunately, at the moment there are not clear guidelines for establishing and successfully operating business incubators. This also includes lack of consensus on what are the critical success factors and key performance indicators of incubators [14]. Some type of metrics however must be defined and implemented, because there is increasing evidence that despite many public policies supporting business incubation and significant financial investments, many of accelerators cannot be declared successful and most of them can be challenged on the general effectiveness.

Forming public-private partnership could be the best compromise that would combine best of both worlds. Business incubation is proven strategy for enhancing an entrepreneurial climate and retaining businesses, creating jobs, accelerating growth and diversifying local economy, it is a perfect fit for a public-private partnership. In order to be successful, the partnership must result in a public benefit, driven by both the public and private sectors [15].

We can finally conclude that type of ownership is relevant when it comes to defining key metrics that will provide incubators and their tenants with success. That being said, choice of right KPI is also heavily influenced by purpose of incubator, its stage of development and external circumstances generated by the factors and policies of the country where it's founded. From everything that has been said so far, we can clearly see that there is a requirement for assessing different needs of incubators depending primarily on phase they are in (what kind of services are they offering), and external factors they are facing (mainly defined by the current situation on the market) [16].

3 Current Situation of Business Incubators in Serbia

Chronologically, during the first wave of development, from macro perspective, programs aimed at economic restructuring and job creation. This means that tenants are mainly offered affordable space along with some kind of basic shared services. The second wave is characterized by the technological research and usually takes form of the science park model that operates as a network commercialization enabler. This basically means that SMEs and entrepreneurs seek values through networking, consulting and skill expansion. The third wave brings multi-purpose models portrayed by complex innovation centers, accelerators and specialized types of incubators.

Previous studies done on business incubators in Serbia are showing us that we can overall identify three main phases or generations in development of the models. It has been noted that majority of incubators belong between first and second generation which means that their main focus is on providing working place and enabling interactions between clients [4].

In order to define key performance indicators that will enable long-term prosperity for all parties involved, we first must determine what are the critical success factors or requirements that must be met to ensure well-being of business actors included. We can start by taking another look at previously mentioned study and this time focus on highlighted obstacles in current operations. This will provide necessary information on what should we measure to induce needed changes and also adapt key metrics to specific circumstances [2].

Some of the identified areas that can be improved are listed below:

- Even though there is an identified need for working space, many of the special purpose designed spaces are unoccupied by tenants and waiting lists are almost non-existent;
- Managers of incubators are indicating that their clients are in need for more specialized services and that current models of business incubators are not equipped to provide that, since that would mean hiring more experts;
- There seems to be a lack of technology-driven tenants in incubators;
- Involvement of regional and municipal governance was identified as insufficient. There seems to be lack of understanding of the role the business incubators could play in the promotion of business and entrepreneurship in the municipalities and the regions;
- Business enhancement activities (hackathons, boot-camps, business success and failure events, public presentations of business stories and cases, etc.), that are perceived as engaging and interesting, need to become a regular part of incubator operations as well as promotional activities among schools, universities, communities, etc.;
- Most of the managers of the more advanced incubators point at the lack of financial support for the development of the tenant companies, and this issue indicates the important asymmetry in financial risk taking between the State (with public money) and the private entrepreneur [4].

Business incubators are most definitely crucial part of national entrepreneurial ecosystem and its future development. They could play key role in economical prosperity of the whole country if sufficient levels of efficiency are ensured. This means that harmony among national policies, legal entities and institutional and organizational elements must be achieved together with understandable values and leadership. It's needless to say that this is only possible with strong national and local governmental support.

Effective performance management and monitoring system on the level of individual incubators and the level of business incubator ecosystem is needed in order to provide continuous improvement of business incubator operations. Efficient functioning of business incubator ecosystem that will provide support for entrepreneurs and SMEs is dependent on well balanced and precisely coordinated national business regulative, involvement of municipal and regional governing entities and competencies of business incubator managers [17].

4 Business Incubation Finance Management

Business incubators can become very successful commercial initiatives but in order to achieve this stage, they generally require sacrifices of short-term results in exchange for the long-term development.

4.1 Capital Expenditures, Operating Costs and Reinvestments

Incubators established on the basis of a public-private partnership are large capital investments. Substantial part of capital spending falls on land acquisition cost. Depending on the ownership of the site parcels and level of construction work needed, these costs can be significant, so one of the important criteria in evaluating different sites and locations can be ownership. Once that location is set there may be a need for further development such as site clearing, leveling, landscaping, development of the internal road network and putting in facilities and services such as energy, water and telecommunications or off-site development costs such as new access roads and traffic junctions [18].

In order to ensure efficiency of operations, incubator manager must be able to properly monitor investments that are made, without any overlapping but with enough detail. This can be accomplished by dividing financial information regarding the uses and disbursement periods for investments needed to develop an incubator into following categories:

Pre-operating Expenditures (consist of feasibility study done prior to the incubator's establishment on site, real estate agency costs, legal services, licenses and permits or consultancy fees); Physical Facilities (building purchase—loan or lease payment, building decorating or remodeling, smaller fixtures and fittings); Equipment (purchase of items that are necessary to start-up and continue the activities of the incubator, such as furniture, computers and printers, desks, seating or telecommunication equipment—Ethernet, routers, fax, photocopier); Human Resources (all the expenditures in relation to its staff, including possible recruitment costs and consultancy fees); General Expenditures (advertising, insurance, maintenance, telecommunications, supplies, banking services, utilities, etc.) and Reserves (amount set aside to build up cash reserves in case unexpected expenditures are required or if revenue falls short of projections) [19].

4.2 Revenue and Cash Flow

We have previously concluded that business incubators may generate significant amount of expenditure, but in return they can also help generate stable stream of revenue. This can be achieved through different sources such as rents and services that clients pay for and it can be used to pay off any debt incurred in developing incubators [20].

European researches have come to realize that is not unusual for business incubators' assets to become more and more valuable once they are proven to be successful and become more desirable. Property value increases as accelerator gains momentum, and can be sold to various occupants. Some incubators have a reputation of very successful commercial ventures. Owners are left with the decision of whether they want to continue owning and managing the operations or whether they want to sell it on to another body such as a commercial organization. However, it should be noted that in this case the objectives which prompted establishing of incubator may be jeopardized, as organization taking over purely for commercial objectives may not be as prepared to pursue the wider goals [21].

When we take another look into Study on the current situation of BIs in the Republic of Serbia [4], we can see that almost 40% of current high-tech incubators are stating that primary funding of their operations comes from rental incomes, projects and other related activities. Only 17% reported that they provide services that can generate income, and on the other hand, some of them have reported that up to 80% of their operations are financed by the local municipalities. Most of them receive cash operating subsidies and if this funding was stopped, they would have to stop or reduce their activities significantly (see Fig. 1).

There is only a limited number of incubators who do not receive any subsidies, which leads to the conclusion that most of the high-tech business incubators are dependent on external subsidies and that they are still far away from developing sustainable models of operations [22].

If we now focus on business incubators that mainly support self-employment, more than 75% of them report that their operations are funded by projects. Rent from tenants is here the second most important funding source with 65%, followed by subsidies from the municipalities. All of them are supported by at least two different sources, but most of them typically have three or four different sources of funding. If cash operating subsidies were to be stopped, 60% responded that their activities would have to be extensively reduced (see Fig. 2).

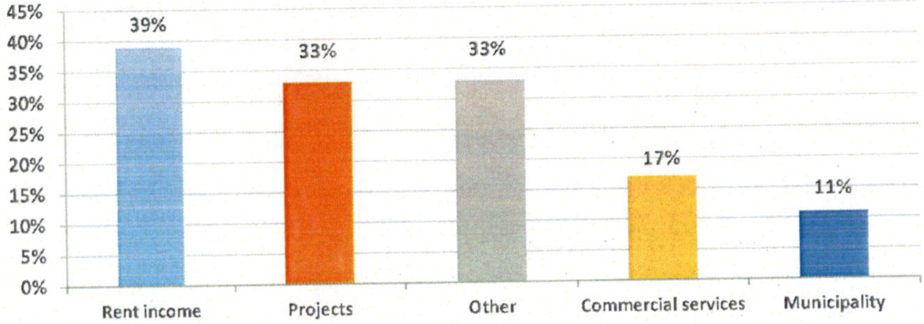

Fig. 1. Sources of income for high-tech supporting business incubators in RS

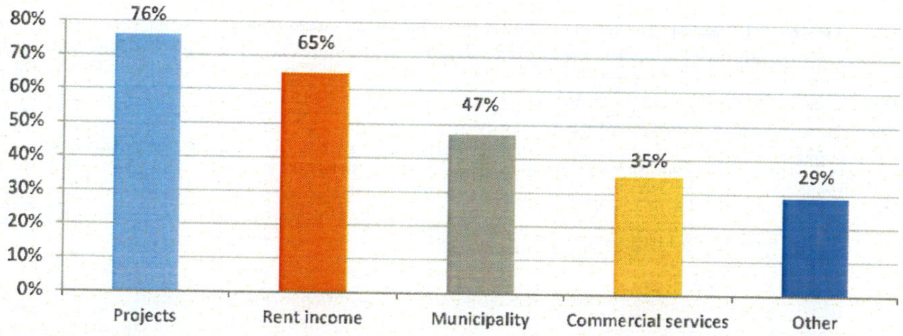

Fig. 2. Sources of income for self-employment supporting business incubators in RS

Generated income of self-employment business incubators is on average very low, at around EUR 24.000 per year. However, there is a wide variation in income among this group of incubators. A few have reported annual incomes between EUR 13.600 and 17.000, whilst other had incomes reported between EUR 65.000 and 88.000 per year [4].

4.3 Pro Forma Statements

Below are listed some items that could be tracked through annual reports that consist of simplified data that could be found in an Income Statement. There are also some other numerical indicators that could be helpful but do not belong in formal Profit and Loss Statement [17] (Tables 1, 2 and 3).

5 Performance Management of Business Incubator

The main task is defining what should be tracked and on what level in order to ensure efficiency and effectiveness of operations. One of the possible solutions is to follow incubator-specific and tenants-specific hard and soft indicators of performance. They should be in relation of their underlining goals that were previously mentioned while also taking into consideration problems in current operations that were observed and that could be closely followed with formally defined metrics [23].

Hard measures for business incubators could include number of clients, number of business graduating, meeting financial targets, and continued operations, where as soft could be growth in staff expertise and experience, recognition by enterprise support community, continued support from stakeholders such as educational institutions, municipals or local communities. Tenants-specific hard measures could be

Table 1. Important metrics—business incubator ecosystem as a whole

REVENUE
Rental Charges
Fees for services
Sponsorships
Appreciating asset values
Total Revenue
EXPENSE
Marketing and PR
Land acquisition cost
Site clearing, leveling, landscaping
Bank Charges
Development of the internal road network
Commissions
Contract Labor
Putting in facilities and services such as energy, water and telecommunications
Off-site development costs such as new access roads and traffic junctions
Insurance (property and liability)
Interest
Legal and Professional Fees
Licenses and Fees
Purchase of items that are necessary to start-up and continue the activities of the incubator (furniture, computers and printers, desks, seating or telecommunication equipment)
Repairs and Maintenance
Reserves
Net Income

Other Important Metrics
Learning and Growth
Number of software SMEs started
Municipal Entities Included
of New high-tech patents
Customer Perspective
Number of Business Incubators
# of New Incentives Introduced
Job opportunities created
Number of applications accepted
Internal Business Processes
# of international partners included
# of international sponsors included
Average time for Serbian tech startup to transform into large company
Percentage of educated locals remaining in the region
Financial Perspective
IT contribution to GDP
Investment rate in IT sector
Amount of Money Invested

sales turnover, increase/decrease in profitability, growth rate or graduation to independent trading. As for the soft ones, they should be oriented on increased professionalism towards clients, improved business skills and increased quality networking with peers, cost savings due to incubator resources usage, positive publicity and overall increased client knowledge.

One of the most frequently mentioned obstacles by managers and business incubator tenants was insufficiency of quality and diversity of consulting services given by incubator staff. One way of putting greater emphasis on this issue is by introducing special set of performance indicators that will focus on percentage of staff with high education degree, percentage of turnover invested in training or percentage of turnover invested in research and development [24].

Aligning of specific metrics among three main stakeholders (client startups, incubation programs and the whole ecosystem), can be done by taking deeper look into the values they should deliver [25].

On national level we can talk about value for the whole ecosystem, and we can expect economy enhancement and talent retention to be most obvious added benefits.

Table 2. Important metrics—high tech startup level (possible example)

REVENUE
Fee for service
Commerce
Subscription
Data manipulation
Productize a service
Transaction Fee
Marketplace
Advertising
Total Revenue
EXPENSE
Salaries
Equipment and communication
Software, servers and hosting
Taxes
Co-working spaces (BI fees)
Commissions
Supplies
Utilities (electric, gas, water, sewer)
Dues and Subscriptions
Vehicle Expenses
Employee Benefit Programs
Insurance (property and liability)
Interest
Legal and Professional Fees
Licenses and Fees
Travel
Office Supplies Expense
Net Income

Other Important Metrics
Learning and Growth
technology, equipment and systems development
innovation pipeline strength
number of contacts with universities and research centers
return on innovation investment
innovation enhancement
number of events attended (such as hackathons, boot-camps)
employee satisfaction index
human capital value
employee churn rate
Customer Perspective
customer retention
customer satisfaction
customer profitability
customer lifetime value
customer engagement
churn rate
client complaints
quality index
brand equity
Internal Business Processes
time to market index
capacity utilization rate
project schedule variance
number of contacts with corporations
number of national partners
customer acquisition cost
conversion rate
quality of assets
salary competitiveness ratio
time to hire
Online metrics
Financial Perspective
earned value analysis
project cost variance
revenue per employee

With this in mind we can define several key performance indicators that will help us evaluate success of these two main goals, but also those metrics can be used as a tool for steering operations in right direction so efficiency can be achieved. We can track and encourage economy enhancement by setting following indicators:

- number of jobs created (5 year period, high weight indicator)
- sales revenue (5 year period, high weight indicator)
- number of international partners included (1 year period, medium weight indicator)

Table 3. Important metrics—business incubator level

REVENUE
Lease revenue
Fees for services
Projects
Sponsorships
Grants
Donations of goods or services
Total Revenue
EXPENSE
Advertising, Marketing and PR
Amortization
Bad Debts
Bank Charges
Charitable Contributions
Commissions
Contract Labor
Depreciation
Dues and Subscriptions
Education and training
Employee Benefit Programs
Entertainment events
Insurance (property and liability)
Interest
Legal and Professional Fees
Licenses and Fees
Miscellaneous
Office Supplies Expense
Payroll Taxes
Postage
Professional services
Programming costs
Rent / Facility lease
Repairs and Maintenance
Supplies
Telecommunications services
Travel
Utilities (electric, gas, water, sewer)
Vehicle Expenses
Wages
Web development and maintenance
Net Income

Other Important Metrics
Learning and Growth
number of university staff
coaching and mentoring hours
number of coaches and mentors
number of contacts with universities and research centers
number of national partners
Customer Perspective
# of application received
percentage of surviving clients
percentage of terminated clients
percentage of profitable clients
percentage of growing clients
tenants' retention rate
tenants' turnover rate
clients utilizing coaching and mentoring
number of training modules
Internal Business Processes
number of graduates
number of IPOs
percentage of acquired clients
number of investors covered by network
number of contacts with investors
number of contacts with municipal representatives
number of alumni
Financial Perspective
number of investors that invested
percentage of unfunded clients
amount of investment attracted
average size of investment

- number of international sponsors included (1 year period, medium weight indicator)
- self-generated revenue (1 year period, medium weight indicator).

Other important factor at this hierarchy level is talent retention, which can be related to:

- percentage of educated locals remaining in the region (1 year period, high weight indicator)
- number of applications accepted (1 year period, medium weight indicator)
- percentage of non-local domestic applications (1 year period, medium weight indicator).

Value of incubation program is crucial for the growth of the ecosystem and emphasizes its function as a facilitator of community and network building through incubation offer and post incubation performance. These factors can be portrayed by eleven following indicators (first five are closely connected to incubation and next six to post incubation):

- incubator investment (1 year period, high weight indicator)
- number of applications received (1 year period, medium weight indicator)
- number of university staff (1 year period, medium weight indicator)
- business program adaptation (1 year period, medium weight indicator)
- number of graduates (1 year period, medium weight indicator)
- number of IPOs (10 year period, high weight indicator)
- percentage of acquired clients (1 year period, medium weight indicator)
- percentage of profitable clients (5 year period, medium weight indicator)
- percentage of growing clients (5 year period, medium weight indicator)
- percentage of surviving clients (1 year period, medium weight indicator)
- percentage of terminated clients (1 year period, low weight indicator).

Value for tenants shows benefits that incubator clients derive from utilizing the incubation programs' services. Quantity and quality of services provided is a crucial indicator of long-term tenants' success and by association indicator of long-term ecosystems' success [26]. Three categories of value creation consist of competence development, access to funds and access to network, and those can be derived into indicators listed below.

Competence development:

- coaching and mentoring hours (1 year period, medium weight indicator)
- number of clients utilizing coaching and mentoring (1 year period, medium weight indicator)
- number of coaches and mentors (1 year period, low weight indicator)
- number of training modules (1 year period, medium weight indicator).

Access to funds:

- amount of investment attracted (5 year period, high weight indicator)
- average size of investment (1 year period, high weight indicator)
- number of investors covered by network (1 year period, high weight indicator)
- number of investors that invested (1 year period, high weight indicator)
- number of contacts with investors (1 year period, medium weight indicator)
- percentage of unfunded clients (1 year period, medium weight indicator).

Access to network:

- number of contacts with corporations (1 year period, high weight indicator)
- number of events such as hackathons, boot-camps, etc. (1 year period, high weight indicator)
- number of national partners (1 year period, high weight indicator)
- number of alumni (1 year period, medium weight indicator)
- number of contacts with universities and research centers (1 year period, low weight indicator)
- number of contacts with municipal representatives (1 year period, medium weight indicator) [27].

Since the intention is to identify the key performance indicators that incubators and tenants must measure, and the interdependencies between incubators and tenants to achieve their goals, we can take into consideration a method which is based on cascading objectives among different hierarchy levels but also highlights balance between financial and non-financial metrics—Balanced Score Card (BSC).

This framework has four perspectives: (1) Customer, (2) Financial, (3) Internal Business Processes and (4) Learning and Growth. It is not necessary for incubators to adopt it in order to be prosperous, but it can be useful in understanding structure and relative impact of all selected metrics together [28].

The Customer Perspective infers about the relationship with customers, explaining how to attract and retain them. It envelops goals like customer retention, customer satisfaction, customer profitability, customer lifetime value and customer engagement, which can be tracked by KPIs like: tenants' retention rate, tenants' turnover rate, churn rate, satisfaction index, client complaints, tenants profitability score, customer lifetime value, ratio of customer acquisition cost and customer lifetime value, tenants engagement rate, net promoter score.

The Financial Perspective contains objectives and financial performance measures, including operational expenses, incomes, other expenditures and investment measurements. This BSC perspective is crucial to evaluate how the strategy is performing. For accenting revenue, profit and margin, we can use net profit, net profit margin, gross profit margin or revenue growth rate, and if we want to highlight investment and expenditure we have ROI, ROA, ROE, CAPEX to sales ratio, working capital ratio, operating profit margin, operating expense ratio or economic value added (EVA). We can also keep track of funding or burn rate.

The Internal Business Processes Perspective identifies the management's objectives of operations and general processes in the company. It is focused to improve customers' value and productivity. Potential KPIs could include quality index, capacity utilization rate, project schedule variance, project cost variance, earned value analysis, customer acquisition cost, conversion rate, time to market index, brand equity, online metrics (search engine rankings, page views and bounce rate, social networking footprint), operational equipment effectiveness.

Finally, the Learning and Growth Perspective is focused on identifying the objectives and necessities of human resources know-how, skills, organizational culture and the company's IT systems. This is done through various measures and it can help in reducing non-technical risk factors associated with social and behavioral aspects [29].

Education and training (training return on investment); technology, equipment and systems development (IT costs as percentage of revenue); innovation enhancement (innovation pipeline strength, return on innovation investment, quality of assets); actual jobs, new jobs and positions retained (employee churn rate, average employee tenure, time to hire, absenteeism, salary competitiveness ratio); human capital value (human capital value added, revenue per employee, 360° feedback score); employee satisfaction and engagement (employee satisfaction index, employee engagement level, staff advocacy score) [30].

There are some recent recommendations regarding standard set of business incubator performance indicators for all business incubators to use in Republic of Serbia, based on the research done in December 2019. To ensure quality monitoring, total of 16 key performance indicators were suggested, broken down into three categories (input efficiency, output effectiveness, results and impact), with an additional four sub-indicators breaking down the operating costs, and seven sub-indicators covering the assistance provided to tenants and clients in greater detail [4].

Input efficiency should portray the overall operational characteristics and performance of operations and should be tracked by measuring following indicators on a yearly basis: average annual operating costs (that can be further decomposed with sub-indicators such as total wages and payroll as percentage of total operating costs, building maintenance as percentage of total, business services to tenants and other as percentage of total), percentage of revenue from public subsidies, business incubator space, number of staff, number of incubator tenants and ratio of tenants and staff.

Output effectiveness indicators addresses how well the business incubator is providing facilities and services to tenants and clients. Delivered value could be measured by occupancy rates, average length of tenancy, percentage of managers' time spent advising clients and number of SMEs assisted (average over past 3 years). Information on SMEs assisted can be broken down by form of assistance through seven sub-indicators: number of clients receiving information and business advisory support, number of clients receiving training, number of clients receiving support in accessing finance, number of clients using business basic incubator facilities, number of pre-incubation events organized, number of new enquiries from potential tenants and clients, percentage of enquiries converted into tenants per year.

Results and impact demonstrate the results of the activities and impact on the local entrepreneurship ecosystem. Six indicators are included, covering business start-up, survival, growth and employment, as well as successful transition from needing business incubator support: number of spin offs/startups created (per year, 3 year average), percentage of survival rates, average size of tenant (number of employees), average number of new jobs/employment created, sales growth (%) and proportion of graduating tenants per year (successfully exited from incubator).

Following model combines different suggestions from various sources mentioned above and can be used as a rough tool since it gives some options for KPIs and benchmark values (Table 4).

Table 4. Suggested KPIs for BIs performance management

Category	KPIs	Benchmark Value	Suggested Method
Results and impact	# Spin Offs/Startups Created (per year, 3 year average)	15	Data on businesses registered (APR)
	# of IPOs (10 year period)	11	Public data
	Survival Rates (%)	80	Annual review with tenant companies / questionnaire
	Average size of Tenant (number of employees)	3-4	
	Time to Market Index (Months)	<18	
	Average number of new Jobs/Employment Created	50	
	Sales Growth (%)	25	
	Proportion of Graduating Tenants per year	33	Post-exit tracking of BI tenants (questionnaire)

Category	KPIs	Benchmark Value	Suggested Method
Output effectiveness	Occupancy Rates (%)	80	Number of rental units / number of tenants
	Average Length of BI Tenancy (Months)	36	Tenancy contracts and rental payments
	# Clients Assisted (Average Over Past 3 Years)	150	Contracts
	# Clients Receiving Information and Business Advisory Support	Benchmarks for the sub-indicators to be set locally, from current baseline, to show year-on-year improvement	Records of meetings and agreed support
	# Clients Receiving Training		Training attendance records
	# Clients Receiving Support in Accessing Finance		Records of meetings and questionnaire
	# Clients Using Business Basic Incubator Facilities		Records of facility bookings
	Number of Pre-Incubation Events Organized		Calendar of events
	Online Metrics (SEO, page views and bounce rate)		Digital Marketing Tools
	Number of Investor that Invested		Financial reports & data, Contracts
	Average Size of Investment		Financial reports & data, Contracts
	Number of New Enquiries from Potential Tenants and Clients	50	Recording of new inquiries (CRM)
	% Of Enquiries Converted into Tenants Per Year	10	Contracts signed
	% Of BI Managers' Time Spent Advising Clients	50	Staff records, diary timekeeping and records of meetings

(continued)

At the end it's important to note that there is no perfect combination of key performance indicators that will provide business incubator managers with perfect formula that will work for every type of accelerator and in every situation. That being said, above mentioned metrics should act as a tool that will be successful for vast majority of incubators and will be good enough to ensure at least average results. The best way to implement selected measures is to approach them critically and adapt them to specific circumstances [31].

Table 4. (continued)

Category	KPIs	Benchmark Value	Suggested Method
Input efficiency	Average Annual Operating Costs (€K)	500	Financial reports & data
	Total Wages and Payroll (% Of Total Operating Costs)	Local benchmarks for sub-indicators, showing increase in share of costs on providing value-added services to tenants	Financial reports & data
	Building Maintenance (% Of Total)		Financial reports & data
	Business Services to Tenants (% Of Total)		Financial reports & data
	Technology, equipment and system development (IT costs as percentage of revenue)		Financial reports & data
	Human Capital development (percentage of turnover invested in training)		Financial reports & data
	Other Operating Costs (% Of Total)		Financial reports & data
	% Of Revenue From Public Subsidies	<50%	Financial reports & data, Contracts
	Business Incubator Space (m²)	800	Building records
	Number of Staff	2 minimum	Staff records
	Percentage of Staff with High Education Degree	>80%	Staff records
	Number of Incubator Tenants	30	Contracts with tenants
	Ratio of Tenants/Staff	10:1	Number of tenants / number of staff
	Ratio of Client Acquisition Cost and Client Lifetime Value	<33%	CAC/LTV

6 Accounting and Reporting the Results

Importance of relevant reporting can be seen in fact that financial statements should be able to ensure full confidence for both governing bodies and investors. Even more important is the need for executives to rely on the numbers in order to make intelligent estimates regarding volume, timing and uncertainty of future cash flows. Adequate reporting mechanisms could also provide stable base for future decisions involving allocation of resources [32].

In order to be consider successful, accounting process for BI's should identify, record, organize and summarize information about economic events, usually in monetary terms. It is also important that this process communicates that information to stakeholders. The communication aspect of accounting involves the reporting of information about business to interested parties, such as managers of BI's. Efficiency and effectiveness are based on the ability of results of all transactions over a period of time, to be properly summarized, presented and interpreted at any given date.

It is common for accounting period to be a yearlong and for businesses to produce sets of results annually, as these are required for specific purposes (taxation, filing with a regulatory authority, etc.). However, in practice, an accounting period can be any length of time in order to provide management with information and guidance for decision making. In certain, well defined circumstances, reporting sets of results for periods other than a year can be accepted for specific purposes.

If well designed, this monitoring and reporting model can be used as transformative process in that it turns the raw data recorded in bookkeeping into useful information. Collected data will only become useful information when they are combined with guide that enables managers of BI's to assess them within a relevant context.

Annual reports of BI's must include mandatory documentation requested by Law on Accounting ("Off. Gazette RS", no. 62/2013 and 30/2018):

1. Reports for statistical purposes
2. Financial statements for public disclosure
    a. Regular annual financial reports
    b. Consolidated annual financial statements
    c. Extraordinary financial statements
3. Associate documentation with financial statements
    a. Decision on the adoption of financial statements
    b. Decision on profit distribution or coverage of losses according to the regular annual financial report
    c. Annual business report
    d. An audit report with the financial statements that were the subject of the audit.

Since a BI's main goal is to generate technology-based firms that are independent and profitable, and are able to leave the program financially viable and freestanding, usually in two to three years, there are reporting tools that can help managers to successfully run BI's, beside above-mentioned scope of documents. Tenants that are able to achieve those targets have potential to commercialize new technologies, while producing more job opportunities, revitalizing neighborhoods and empowering local and national economies. That is why there is a crucial need for applicable set of indicators that can be tracked and easily interpreted [33].

On yearly basis, executives should conduct simple questionnaire to document activities for the current reporting period and keep track of following macro data: tenant capacity, small businesses that are currently leasing space in incubators, type of business targeted, type of location that the firm operates on, full-time jobs provided by incubator tenants, support services offered and provided (Table 5).

On the BI level, notable reports, besides mandatory ones (that have already been mentioned), could be classified into the following categories:

- Program and Tenants' Report
- Impact Report
- Finance and Risk.

Table 5. Annual program monitoring of business incubators

Incubator	Tenant capacity	Current number of tenants	*Type of firm targeted	Support services offered					
				Computer Network	Phone service	Business planning	Marketing assistance	Conference room	Other
Incubator1									
Incubator2									
Incubator3									

Table 6. Recommended indicators for Tenants' Report and corresponding monitoring period of time

Indicator to track	Period of time
Business Incubator Space (m^2)	End of fiscal year
Number of Staff	End of fiscal year
Ratio of Tenants/Staff	End of fiscal year
Number of Incubator Tenants	Every Quarter
Occupancy Rates (%)	Every Quarter
Average Length of BI Tenancy (Months)	End of fiscal year
# Clients Assisted	Every Quarter
# Clients Receiving Information and Business Advisory Support	Every Quarter
# Clients Receiving Training	Every Quarter
# Clients Receiving Support in Accessing Finance	Every Quarter
% Of Enquiries Converted into Tenants Per Year	Every Quarter
% Of BI Managers' Time Spent Advising Clients	Every Quarter
Survival Rates (%)	End of fiscal year
Average size of Tenant (number of employees)	Half-year Report (Every 6 Months)
Proportion of Graduating Tenants per year	Half-year Report (Every 6 Months)

6.1 Tenants' Report

This report should consist of data regarding the targets set for tenants, previously mentioned here, and should be generated every six months for most of the indicators, since developing of independent tech firms is main goal for efficiently operating BI. This would provide strong base for timely reaction in case projected targets are not met. It illustrates how well internal systems of BI operating are working, and portrays how well BI is managed [34].

It should review defined objectives and whether or not they are met, main activities and achievements and any research done in this area and relevant results (Table 6).

6.2 Impact Report

This part of reporting should capture impact that BI has on community goals as previously described and should provide information about benefits of having incubators in neighborhood [35] (Table 7).

Table 7. Recommended indicators for impact report and corresponding monitoring period of time

Indicator to track	Period of time
# Clients Using Business Basic Incubator Facilities	End of fiscal year
Number Of Pre-Incubation Events Organised	Half-year Report (Every 6 Months)
# Spin Offs/Start ups Created (per year, 3 year average)	End of fiscal year
Average number of new Jobs/Employment Created	End of fiscal year

Table 8. Recommended financial indicators and corresponding monitoring period of time

Indicator to track	Period of time
Average Annual Operating Costs (€K)	End of fiscal year
Total Wages And Payroll (% Of Total Operating Costs)	End of fiscal year
Other Operating Costs (% Of Total)	End of fiscal year
% Of Revenue From Public Subsidies	End of fiscal year
Sales Growth (%)	End of fiscal year

6.3 Finance and Risk

Financial review must focus on monetary data (Table 8).

7 Conclusion

There is no denying that the major function of nurturing high-tech start-ups should be beneficial to Serbian tech landscape and economy as a whole and therefore it is essential to enable proper governance of business incubators dedicated to developing them. Wide range of literature does give a lot of insights on current business incubation ecosystem, but naturally cannot offer a unified solution when it comes to recommended instruments of managing incubators, especially because these vary in phase they are in, services they are providing and external factors they are facing. We can definitely conclude that the smooth functioning of business incubators depends on symbiotic relationship between national business regulative, involvement of municipal and regional governing entities and competencies of business incubator managers.

The overall benefits of investing in high-tech tenants enlist economy enhancement, encouragement of talent retention, technological expansion and many other financial and non-financial advantages. By thoroughly inspecting efficiency and effectiveness of accelerators through large set of KPIs, we can rationalize the need for strategic governance and systematic support. If managed properly, incubators can become a key component of national development strategy since they can help high-tech companies to survive and develop in most critical initial phase, by providing wide range of services such as management support, financial resources, technical services or simply offering adequate office space.

At the end, business incubator should be a sustainable business. Managers must lead their operations in a way that will provide desirable rate of return on invested capital in the long run or at least ensure survival by generating enough income from its property and service delivery. This entails involvement of private sector, especially in supporting progress of commercial technology or development of new assets. However, the role of public sector remains crucial, particularly in the early stages when the concept needs proving within its local environment.

References

1. Petrović, D., Mihić, M., Stošić, B.: Strategic IT portfolio management for development of innovative competences. In: Tan, Kwan, A.W. (Eds.) Strategic Information Technology and Portfolio Management, pp. 150–169. IGI Publishing, Hershey, USA (2009)
2. Salem, M.I.: The role of business incubators in the economic development of Saudi Arabia. Int. Bus. Econ. Res. J. **13**(4), 853–860 (2014)
3. Adnan, A.I., Yildirim, S.E., Eyamba, P.: Impact of business incubators' facilities on entrepreneurial ecosystem creation: a case of business incubator in Jordan. J. Manage. Sustain. **10**, 189–196 (2020)
4. Jach, K.-H., Černevičius, T., Kovačević, M., Batston, S., Mandić, V., Marinković, M., Uus, I.: Study on the current situation of BIs in the Republic of Serbia. GFA Consulting Group GmbH (Belgrade: Maxnova) (2020)
5. Statistical office of the Republic of Serbia: Godišnji nacionalni računi. Statistical office of the Republic of Serbia (2019). https://publikacije.stat.gov.rs/G2019/Pdf/G20191345.pdf. Accessed 16 Aug 2020
6. Mihic, M., Arsic, S., Arsic, M.: Impacts of entrepreneurs' stress and family members on SMEs' business success in Serbian family-owned firms. J. East Eur. Manage. Stud. **20**, 452–483 (2015)
7. APMG International. Appraising PPP projects. World Bank Group (2020)
8. Matijević, M., Šolaja, M.: ICT in Serbia—At a Glance, 2020. Vojvodina ICT Cluster, Novi Sad, Serbia (2020)
9. Kukić, Z., Kuzman, T., Kovač, A.: Startup Scanner. Digital Serbia Initiative, Belgrade (2020)
10. World Bank Group (2010). http://www.infodev.org/businessincubation. https://www.infodev.org/sites/default/files/m7_traineemanual_part1_20101029.pdf. Accessed 17 July 2020
11. Milutinović, R., Stošić, B., Mihić, M.: Concepts and importance of strategic innovation in SMEs: evidence from Serbia. Management **77**, 35–42 (2015)
12. Grimaldi, R., Grand, A.: Business incubators and new venture creation: an assessment of incubating models. Technovation **25**, 111–121 (2005)
13. Dietrich, F., Harley, B., Langbein, J.: Development Guidelines for Technology Business Incubators. InWEnt – Internationale Weiterbildung und Entwicklung, Bonn (2010)
14. Stosic, B., Mihic, M., Milutinovic, R., Isljamovic, S.: Risk identification in product innovation projects: new perspectives and lessons learned. Technol. Anal. Strat. Manage. **29**(2), 133–148 (2017)
15. Putarek, J.: Comparative Analysis of Public and Private Start-up Incubators (From a Users' Perspective). Rochester Institute of Technology (2018)
16. ADB, EBRD, IDB, IsDB, and WBG. *The APMG Public - Private Partnership (PPP) Certification Guide*. PPIAF. (2016).

17. European Commission Enterprise Directorate General. Benchmarking of Business Incubators. Centre for Strategy & Evaluation Services (CSES), United Kingdom (2002)
18. World Bank Group. (2008, October). Financing an Incubator. https://www.infodev.org/. https://www.infodev.org/infodev-files/m5_traineemanual_part2_20101029_0.pdf. Accessed 15 Aug 2020
19. World Bank Group: Identifying the Required Financing Towards Breakeven (2008). https://www.infodev.org/businessincubation. https://www.infodev.org/sites/default/files/m5_traineemanual_part1_20101029.pdf. Accessed 17 Aug 2020
20. COWI. Economic and Financial Analysis. Nordic Energy Research, Denmark (2016)
21. Commission, E.: Setting Up, Managing and Evaluating EU STP. Publication Office of the European Union, Luxembourg (2013)
22. Klofsten, M., Jones-Evans, D.: Stimulation of technology-based small firms—a case study of university-industry cooperation. Technovation **16**, 187–213 (1996)
23. Erasmus+ Programme of the European Union. Critical Success and Failure Factors of Business Incubation in HEI (2018)
24. European Commission—Joint Research Centre. Knowledge Transfer Metrics. Publications Office of the European Union, Luxembourg (2020)
25. Hackett, S.M., Dilts, D.M.: A systematic review of business incubation research. J. Technol. Transf. **29**, 55–82 (2004)
26. Aernoudt, R.: Incubators: tool for entrepreneurship? Small Bus. Econ. **23**, 127–135 (2004)
27. Holger, M.: Measure for Success: Data-driven Innovation Ecosystem Development. UBI Global, Stockholm (2015)
28. Llorach, C., Ottosson, E.: The balanced scorecard during the early stages of a tech firm. Uppsala University—Department of Business Studies, Sweden (2016)
29. Mihic, M., Dodevska, Z., Todorovic, M., Obradovic, V., Petrovic, D.: Reducing risks in energy innovation projects: complexity theory perspective. Sustainability **10**(9) (2018)
30. Cavalcanti, A.: A Method for Documenting the Goals and KPIs of Accelerators and Startups. Carleton University, Ottawa, Ontario, Canada (2018)
31. Torun, M., Peconick, L., Sobreiro, V., Kimura, H., Pique, J.: Assessing business incubation: a review on benchmarking. Int. J. Innov. Stud. **2**, 91–100 (2018)
32. Arsić, S., Mihic, M.: Integral model of management support: review of quantitative management techniques. In: Mitrovic, N., Milosevic, M., Mladenovic, G (Eds.) Advances in Computational and Experimental Approaches in Materials Science and Engineering, pp. 250–264. Springer, Cham (2020)
33. Department of Commerce. Small Business Incubator Certification Program. Department of Commerce, Oklahoma City (2019)
34. Prasetyawan, Y., Agustiani, E., Jumayla, S.: Aligning business strategy of incubator center and tenants. In: AIP Conference Proceedings 1855. AIP Publishing (2017)
35. Barbero, J.L., Casillas, J.C., Wright, M., et al.: Do different types of incubators produce different types of innovations? J. Technol. Transf. **39**, 151–168 (2014)

Development and Validation of an Open-Source Finite-Volume Method Solver for Viscoplastic Flows

Nikola Mirkov, Seif Eddine Ouyahia, Sara Lahlou, Milada Pezo, and Rastko Jovanović

Abstract In the present paper, we discuss implementation details of a free and open-source numerical solver based on the finite volume method for numerical simulation of viscoplastic non-Newtonian fluids. In addition to the fact that they are involved in many industrial applications, both their physical properties and their rheological behavior make them challenging for numerical simulation. Viscoplastic fluids are known to behave as solid unless the shear stress reaches a critical level, known as yield-stress, beyond which they behave as liquid. In most cases, both yielded and unyielded regions coexist in the fluid domain. In mathematical model of viscoplastic fluid, the constitutive equation is a non-differentiable function. This is often overcome by using the approximate constitutive equation that has a regularized form, e.g. the Papanastasiou regularization model. Using the same approach, we assess the influence of regularization parameters on simulation convergence and results accuracy. In this study, we give implementation details of viscoplastic fluid models in *freeCappuccino* open-source Computational Fluid Dynamics code. Moreover, we perform validation on several well known benchmark cases and compare proposed approach with those existing in published literature. We also perform a parametric analysis and show the effect of Reynolds and Bingham numbers on the extent of the yielded regions. Conclusions of the study have relevance in practical application of computational fluid dynamics to viscoplastic fluids in particular and to non-Newtonian fluids in general.

N. Mirkov (✉) · M. Pezo · R. Jovanović
VINČA Institute of Nuclear Sciences-National Institute of the Republic of Serbia, University of Belgrade, Mike Petrovića Alasa 12-14, 11351 Belgrade, Serbia
e-mail: nmirkov@vin.bg.ac.rs

S. E. Ouyahia · S. Lahlou
Laboratory of Transport Phenomena, Faculty of Mechanical and Process Engineering, USTHB B.P. 32 El-Alia Bab-Ezzouar, 16111 Algiers, Algeria

S. E. Ouyahia
Sonatrach–Direction Centrale Recherche et Développement, Avenue du 1er novembre, 35000 Boumerdes, Algeria

N. Mitrovic et al. (eds.), *Current Problems in Experimental and Computational Engineering*, Lecture Notes in Networks and Systems 323,
https://doi.org/10.1007/978-3-030-86009-7_12

Keywords Computational fluid dynamics · Finite volume method · Non-newtonian fluids · Bingham fluid · Papanastasiou regularization.

1 Introduction

In the present article we are interested in numerical simulation of the flow for a specific class of fluids, finding large technological application, for which linear constitutive relation doesn't hold i.e. the non-Newtonian fluids. Most of their characteristic features are elastic properties, time-dependent viscosity, strain-dependent viscosity and yield stress. To account for these behaviors, a constitutive equation is needed to relate the extra stress tensor $\boldsymbol{\tau}$ to the rate of strain tensor $\mathbf{D}$, and to the apparent viscosity $\eta(\dot{\gamma})$ that depends on the second invariant $(\dot{\gamma})$ of the rate of strain tensor $\mathbf{D}$. Constitutive equations are not derived from any physical principles, although they are subject to obeying certain rules. They are rather mathematical models of the material's rheological behavior, that are validated against experimental results [1]. A useful and widely employed constitutive equation for non-Newtonian fluids is the Generalized non-Newtonian fluid concept [2] which may be expressed by, $\boldsymbol{\tau} = 2\eta(\dot{\gamma})\mathbf{D}$.

An important class of non-Newtonian inelastic fluids is the viscoplastic fluids. Their main characteristic feature is that they exhibit a yield stress τ_y, below which they present a solid-like behavior with no noticeable deformation. Above the yield stress, a viscosity decay occurs allowing the material to flow like a liquid. A wide range of industrial applications and scientific areas involve flowing viscoplastic fluids, such as in oil and gas industry, for crude oil transport and drilling operations, nuclear energy applications, food processing, geophysical flows, etc. [3].

Many rheological models for viscoplastic fluids exist in literature. The most general one is the Herschel-Bulkley model, which reverts to five fluid category limiting cases. Its constitutive equation takes the following form,

$$\begin{cases} \tau = \tau_y + K\dot{\gamma}^n, & \tau > \tau_y \\ \dot{\gamma} = 0, & \tau \leq \tau_y \end{cases},$$

where $\tau = \left(\frac{1}{2}\boldsymbol{\tau} : \boldsymbol{\tau}\right)^{1/2}$ is the magnitude of the extra stress tensor $\boldsymbol{\tau}$ defined above, $\dot{\gamma}$ is the shear rate, τ_y the yield stress, K the consistency and n is the power law index. When $0 < n < 1$, the fluid exhibits a shear-thinning behavior after yielding and when $n > 1$, a shear-thickening behavior after yielding. Remark that in the case of zero yield stress, the Herschel-Bulkley model reduces to the Power law model,

$$\tau = K\dot{\gamma}^n,$$

with shear-thinning or shear-thickening behavior. If the power law index $n = 1$, the Herschel-Bulkley model reduces to the well-known Bingham model,

$$\begin{cases} \tau = \tau_y + \mu_p \dot{\gamma}, & \tau > \tau_y \\ \dot{\gamma} = 0, & \tau \leq \tau_y \end{cases},$$

where the consistency K becomes the Bingham plastic viscosity μ_p. And finally, if $\tau_y = 0$ and $n = 1$, the model corresponds to the Newtonian fluid's one and K reduces to the constant Newtonian viscosity μ. In this work, the discussions are limited to the case of Bingham's fluids for which, the generalized rheological constitutive equation is expressed in terms of the apparent viscosity as:

$$\begin{cases} \eta(\dot{\gamma}) = \frac{\tau_y}{\dot{\gamma}} + \mu_p, & \tau > \tau_y \\ \dot{\gamma} = 0, & \tau \leq \tau_y \end{cases}.$$

Computing viscoplastic flows is a challenging task. As it can be seen, one of the main arising difficulties is the inherent discontinuous nature of the viscoplastic models, that prevent the constitutive equation to be directly implemented. To overcome this, two main approaches exist in literature. The first one is based on the variational inequalities theory, and involves using optimization techniques as the Augmented Langrangian method [4]. Numerical works on viscoplastic fluids using this method can be found in literature, e.g. [5]. Despite that it accounts for the exact solution and its accuracy to identify unyielded regions, the method might not be easily implemented [6]. The second, and most frequently adopted approach, is the regularization technique. Essentially, it consists of approaching the discontinuous constitutive equation by an approximated continuous and differentiable one. This involves writing the constitutive equation in terms of apparent viscosity and to smooth the yielding process such that the computation be feasible and the obtained solution be close enough to the ideal one [7]. To do so, an additional artificial parameter is necessarily included to control the smoothing process. Its value can be adjusted so that the approximation approaches as possible the exact solution. This second approach is more physically acceptable, since the existence of a true yield stress has long been very controversial [8]. In addition, most commercial and open-source CFD software use regularization techniques in non-Newtonian modeling.

Many regularization models have been proposed. The first attempt was made by Bercovier and Engelman [9] who write apparent viscosity as,

$$\eta(\dot{\gamma}) = \mu_p + \frac{\tau_y}{\sqrt{\dot{\gamma}^2 + \delta^2}}.$$

The regularization parameter δ must be small compared to the characteristic shear rate of the flow $\dot{\gamma}$, but large enough to avoid numerical disturbances. The second contribution was made by [10], who proposed the so-called bi-viscosity model to approximate the Bingham fluid case. The regularization allows the model having two Newtonian regions with finite viscosity slopes,

$$\begin{cases} \eta(\dot{\gamma}) = \mu_p + \frac{\tau_y}{\dot{\gamma}}\left[1 - \frac{\mu_p}{\mu_y}\right], & \dot{\gamma} > \frac{\tau_y}{\mu_y} \\ \eta = \mu_y, & \dot{\gamma} \le \frac{\tau_y}{\mu_y} \end{cases}.$$

This model replaces the solid material by a fluid of high viscosity, where is the yield viscosity and the plastic viscosity. O'Donovan and Tanner [10] reported that a value of $\mu_y = 10^3\mu_p$ mimics satisfactorily the ideal Bingham model. Taking into account earlier works, Papanastasiou [11] proposed an exponential regularization of constitutive equation of the Bigham model, valid for both yielded and unyielded regions, in the form,

$$\eta(\dot{\gamma}) = \mu_p + \frac{\tau_y}{\dot{\gamma}}\left[1 - exp(-m\dot{\gamma})\right].$$

He introduces the regularization parameter m which controls the stress growth exponent such that, below τ_y, the model allows vanishingly small shear rates. In the limit of $m = 0$ the Newtonian fluid is recovered, while as $m \to \infty$, the model approaches the exact Bingham solution. However, the greater is the value of m, the steeper is the progress from unyielded to yielded regions, which can cause numerical convergence problems [12].

Another main challenging feature in computing viscoplastic fluids is the yield surface determination. The main consequence is that the flow, experiencing very small shear rates in these regions, will not be deformed and form a plug. Therefore, an accurate prediction of their extent is a key asset, particularly in heat transfer applications, mixing and pumping processes. In theory, unyielded regions are delimited with the criterion. But as long as regularized models are involved, the apparent viscosity is extremely high but finite, and this criterion does not hold anywhere in the domain. Rather, Mitsoulis and Tsamopoulos [13] proposed the criterion that the material flows only when the magnitude of the extra stress tensor exceeds the yield stress, i.e.,

$$\begin{cases} Yielded: & \tau > \tau_y \\ Unyielded: & \tau \le \tau_y, \end{cases},$$

which was widely adopted in literature.

The area of viscoplastic flows has been an area of interest for a long time. The square enclosure has extensively been investigated numerically, in natural and forced convection [14, 15], in lid-driven [16] benchmark cases, and also experimentally [17]. Various geometries have also been investigated such as flows in channels and ducts [22], in the so-called Backward Facing step [23], through contraction-expansions [18], around cylinders [19], in T-Branch configurations [20], in peristaltic flows [21], etc.

Numerical methods are also constantly improving, therefore in this article we present a general purpose viscoplastic flow solver, based on *freeCappuccino*[1] CFD

[1] Available at https://www.github.com/nikola-m/freeCappuccino-dev/.

code and a library for computational continuum mechanics, aimed at numerical simulations in complex domains using unstructured meshes and based on Finite Volume Method. In this article the viscoplastic flow solver is evaluated on three well known benchmark cases of lid-driven cavity, planar sudden expansion and planar T-branch flow and its performance is compared to that of competing methods and software tools, both in-house and commercial.

2 Methodology

In this section we give the mathematical formulation of the implemented model for viscoplastic fluids and computational aspects of the non-Newtonian solver in *freeCappuccino*.

2.1 Governing Equations

In this study the isothermal incompressible flow of viscoplastic fluids is considered, which is governed by the equation of continuity,

$$\nabla \cdot \mathbf{u} = 0, \tag{1}$$

and by momentum conservation equation,

$$\frac{\partial (\rho \mathbf{u})}{\partial t} + \nabla \cdot (\rho \mathbf{u}\mathbf{u}) = -\nabla p + \nabla \cdot \tau, \tag{2}$$

where ρ is fluid density, $\mathbf{u}$ the velocity, p is the pressure, and $\boldsymbol{\tau}$ is the stress tensor. The extra stress tensor $\boldsymbol{\tau}$, is formulated by the generalized Newtonian constitutive equation,

$$\boldsymbol{\tau} = 2\eta(\dot{\gamma})\mathbf{D}. \tag{3}$$

The apparent viscosity of the Bingham fluid, regularized with the Papanastasiou model is written as,

$$\eta(\dot{\gamma}) = \mu_p + \frac{\tau_y}{\dot{\gamma}} \left[1 - exp(-m|\dot{\gamma}|)\right], \tag{4}$$

where μ_p is the plastic viscosity, τ_y the yield stress, $\dot{\gamma}$ the shear rate and m the regularization parameter. In the above equations, the shear rate is expressed as, $\dot{\gamma} = \sqrt{2\mathbf{D} : \mathbf{D}}$, where $\mathbf{D}$ is the deformation rate tensor, defined as,

$$\mathbf{D} = \frac{1}{2}\left(\nabla \mathbf{u} + (\nabla \mathbf{u})^T\right). \tag{5}$$

2.2 Numerical Algorithm

The CFD (*Computational Fluid Dynamics*) code *freeCappuccino* [27–29] used in the present study is based on the second-order finite volume method for fully unstructured numerical meshes. The variable arrangement is cell-centered, pressure and velocities are collocated, and it is especially adapted for grids consisting of highly distorted computational cells. The distinguishing characteristic of the present algorithm is consistent application of corrections taking into account the intersection-point offset in non-orthogonal grids. The offset is determined by distance between two points on cell face, one found at the intersection of line connecting two cell centers and the cell face, and the cell face center. This type of correction is, as shown previously in [27] crucial for higher accuracy of solution. Accuracy of the calculation procedure strongly depends on the reconstruction of cell-centered gradients. In the present algorithm, a highly accurate least-squares gradient reconstruction procedure is used. It is based on the QR decomposition and devised to minimize the number of arithmetic operations needed in gradient field update, repeated several times during simulation, for all variables (pressure, velocity components, and scalars). Convection term discretization is defined using the mid-point rule for approximation of the convection terms in integral form which lead to two unknowns, cell-face mass flux and cell-face interpolated value of dependent variable. The iterative process which follows the decoupling of pressure and velocity fields in SIMPLE algorithm, allows the use of mass flux from the previous iteration, which introduces error due to lagging of mass flux value, which vanishes once the SIMPLE iterations converge. Mass fluxes on cell faces are calculated using the Rhie-Chow interpolation, usual for collocated cell arrangement [27]. The cell-face interpolated value is found using the expression taking into account grid non-orthogonality and uses surrounding cell-center values of dependent variables, as well as their gradients, computed using the previously described least-squares procedure. The options for convection fluxes include flux-limited interpolation, adapted for non-orthogonal grids, which is described in detail in [27]. There is a variety of implemented flux-limiter schemes, such as MUSCL, SMART, and UMIST, etc. The second order approximation of the diffusion term using the mid-point rule leads to cell-face value of the gradient of a dependent variable. The approximation of diffusion term therefore relies on accurate approximation of this quantity. The cell-face centered gradient is found by interpolation from the neighboring cell-centers in an original form which takes into account the grid non-orthogonality. The procedure, defined as *power of relaxation* approach, is described in detail in [27]. It is a generalization of approaches including minimal correction approach and over-relaxed approach, which controls the effect on explicitly treated non-orthogonal correction to iterative process by a parameter depending on the angle but in contrast to previous approaches takes into account the intersection-point offset. The volumetric source terms are approximated using the mid-point rule. The pressure and velocity fields are updated in a segregated manner within the SIMPLE iterative algorithm. The distinctive characteristic of the present algorithm is multiple pressure-correction solution to attenuate adverse effects of non-orthogonality to solution pro cedure. In additional pressure correction solutions, interpolations are conducted using the least-square gradients, not sensitive to loss of accuracy due to grid distortions. Linear system of equations, resulting from discretization process is solved by a variety of iterative solution methods. Most usually the pressure-correction equations are solved using Incomplete-Cholesky preconditioned Conjugate Gradient algorithm (PCG(IC(0)) due to its symmetry and positive definiteness, while other non-symmetric linear systems are solved using ILU preconditioned BiCGStab (BiCGStab(ILU(0))) algorithm.

In every SIMPLE iteration the flow dependent value of apparent viscosity is updated according to a chosen rheological model and is based on explicit field manipulation technique that is described next.

Tensor field manipulation The *freeCappuccino* CFD code and library for computational continuum mechanics has incorporated derived data types representing volume scalar fields, volume vector fields, volume tensor fields as well as equivalent types defined for cell faces instead of cell volumes. A typical manipulation sequence for updating the apparent viscosity, e.g. for Bingham rheological model, in single step of SIMPLE iteration amounts to following commands,

```
type(volVectorField) :: Uvec
type(volTensorField) :: gradU
type(volScalarField) :: shear

Uvec = volVectorField( "VelocityVec",     &
                        U, &
                        V, &
                        W  &
                      )
gradU = Grad( Uvec )
shear = sqrt2*(.mag.(.symm.( gradU ) ))

! Now return to array based storage of variables
vis(1:numCells) = urfVis * &
 ( muplastic + Tau_0/( shear%mag(1:numCells) + 1e-20 )   ) &
                + (1.0_dp-urfVis)*vis(1:numCells)
```

Where first the derived data types for rank-0, rank-1, and rank-2 tensor fields are declared, velocity vector is assembled from its components stored as arrays, velocity gradient tensor calculation and shear magnitude as a scalar field calculation using tensor field manipulations and finally calculation of apparent viscosity using the Bingham model and its under-relaxation common for SIMPLE pressure-velocity coupling. In the last command many ordinary operations of addition and multiplication are overloaded, meaning their ordinary use is seamlessly transferred by *freeCappuccino* library to equivalent operations over tensor fields.

For Bingham-Papanastasiou model the viscosity update is conducted in a following way (showing only relevant part of the code which differs from the one above),

```
! Now return to array based storage of variables
vis(1:numCells) = urfVis * &
 ( muplastic + Tau_0/( shear%mag(1:numCells) + 1e-20 ) * &
       ( 1.0_dp - exp( -megp*shear%mag(1:numCells) ) ) ) &
                + (1.0_dp-urfVis)*vis(1:numCells)
```

Another often used is Herschel-Bulkley-Papanastasiou constitutive model coded in *freeCappuccino*, represented operationally in following way,

```
vis(1:numCells) = urfVis * min( mumin, &
  ( Tau_0*( 1.0_dp - exp( -megp*shear%mag(1:numCells) ) ) + &
    Consst * shear%mag(1:numCells) ** npow ) / &
    ( shear%mag(1:numCells) + 1e-30 ) ) &
                + (1.0_dp-urfVis)*vis(1:numCells)
```

In all cases we used tensor field operation for obtaining the symmetric part of the velocity gradient tensor based on the following library function:

```
function symm(T)  result(D)
    implicit none
    type(volTensorField), intent(in) :: T
    type(volTensorField)             :: D

!              overloaded operator - here '*' multiplies
!              | tensor fields by a constant scalar
!              |     overloaded operator - here '+' adds
               |     | two tensor fields
!              |     |
    D = 0.5_dp * ( T + .trans.T )

end function
```

And magnitude of the symmetric tensor field is obtained using following procedure, called after code recognizes symmetric rank two tensor field as the argument of unary *.mag*. operation:

```
function magSymmetricTensorField(S) result(scalar)
    implicit none
    type(volSymmetricTensorField), intent(in) :: S
    type(volScalarField)                      :: scalar

        scalar = S**S
        scalar%mag = sqrt(scalar%mag)

end function
```

Where two stars symbol, ** denotes contraction over two indices of rank two tensor fields appearing as the argument. In the example of *symm* operation for obtaining symmetric part of a general tensor field, above, operator overloading is crucial for compact and elegant formulation, and this is stressed in the code segment itself. More details on implementation of discrete tensor fields and their manipulation in *freeCappuccino* library, can be found in [29].

3 Results and Discussion

In order to assess the capability and reliability of *freeCappuccino* flow solver on non-Newtonian flows we present in this section validations of some benchmark cases from the literature, dealing with the flow of a non-Newtonian Bingham fluid. The cases selected are the lid-driven cavity, flow through a sudden expansion and flow in a T-junction.

Since we are dealing with the cases of Bingham fluids, the two dimensionless numbers the flow field is characterised by in following examples and discussion are the Reynolds number, which is defined in terms of the plastic viscosity μ,

$$Re = \frac{\rho U L}{\mu}, \tag{6}$$

and the Bingham number Bn, defined in a following way,

$$Bn = \frac{\tau_y L}{\mu U}. \tag{7}$$

All the example cases are tested in a range of these non-dimensional numbers, reflecting different regimes the validation is focused on.

3.1 *Flow of Bingham Plastics in Lid-Driven Square Cavity*

The first case selected to test *freeCappuccino* code is the lid-driven problem. It is a significant and widely used benchmark test for computing non-Newtonian flows, and plenty of references are available in the literature. The paper selected for our validation is the square lid-driven cavity flow of a Bingham fluid, performed by Syrakos et al. [24]. The computational domain is a square cavity of unit length, bounded by solid walls and the upper one (lid) moves toward the right with a uniform horizontal velocity. To obtain the results in reference study of Syrakos et al., governing equations are discretized using the finite volume method, on a cartesian grid, and the pressure-velocity coupling was solved by the SIMPLE pressure-correction algorithm. For simulation they used their in-house code. Authors performed simulations

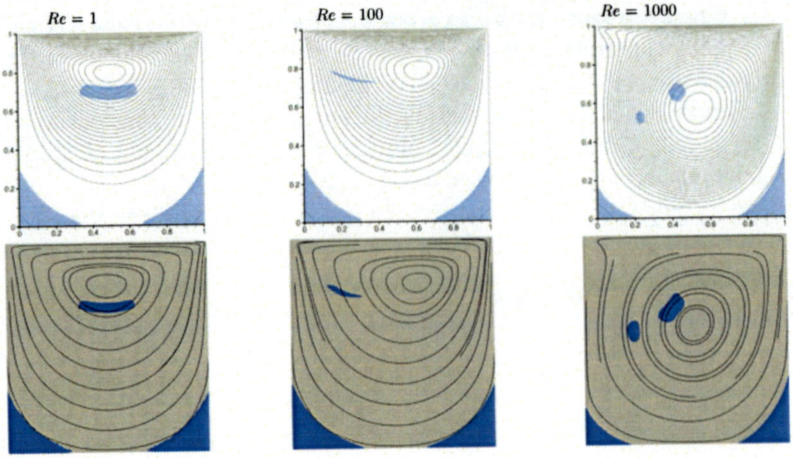

Fig. 1 Lid-driven square cavity flow—comparison of the streamlines and unyielded areas for $Bn = 1$ and three different Reynolds numbers $Re = 1$, 100 and 1000. Syrakos et al. [24] (upper row), and present results (bottom row)

on a 512x512 grid. The problem was solved for a wide range of Reynolds and Bingham numbers. The Papanastasiou regularization was adopted for the Bingham constitutive model, and the stress growth number was set to $m = 400$, unless for the highest values of Bn and Re, for which a lower value of $m = 200$ was used to avoid convergence difficulties.

Figure 1 displays the streamlines and unyielded areas obtained with *freeCappuccino*, at $Bn = 1$ and three values of the Reynolds number, including the results from the aforementioned reference. We reiterate author's findings. The main characteristic of the flow field is a recirculation cell at the upper central region of the cavity. Three distinct unyielded regions form: one at each bottom corner of the cavity and one just below the recirculation cell. The flow field is symmetric with respect to the vertical centerline and, as Re increases, the vertex expands and its centerline moves toward the right while the unyielded region located just below moves to the left. As it can be seen, our results agree satisfactorily well with those of Syrakos et al. [24].

3.2 Flow in a Sudden Expansion (1:2 Ratio)

The second test case is the entry flow through an abrupt expansion. The selected study to validate, is the one performed by Mitsoulis and Huilgol [25], on the laminar flow of a Bingham fluid in a sudden expansion. Authors analyzed the effect of inertia and yield stress on the flow pattern and the yielded/unyielded zones.

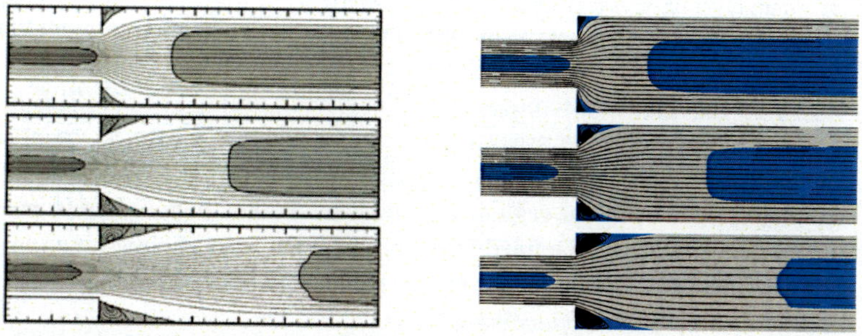

Fig. 2 Flow in a sudden expansion (1:2 ratio)—comparison of the streamlines and unyielded areas for $Bn = 3.9$ and three different Reynolds numbers $Re = 0$, 100 and 200. Mitsoulis [25] (left column), and present results (right column). Reynolds numberss increase from top to bottom

The geometry of the problem is a plane channel, characterized by an expansion with an expansion ratio of 2:1 between the outlet and inlet height, as depicted in Fig. 2. To guarantee the fully developed velocity profile boundary condition, the inflow has been set at -4H and the outflow at +52H from the expansion, H being the half channel width, while no slip boundary condition was imposed at the walls. In [25] the numerical solution was obtained by the Finite Element Method, on a quadrilateral Lagrangian mesh of 2220 elements. The discontinuity of the Bingham model was carried by the Papanastasiou regularization, with a stress growth exponent value of $m = 1000$. Present calculation used multiblock mesh of comparatively same size, containing 2200 cells.

The flow pattern in such configuration is characterized by a symmetric flow with two recirculation zones of equal size in the expansion corners. Unyielded regions are mainly located at the core of the channel and in the corners of the expansion. Figure 2 shows the streamlines and unyielded regions for different values of the Re number, at $Bn = 3.9$. Results from the reference and those from our code are presented side by side, for comparison purposes. It can be seen that with the increasing Reynolds number, the effect of inertia increases the recirculation zones length and elongate the unyielded regions farther, while the unyielded region upstream is retracted. These results are expected since inertia pushes the main flow further away in the expansion, leaving bigger dead spaces behind, which appear as unyielded near the walls at the separation point, see [25] for more details. In qualitative comparison, we can see that results obtained by *freeCappuccino* non-Newtonian solver are almost identical with the original reference.

3.3 *Flow in a T-Branch*

The third benchmark case is a flow in two dimensional T-branch. For validation the numerical simulations performed by Inácio et al. [26] are considered, performed using commercial software. The program uses the finite volume method (FVM) to discretize the conservation equations based on a cell-centered formulation and a second order scheme for discretization of the viscous terms. The SIMPLE method was adopted to approach the pressure-velocity coupling. Collocated variable arrangement is enabled using the Rhie-Chow interpolation scheme. The viscosity function was determined using the regularized Herschel-Bulkey-Papanastasiou model implemented via UDF (User Defined Function) to extend the original functionality of the original software. This being said, the referenced study uses the same method as implemented in *freeCappuccino* CFD code.

The flow requires a long inlet section, therefore it is interesting for validating the streamwise velocity component in the channel as the first step. The comparison of the two profiles, one from [26] and present simulation, for the profile at same cross-section position, at $x = 15H$, where H is the height of the channel, is given in Fig. 3. Viscoplastic fluids have a characteristic flattened profile relative to laminar, Newtonian fluid case, which is visible in this figure. The two profiles are in excellent agreement.

Following this we compare streamlines for $Bn = 1$ and three Reynolds numbers $Re = 0, 30, 50$ in Fig. 4. At $Re = 50$ we see symmetrical recirculation regions forming following the branching of the flow. Both cases agree in great detail. Finally, and most importantly, the yielded and unyielded regions in the planar T-branch flow for the same range of non-dimensional numbers is shown in Fig. 5. Figures in the top row identify characteristic regions by circles, which are repeated in present simulations almost exactly.

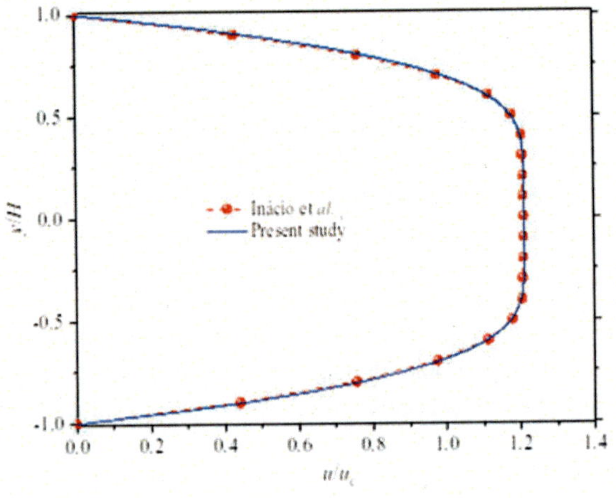

Fig. 3 Flow in a two-dimensional T-branch—Comparison of the fully developed axial velocity in the inlet channel (at x = 15*H*) *Re* = 20 and *Bn* = 1. Shown are Inacio et al. [26], and present results

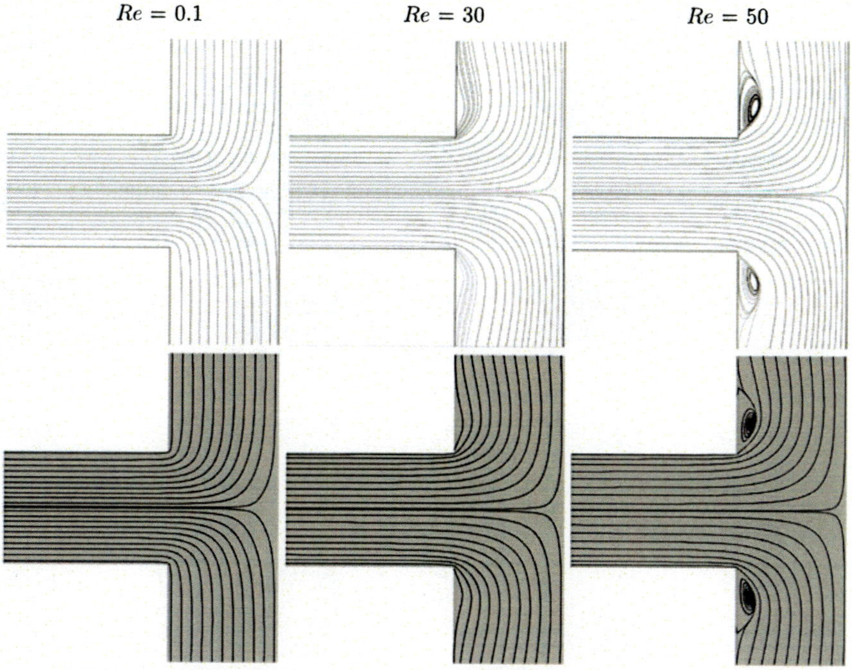

Fig. 4 Flow in a two-dimensional T-branch - comparison of the streamlines and unyielded areas for $Bn = 1.0$ and three different Reynolds numbers $Re = 0$, 30 and 50. Inacio et al. [26] (top row), and present results (bottom row)

3.4 Conclusion

We have seen how a general purpose, open-source, unstructured finite volume code *freeCappuccino* was extended for the solution of viscoplastic non-Newtonian flows. The computational challenges arising with such fluids is treated in literature with various computational approaches, ranging from Finite-Volume Method, Finite Element Method, Lattice Boltzmann Method, etc. In this study, for validation purposes, we have chosen references where both the in-house and the commercial codes were used. In the focus of the validation task was proper identification of yielded and unielded regions in the flow domain, for characteristic flow cases and for a range of Reynolds and Bingham numbers. All the reported results were reproduced in almost identical manner. Although the reported cases were exclusively two-dimensional, the *freeCappuccino* code is capable of simulating flow cases in highly complex three-dimensional domains. We have restricted our attention to strain-dependent viscosity cases in this article, and have left temperature dependent viscosity in flow cases including heat transfer for future validation studies. Our next goal will be extension of the solver capability for viscoelastic fluids as well.

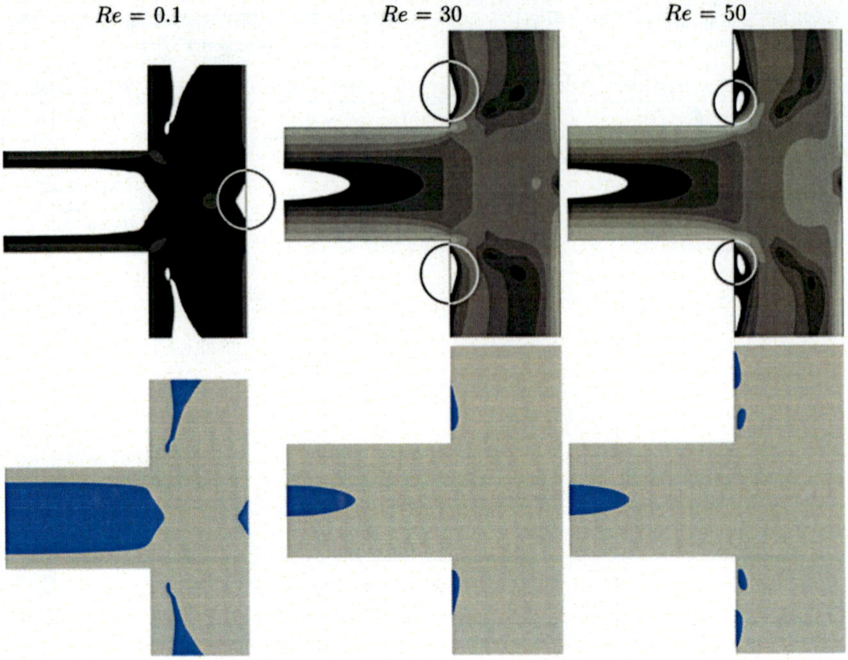

Fig. 5 Flow in a two-dimensional T-branch - comparison of the unyielded areas for $Bn = 1.0$ and three different Reynolds numbers $Re = 0$, 30 and 50. Inacio et al. [26] (top row), and present results (bottom row)

Acknowledgements The research of Serbian co-authors was funded by the Ministry of Education, Science and Technological Development of the Republic of Serbia.

References

1. Reddy, J.N.: Constitutive equations. In: An Introduction to Continuum Mechanics. 2nd edn., pp. 221–264. Cambridge University Press, Cambridge (2013)
2. Bird, R.B., Dai, G.C., Yarusso, B.J.: The Rheology and flow of viscoplastic materials. Rev. Chem. Eng. **1**(1), 1–70 (1983)
3. Wilson, D.I.: Industrial applications of yield stress fluids. In: Mechanics, G., Ovarlez, Hormozi, S. (eds.) Lectures on Visco-Plastic Fluid, pp.195–259. Springer International Publishing, Cham (2019)
4. Fortin, M., Glowinski, R.: Augmented Lagrangian Methods: Applications to the Numerical Solution of Boundary-Value Problems. Elsevier (2000)
5. Huilgol, R.R., You, Z.: Application of the augmented Lagrangian method to steady pipe flows of Bingham, Casson and Herschel-Bulkley fluids. J. Nonnewton. Fluid Mech. **128**(2–3), 126–143 (2005). https://doi.org/10.1016/j.jnnfm.2005.04.004
6. Ahmadi, A., Karimfazli, I.: "A quantitative evaluation of viscosity regularization in predicting transient flows of viscoplastic fluids. J. Nonnewton. Fluid Mech. **287**, 104429 (2021). https://doi.org/10.1016/j.jnnfm.2020.104429

7. Wachs, A.: Computational methods for viscoplastic fluid flows. In: Ovarlez, G., Hormozi, S. (eds.) Lectures on Visco-Plastic Fluid Mechanics, pp. 83–125. Springer International Publishing, Cham (2019)
8. Barrnes, H.A.: The Yield Stress-a review or "$\pi\alpha\nu\tau\alpha\ \rho\epsilon\iota$"'-everything flows. J. Non-Newtonian Fluid Mech **81**, 133–178 (1999)
9. Bercovier, M., Engelman, M.: A finite-element method for incompressible non-Newtonian flows. J. Comput. Phys. **36**(3), 313–326 (1980). https://doi.org/10.1016/0021-9991(80)90163-1
10. O'Donovan, E.J., Tanner, R.I.: Numerical study of the Bingham squeeze film problem. J. Nonnewton. Fluid Mech. **15**(1), 75–83 (1984)
11. Papanastasiou, T.C., Boudouvis, A.G.: Flows of viscoplastic materials: Models and computations. Comput. Struct. **64**(1–4), 677–694 (1997). https://doi.org/10.1016/S0045-7949(96)00167-8
12. Mitsoulis, E.: Fountain flow of pseudoplastic and viscoplastic fluids. J. Nonnewton. Fluid Mech., **165**, 45–55 (2010). https://doi.org/10.1016/j.jnnfm.2009.09.001
13. Mitsoulis, E., Tsamopoulos, J.: Numerical simulations of complex yield-stress fluid flows. Rheol. Acta **56**(3), 231–258 (2017). https://doi.org/10.1007/s00397-016-0981-0
14. Ouyahia, S.-E., Benkahla, Y. K., Berabou, W. and Boudiaf, A.: Numerical study of the flow in a square cavity filled with Carbopol-TiO 2 nanofluid. Powder Technol., **311**, 101–111 (2017). https://doi.org/10.1016/j.powtec.2017.01.026
15. Lahlou, S., Labsi, N., Benkahla, Y. K., Boudiaf, A., and Ouyahia,S.-E.: Flow of viscoplastic fluids containing hybrid nanoparticles: extended Buongiorno's model. J. Nonnewton. Fluid Mech. 104308 (2020). https://doi.org/10.1016/j.jnnfm.2020.104308
16. Mitsoulis, E., Zisis, T.: Flow of Bingham plastics in a lid-driven square cavity. J. Non-Newtonian Fluid Mech **101**, 173–180 (2001)
17. Hassan, M.A., Pathak, M., Khan, M.K.: Rayleigh-Benard convection in Herschel-Bulkley fluid. J. Nonnewton. Fluid Mech. **226**, 32–45 (2015). https://doi.org/10.1016/j.jnnfm.2015.10.003
18. Soto, H.P., Martins-Costa, M.L., Fonseca, C., Frey, S.: A numerical investigation of inertia flows of Bingham-Papanastasiou fluids by an extra stress-pressure-velocity galerkin least-squares method. J. Brazilian Soc. Mech. Sci. Eng. **32**, 450–460 (2010). https://doi.org/10.1590/S1678-58782010000500004
19. Mossaz, S., Jay, P., Magnin, A.: Non-recirculating and recirculating inertial flows of a viscoplastic fluid around a cylinder. J. Nonnewton. Fluid Mech. **177–178**, 64–75 (2012). https://doi.org/10.1016/j.jnnfm.2012.04.008
20. Maurya, A., Tiwari, N., Chhabra, R.P.: Effect of a rotating cylinder on the flow of a Bingham plastic fluid in T-junction: Momentum and heat transfer characteristics. Int. J. Heat Mass Transf. **143**, 118506 (2019). https://doi.org/10.1016/j.ijheatmasstransfer.2019.118506
21. Khabazi, N.P., Taghavi, S.M., Sadeghy, K.: Peristaltic flow of Bingham fluids at large Reynolds numbers: A numerical study. J. Nonnewton. Fluid Mech. **227**, 30–44 (2016). https://doi.org/10.1016/j.jnnfm.2015.11.004
22. Danane, F., Boudiaf, A., Boutra, A., Labsi, N., Ouyahia, S.-E., and Benkahla, Y. K.: 3D analysis of the combined effects of thermal buoyancy and viscous dissipation on the mixed convection of Bingham plastic fluid in a rectangular channel. J. Brazilian Soc. Mech. Sci. Eng., **40**(3), (2018) 10.1007/s40430-018-1048-1
23. Danane, F., Boudiaf, A., Mahfoud, O., Ouyahia, S.-E., Labsi, N., and Benkahla, Y. K.: Effect of backward facing step shape on 3D mixed convection of Bingham fluid. Int. J. Therm. Sci., **147**(September 2019), p. 106116 (2020) 10.1016/j.ijthermalsci.2019.106116
24. Syrakos, A., Georgiou, G. C., and Alexandrou, A. N.: Performance of the finite volume method in solving regularised Bingham flows: Inertia effects in the lid-driven cavity flow. J. Nonnewton. Fluid Mech., **195**, 19–31 (2013) 10.1016/j.jnnfm.2012.12.008
25. Mitsoulis, E., Huilgol, R.R.: Entry flows of Bingham plastics in expansions. J. Nonnewton. Fluid Mech. **122**(1–3), 45–54 (2004). https://doi.org/10.1016/j.jnnfm.2003.10.007
26. Inácio, G.R., Tomio, J.C., Vaz, M., Zdanski, P.S.B.: Numerical study of viscoplastic flow in a T-bifurcation: Identification of stagnant regions. Brazilian J. Chem. Eng. **36**(3), 1279–1287 (2019). https://doi.org/10.1590/0104-6632.20190363s20180361

27. Mirkov, N., Rašuo, B., Kenjereš, S.: On the improved finite volume procedure for simulation of turbulent flows over real complex terrains. Journal of Computational Physics 287, 18–45 (2015) 10.1016/j.jcp.2015.02.001
28. Mirkov, N.S., Stevanović, ŽM.: New Non-Orhogonality Treatment for Atmospheric Boundary Flow Simulation Above Highly Non-Uniform Terrains. Thermal Science **20**(Suppl. 1), S223–S233 (2016)
29. Mirkov, N., Vidanović, N., Kastratović, G.: freeCappuccino - An Open Source Software Library for Computational Continuum Mechanics. In: N. Mitrovic et al. (Eds.): CNNTech 2018, LNNS 54, pp. 137–147, Springer Nature Switzerland AG (2019). 10.1007/978-3-319-99620-2_11

Comparison of Tensile Properties of Carbon/Epoxy Composite Materials with Different Fiber Orientation Using Digital Image Correlation

Aleksandra Jelić, Milan Travica, Vukašin Ugrinović, Aleksandra Božić, Marina Stamenović, Dominik Brkić, and Slaviša Putić

Abstract Due to its remarkable qualities, carbon fiber epoxy composite sandwich panels are used in a variety of engineering applications. The goal of this research is to use tensile testing and a full-field non-contact 3D Digital Image Correlation (DIC) method to characterize carbon fiber reinforced composite sandwich panels with varied fiber orientations (0°/90° and ± 45°). The tested materials were composed of carbon fiber prepregs with epoxy resin systems and Aramid synthetic fiber. The properties of the materials were determined using full-field data derived from 3D DIC measurements and a set of experiments by according to ASTM standards. Values of maximum stress and strain at entire areas, break stress and strain, and toughness at entire areas and, modulus of elasticity of both structures were compared. The adherend's full-field, out-of-plane deformation, strain distribution, and strain evolution along the bond line were captured using a digital image correlation method, allowing the fracture mechanism to be visually defined. Since DIC produces the displacement field, the strain field must be deduced from it. The orientation of the fibers had a significant impact on the tensile properties of the tested materials. The results revealed that the specimen with 0°/90° fiber orientation had higher break stress and brittle fracture, whereas the specimen with ± 45° fiber orientation twisted in the fiber direction had higher elongation values while carrying the applied load. In order to complement previously obtained results, scanning electron microscopy (SEM) analysis of the fibers and core, as well as fracture surfaces was performed.

A. Jelić (✉) · A. Božić · M. Stamenović · D. Brkić
Department of Belgrade Polytechnic, The Academy of Applied Technical Studies Belgrade, Belgrade, Serbia
e-mail: ajelic@politehnika.edu.rs

M. Travica
Innovation Center, Faculty of Mechanical Engineering, University of Belgrade, Belgrade, Serbia

V. Ugrinović
Innovation Center, Faculty of Technology and Metallurgy, University of Belgrade, Belgrade, Serbia

S. Putić
Faculty of Technology and Metallurgy, University in Belgrade, Belgrade, Serbia

N. Mitrovic et al. (eds.), *Current Problems in Experimental and Computational Engineering*, Lecture Notes in Networks and Systems 323,
https://doi.org/10.1007/978-3-030-86009-7_13

Keywords Sandwich panels · Tensile testing · Digital image correlation

1 Introduction

Known for its high stiffness to weight ratio and high strength to weight ratio [1–7], composite materials have a broad range of applications in aerospace [8–10], transportation [11, 12], construction [13–15], energy [16, 17], maritime [18, 19], military [20, 21], and civil applications [22–27]. Carbon fiber reinforced polymer (CFRP) has proved to be the most attractive for a wide variety of applications in recent years due to its remarkable properties and plain formability [28, 29].

The form of reinforcement fiber, matrix material, and manufacturing techniques all influence the properties of composite materials. Sandwich composite structures are composed by connecting two thin facings to a lighter core leading to the increase of the flexural rigidity, minimized weight while increasing strength and improving overall thermal insulation properties [30]. Novel sandwich systems now have much more ability to conform to a variety of working environments and design constraints thanks to fiber-reinforced composites [31]. Since these structures can be individually designed in order to improve their properties making them desirable for particular applications, serious attention when evaluating their mechanical properties is required. Since epoxy resins are porous, the strengthening and toughening the effects of the fillers at the same time is demanding [32–36].

Over the last three decades, in order to define the strength of composite sandwich panels, extensive efforts have been made to better understand the impact of design parameters. Shahdin et al. have investigated the mechanical properties of sandwich structures and fiber network used as a new core material and unidirectional carbon pre-impregnated plies as skins. Compression and vibration tests showed low values of structural strength and lesser vibratory levels of lighter specimens, but the authors have concluded that specimens were successfully reproduced [37]. Dinesh et al. examined the mechanical properties of sandwich panels with carbon fiber reinforcement and different core materials like Aluminum Honeycomb, Rohacell and High-Density Polyurethane Foam and epoxy resin. Three-point bending test, tensile and compressive tests were performed leading to a conclusion foam-based sandwich panels have better tensile, and compressive load-bearing capacity [30]. Li et al. performed an experimental analysis of the mechanical behavior of carbon fiber reinforced composite sandwich panels with different pyramidal truss cores at different temperatures using a series of compression tests. The authors concluded that compressive strength and stiffness decreased as the temperature rose and explained the influence of temperature on the failure mode [38]. As a potential substitute to honeycombs, George et al. evaluated the mechanical behavior of carbon fiber composite sandwich panels with pyramidal truss cores [39]. Xu et al. investigated the mechanical performance of carbon/epoxy composite sandwich structures with three-dimensional corrugated cores with various graded parameter values [40]. Kwon et al. studied carbon fiber reinforced plastic sandwich composites utilizing compressive, flexural strength, lap shear strength, and impact tests and concluded that epoxy foam increased sandwich structure interfacial bonding [41]. Zheng et al. created a novel epoxy resin matrix design for carbon fiber reinforced composites with viscoelastic sandwich layers and used static mechanical testing to examine their characteristics. The experimental findings showed that the study encouraged the production of practical composites for vibration and noise reduction [42]. Daniel et al. investigated failure mechanisms and parameters in unidirectional carbon/epoxy facings and aluminum and PVC closed-cell foam cores, then analyzed the accuracy to theoretical predictions. The results indicated the occurrence of face sheet failure followed by adhesive bond failure, core failure, and facing wrinkling [43]. Lu et al. used carbon fibers and epoxy resin to prepare honeycomb sandwich panels. The investigation showed that the new material had higher bending properties in comparison to traditional and Nomex honeycomb sandwich panels [42]. Jean-St-Laurent et al. examined the impact behavior of carbon epoxy composite sandwich panels with Nomex honeycomb cores at various temperatures and verified that temperature has an effect on the damage caused by impact loading. [44]. Kazemhvazi et al. studied corrugated carbon fiber reinforced epoxy sandwich cores under compressive stress utilizing a Kolsky-bar setup, and the findings revealed substantial strength enhancement with increasing loading rate [45].

A critical component of any subsequent work is correctly determining the timing and location of the onset of damage. There are several methodologies for qualifying the failure of carbon epoxy sandwich plates, but only a few attempts have been made to use the Digital Image Correlation (DIC) methodology for qualifying strain and/or displacement fields. DIC is used to record field variables such as deformation and strains throughout the trailing edge segment during the failure series [46–48]. Sutton et al. defined the procedure for performing DIC for the first time in 1983. The authors defined a method for taking digital images of an object before and after deformation and then using the induced light intensity levels to transform the discrete data into a continuous shape using a surface fit [49]. Measuring is done by measuring a sequence of sequential photographs collected while testing at a given period. DIC can be used as a 2D system with a single camera or as a 3D method with two cameras, depending on the criteria given with an ability to test large structures and give highly accurate results [50–54].

Hower et al. performed a single Cantilever Beam test on aluminum honeycomb sandwich panels with carbon fiber reinforced polymer face sheets. The experimental findings showed fiber bridging and crack process zones and were proven using digital image correlation [55]. Shams et al. investigates the effects of scratch damage on the progressive failure of laminated carbon fiber/epoxy composites under tensile loading using theoretical and computational analysis methods. The findings showed that the suggested simulation method for initiation and propagation load levels had a positive association with interlaminar crack propagation. These deformation modes have been validated using digital image correlation assessment [56]. Rolfe et al. have investigated the influence of high-velocity impact and blast loading of carbon and glass composite sandwich panels [57]. Crump et al. performed a research study on the strain accuracy measured using separate correlation patterns of tested secondary sandwich aircraft panels with different ply configurations [58]. Leone used DIC to describe the fracture behavior and structural reaction of large carbon/epoxy prepreg face sheets and a Nomex honeycomb core sandwich composite aircraft fuselage plates up to collapse, quasi-static combined friction, hoop, and axial loading [59]. Using the DIC, Barile et al. investigated the mechanical behavior of two separate carbon reinforced composite sandwich structures, with and without impact, subjected to compressive load in an anti-buckling fixture. The goal was to use a full-field and noncontact technique to calculate out-of-plane displacements, compression test characteristics, in all points of the area of interest [60].

In this study, tensile loading was applied to carbon epoxy composite sandwich panels with different fiber orientations (0°/90° and ± 45°). The findings were used to clarify the effect of fiber orientation on tensile properties. The novel part of this work is thus to address this issue using an unconventional method using the DIC 3D technique that when applied correctly, allows for the detection of local strain gradients in regard to the material's microstructure [53].

2 Experimental

2.1 Materials

The tested carbon fiber reinforced epoxy composite sandwich panel was obtained from Wing d.o.o. (Belgrade, Serbia). The material's specific density was 300 g/m^3, and the fiber diameter was 5 ÷ 7 μm. As top layers for sandwich panels, carbon fiber prepregs with epoxy resin systems are used. The core structure was based on the epoxy resin (D.E.R. 667, Dow Plastics, USA) and Aramid synthetic fibers (Nomex®, E.I. DuPont, USA). The specimens were autoclaved at 175 °C and 7 bar for 1 h, then 190 °C and 7 bar for 4 h, as directed by the manufacturing company, and eventually cooled to 40 °C. Every specimen had a fiber mass fraction of about 65 percent.

WaterJet, a water processing unit, was used to remove the specimens (PTV, Czech Republic). Specimen dimensions were 250 × 25 mm.

Each specimen with 0°/90° fiber orientation (SZ-1) and (SZ-2) were tested using five samples. The tested samples are shown in Fig. 1.

2.2 Characterization

The experiment was carried out with the aid of the 3D optical system Aramis 2 M (GOM, Braunschweig, Germany) and the Tensile testing machine Shimadzu Autograph AGS-X Series with max load of 100kN at 23 °C with a constant cross-head speed of 1 mm/min. 3D Digital Image Correlation provides the capacity to determine

Fig. 1. Tested samples

non-contact 3D coordinates, displacements, and stresses of materials and structures (DIC) [61, 62]. Two optical cameras are included in the system to provide a synchronized stereo view of the specimen. The machine also contains a stand for sensor stabilization, power management, and image storage unit, and a data processing system. The camera's optical sensor is in charge of data generation. A sheet of white paint was added to the measurement surface before recording, accompanied by a layer of finely spaced black points (Kenda Color Acril-207ico, Kenda Farben). The experiment setup for tensile testing and 3D Image Correlation is shown in Fig. 1.

Scanning electron microscopy (SEM) analysis was used to analyze the fracture surfaces of mechanically fractured specimens for the study of fatigue micromechanisms (MIRA3 TESCAN, Tescan, Czech Republic). The fracture surfaces were vapor-coated with a small coating of gold to increase optical visibility during SEM analysis (Fig. 2).

3 Results and Discussion

3.1 Tensile Testing and 3D Digital Image Correlation

The results obtained after tensile testing (maximum stress and strain at entire areas, fracture stress and strain, and toughness at entire areas) with standard deviations are shown in Table 1. The values of fracture stress and fracture strain for SZ-II were not determined since the specimen did not break, but the specimen twisted in the fiber direction.

Stress – strain curves for SZ-I-2 and SZ-II-2 specimens are shown in Fig. 3. SZ-I exhibited catastrophic failure behavior, as predicted, but SZ-II exhibited high ductility. SZ-I exhibited higher values of tensile strength, but SZ-II showed great elongation while tensile testing when specimen twisted in the fiber direction. As

Fig. 2. Experiment setup

Table 1. Tensile properties for SZ-I and SZ-II with standard deviations

Specimen	Max. stress (MPa)	Max. strain (%)	Fracture stress (MPa)	Fracture strain (%)	Toughness (J)
SZ-I	96.431 ± 24.778	1.091 ± 0.319	94.137 ± 24.833	1.102 ± 0.330	3.593 ± 1.978
SZ-II	31.656 ± 0.747	3.287 ± 0.273	–	–	6.290 ± 0.653

a result, the inclusion of carbon fibers with a ± 45-degree orientation might be a viable alternative for improving the very low tensile strain-to-failure of carbon fiber reinforced composite materials [63]. When subjected to tensile loading, fibers in the longitudinal direction bear the majority of the force, and the tensile strength is dictated by the fibers' tensile characteristics [64]. The curves for ± 45-degree oriented fibers architectures reveal a substantial nonlinear transition and an apparent plastic platform, indicating distinct failure causes. The principal load-bearing items are the interfaces between the ± 45-degree oriented fibers, so it was concluded that with

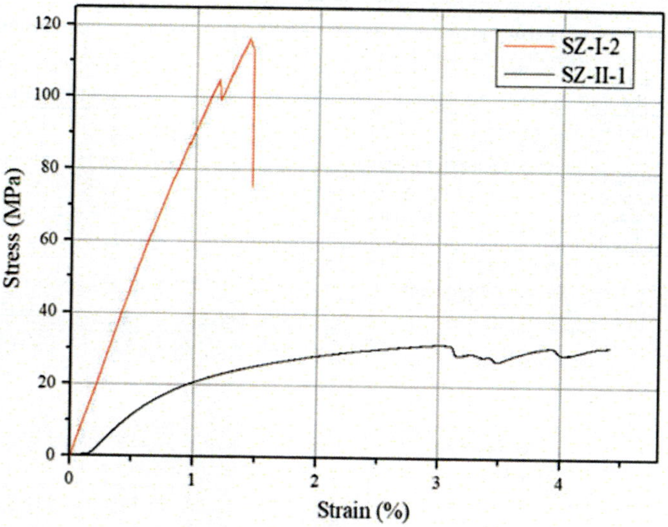

Fig. 3. Stress – strain curves for SZ-I-2 and SZ-II-1 specimen

increasing fiber orientation, the composite materials demonstrated a longer strain performance.

Furthermore, the values of toughness that is considered to be the resistance against failure, were calculated by computing the areas under the stress–strain curves, and the results are also shown in Table 1. As the previous investigations showed, the toughness of materials in general is believed to result in reduction of strength in materials selection [65, 66]. This was confirmed while comparing the results obtained after the experiment: SZ-I showed high tensile strength, but a lower value of toughness (3.593 ± 1.978 J) in comparison to SZ-II that exhibited higher toughness value up to 6.290 ± 0.653 J.

According to the experimental data, values of modulus of elasticity for SZ-I and SZ-II specimens were also calculated. SZ-I exhibited higher modulus of elasticity of 93.66 GPa, i.e. greater rigidity. The fiber becomes gradually stretched as the tension increases, resulting in this effect [67]. On the other hand, SZ-II showed lower values of modulus of elasticity (29.24 GPa) proving the effects of fiber orientation.

In order to explain the stress and strain conditions of various materials, numerical and experimental analysis is required. In this article, an experiment was carried out using the 3D DIC and Aramis software to compute displacement fields. Aramis is the source of the complete efficient Von Mises strain. Since the elastic area of the material in question is too limited in comparison to the plastic region, the effective Von Mises strain is considered to be the local corresponding plastic strain.

Figure 4 depicts the von Mises deformation findings for the SZ-I specimen at a maximal force of 3399 N. Sections 0 and 1 are used to evaluate the deformation area, as well as stage points 0 and 1, as well as stage points 2 and 3 (Fig. 4). Section 0 (black line) is oriented vertically and has a length of 50 mm, while Section 1 (yellow

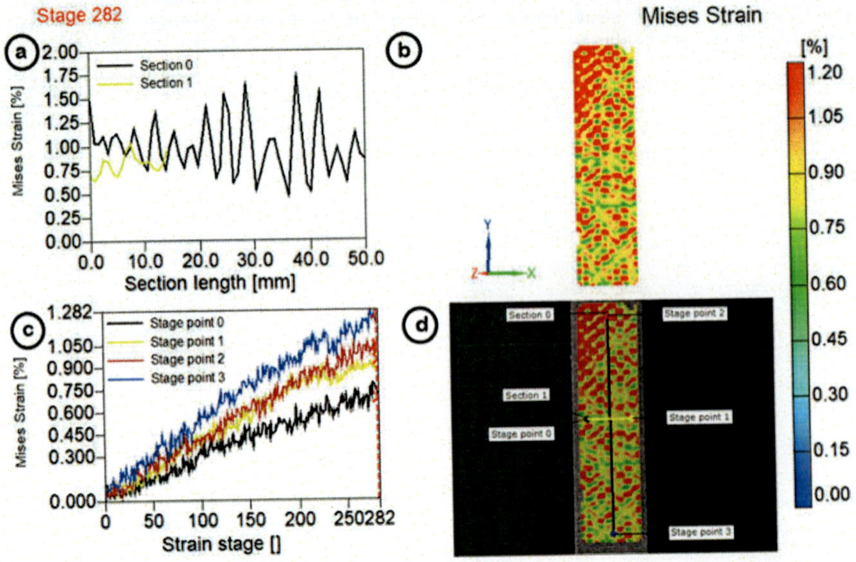

Fig. 4. Experimental von Mises strain for the maximum force of 3399 N **a** von Mises strain - Section length, **b** von Mises strain - Strain stage, **c** von Mises strain field, **d** a sample photograph with the overlaying von Mises strain field

line) is oriented horizontally and has a length of 14 mm. Section 1 is assigned to points 0 (black) and 1 (yellow), while Section 0 is assigned to points 2 (red) and 3 (blue). The highest values (red) are uniformly distributed around the entire surface of the test sample in the 3D von Mises deformation region on the sample surface (Fig. 4c and d). Figure 4a shows the von Mises deformation values as a function of length cross-section, and Fig. 4b shows the observed points as a function of load time represented over the number of images. Sharp tips in Sections 0 and 1, Fig. 4a, exhibit the highest deformation of 0.39 percent in Section 0, and the maximum deformation of 0.27 percent in Section 1. Points 2 and 3, Fig. 4b, show a related deformation growth pattern with rising load, with maximum values of around 1.05 and 1.20 percent, respectively.

The von Mises deformation results for the SZ-II specimen at a maximum force of 1115 N are seen in Fig. 5. The same concept of estimating von Mises deformation and deformation field analysis was used for fiber orientation samples ± 45° as it was for fiber orientation samples 0°/90°. Section 0 (black line) is vertically aligned and has a length of 50 mm, while Section 1 (yellow line) is horizontally oriented and has a length of 14 mm. The highest values (red) are spread on a specific region of the test sample surface in the 3D von Mises deformation region on the sample surface i.e. in the center of the sample in the direction of carbon fiber marked as the material's most loaded part (Fig. 5c and d). Sharp tips in Sections 0 and 1, Fig. 5a, exhibit the highest deformation: 20.10 percent in Section 0 and 23.27 percent in Section 1.

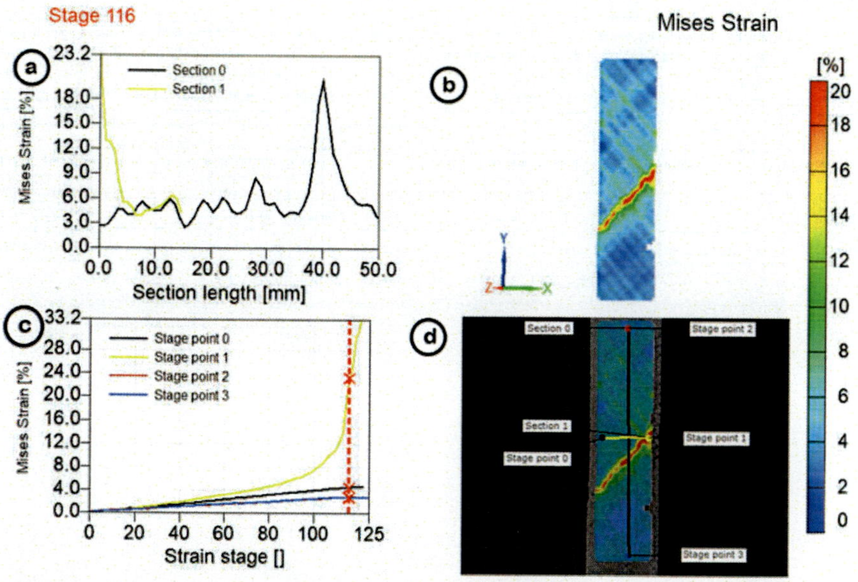

Fig. 5. Experimental von Mises strain for the maximum force of 1115 N **a** von Mises strain - Section length, **b** von Mises strain - Strain stage, **c** von Mises strain field, **d** a sample photograph with the overlaying von Mises strain field

Points 2 and 3, Fig. 5b, show the same trend of deformation growth with increasing load, with maximum values of about 3%.

3.2 *SEM Analysis*

In order to investigate the fracture surface of the tested samples, SEM analysis was performed. Fiber diameter was confirmed to be 5 ÷ 7 μm after SEM, as shown in Fig. 6.

SEM images of the hexagon-shaped honeycomb cells of the core structure are shown in Fig. 7a. Figures 6c and 7b show substantial fiber bridging and fracturing at the face–core interface at a higher magnification.

Figure 8 depicts SZ-I specimen's fracture surface. Severe fiber damage and delamination and core cracking were spotted (Fig. 8a, b, d). The SZ-I fractured when it was undermined by longitudinal fissures and fractures caused by delamination of layer with opposing orientations, permitting appropriate stress transmission among them.

Fiber pull-out and matrix breakage were also noticed as a result of applied tension (Fig. 8c). Therefore, it is apparent that the majority of disruption in sandwich structures is classified as matrix splitting, matrix debonding, fiber pullout, and fiber fracturing. Fiber breakdown within the least effective layer was followed by gradual loss of the fibers in the remaining layers. Based on the images observed, it was possible

Fig. 6. SEM images of carbon fibers

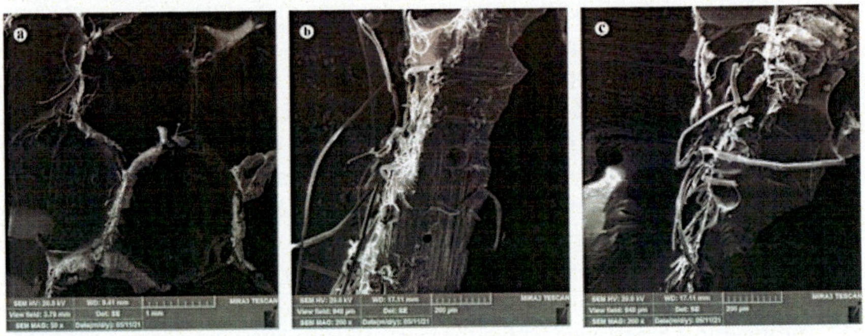

Fig. 7. SEM images of honeycomb structure

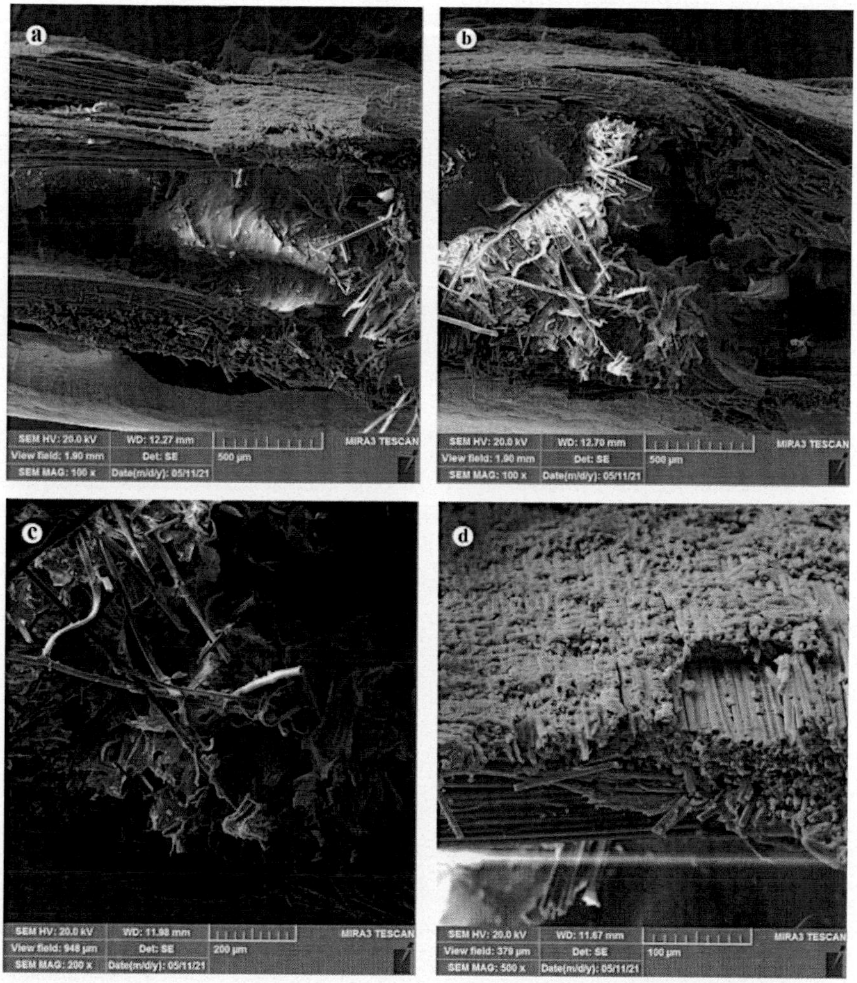

Fig. 8. Z-I fracture surface

to establish that the major crack expansion was caused by the sequential breakage of fibers in the lay-up comprising 0° oriented fibers.

Figure 9 displays the fracture surface of a SZ-II specimen. Based on the obtained images, matrix breaks and visible plastic deformation were seen. Failure occurred prior to the breakage of the majority of the fibers in the layers, as well as the matrix's macroscopically obvious fiber withdraw from the matrix. This was followed by delamination, which reflected the extra failure state seen during tensile testing.

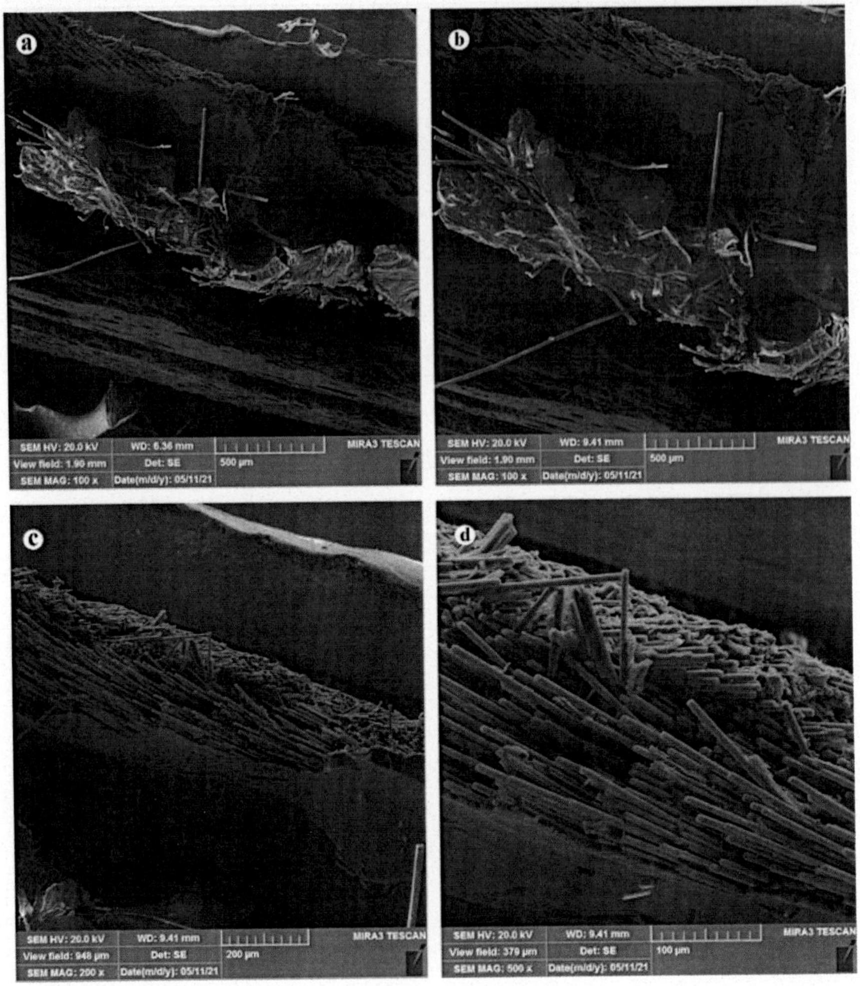

Fig. 9. Z-II fracture surface

4 Conclusions

The purpose of this work was to analyze carbon fiber reinforced composite sandwich panels with varied fiber orientations (0°/90° and ± 45°) and Aramid fiber honeycomb core by employing tensile testing and a full field non-contact 3D DIC. The structure and mechanical behavior of the examined materials was confirmed using SEM images.

Composite structures with 0°/90° oriented fibers exhibited break stress values higher for ~ 30% in comparison to structures with ± 45° fiber orientation. However, this leads to lower strain values, brittle fracture of SZ-I specimen and values of modulus of elasticity ~ 74% higher in comparison to SZ-II specimen. Specimen ± 45° fiber orientation did not break, but twisted in fiber direction, leading to its elongation, strain, and lower value of modulus of elasticity. Because the interfaces between the 45-degree oriented fibers are the primary load-bearing components, it was inferred that increasing fiber orientation resulted in better strain performance for composite materials. This led to the conclusion that incorporating carbon fibers with a 45-degree orientation might be a feasible option for increasing the extremely low tensile strain-to-failure of carbon fiber reinforced composite materials.

The von Mises deformation results for the SZ-I specimen at a maximal force of 3399.20 N showed Section 0 had the greatest distortion of 0.39 percent, whereas Section 1 had the greatest deformation of 0.27 percent. The von Mises deformation data for the SZ-II specimen at a maximum force of 1115.86 N revealed that the most loaded portion of the material was around the center of the sample, in the direction of the carbon fiber.

The micromechanical failure behavior of the studied materials was explained using SEM analysis, which verified previously acquired results. The ultimate fracture od SZ-I was caused by the dominating crack growth in the layers with 0° fiber orientation. In the samples including both structures (SZ-I and SZ-II), the formation of a crack in the layers led its growth in the thickness of the layers, resulting in delamination, fiber pull-out and matrix damage which decreased the sample's strength and caused the ultimate fracture.

References

1. Stamenović, M., Putić, S., Drmanić, S., Rakin, M., Medjo, B.: The influence of service solutions on the longitudinal and circumferential tensile properties of glass-polyester composite pipes. Mater. Sci. **47**, 61–69 (2011). https://doi.org/10.1007/s11003-011-9368-7
2. Putic, S.M., Stamenovic, Petrovic, J., Rakin, M., Medjo, B.: Effect of alkaline solutions on the tensile properties of glass-polyester pipes. Acta Period. Technol. 185–195 (2011). https://doi.org/10.2298/APT1142185P
3. Muneer Ahmed, M., Dhakal, H.N., Zhang, Z.Y., Barouni, A., Zahari, R.: Enhancement of impact toughness and damage behaviour of natural fibre reinforced composites and their hybrids through novel improvement techniques: a critical review. Compos. Struct. **259**, 113496 (2021). https://doi.org/10.1016/j.compstruct.2020.113496
4. Chak, V., Chattopadhyay, H., Dora, T.L.: A review on fabrication methods, reinforcements and mechanical properties of aluminum matrix composites. J. Manuf. Process. **56**, 1059–1074 (2020). https://doi.org/10.1016/j.jmapro.2020.05.042
5. Tasić, A., Rusmirović, J.D., Nikolić, J., Božić, A., Pavlović, V., Marinković, A.D., Uskoković, P.S.: Effect of the vinyl modification of multi-walled carbon nanotubes on the performances of waste poly(ethylene terephthalate)-based nanocomposites. J. Compos. Mater. **51**, 491–505 (2017). https://doi.org/10.1177/0021998316648757
6. Ghali, L., Msahli, S., Zidi, M., Sakli, F.: Effects of fiber weight ratio, structure and fiber modification onto flexural properties of luffa-polyester composites. Adv. Mater. Phys. Chem. **01**, 78–85 (2011). https://doi.org/10.4236/ampc.2011.13013
7. Daniel, I.M., Ishai, O., Engineering Mechanics of Composite Materials, Oxford University Press, Inc., New York, 2006.

8. Eken, S.: Free vibration analysis of composite aircraft wings modeled as thin-walled beams with NACA airfoil sections. Thin-Walled Struct. **139**, 362–371 (2019). https://doi.org/10.1016/j.tws.2019.01.042
9. Marsh, G.: Wing worker for the world. Reinf. Plast. **54**, 24–28 (2010). https://doi.org/10.1016/S0034-3617(10)70109-0
10. Zhao, W., Gupta, A., Regan, C.D., Miglani, J., Kapania, R.K., Seiler, P.J.: Component data assisted finite element model updating of composite flying-wing aircraft using multi-level optimization. Aerosp. Sci. Technol. **95**, 105486 (2019). https://doi.org/10.1016/j.ast.2019.105486
11. Forintos, N., Czigany, T.: Multifunctional application of carbon fiber reinforced polymer composites: electrical properties of the reinforcing carbon fibers – a short review. Compos. Part B Eng. **162**, 331–343 (2019). https://doi.org/10.1016/j.compositesb.2018.10.098
12. Willicombe, K., Elkington, M., Hamerton, I., Ward, C.: Development of novel transportation shells for the rapid automated manufacture of automotive composite parts. Procedia Manuf. **51**, 818–825 (2020). https://doi.org/10.1016/j.promfg.2020.10.115
13. Kogure, T.: Characterisation of halloysite by electron microscopy. In: Developments in Clay Science, 1st edn., pp. 92–114. Elsevier Ltd. (2016). https://doi.org/10.1016/B978-0-08-100293-3.00005-4.
14. Bodea, S., Zechmeister, C., Dambrosio, N., Dörstelmann, M., Menges, A.: Robotic coreless filament winding for hyperboloid tubular composite components in construction. Autom. Constr. **126**, 103649 (2021). https://doi.org/10.1016/j.autcon.202
15. Klyosov, A.A.: Improving wood–polymer composite products: a case study. In: Wood–Polymer Compos. Elsevier, pp. 331–353 (2008). https://doi.org/10.1533/9781845694579.331
16. Dou, S., Tao, L., Wang, R., El Hankari, S., Chen, R., Wang, S.: Plasma-assisted synthesis and surface modification of electrode materials for renewable energy. Adv. Mater. **30**, 1705850 (2018). https://doi.org/10.1002/adma.201705850
17. Gattineni, V., Nathi, V.: Expansion based composite tubes as energy absorbers for improved specific energy absorption: FEA based study. Mater. Today Proc. (2021). https://doi.org/10.1016/j.matpr.2020.12.786
18. Shenoi, R.A., J.M. Dulieu-Barton, Quinn, S., Blake, J.I.R., Boyd, S.W., Composite Materials for Marine Applications: Key Challenges for the Future, in: Compos. Mater., Springer London, London, pp. 69–89 (2011). https://doi.org/10.1007/978-0-85729-166-0_3
19. Tran, P., Nguyen, Q.T., Lau, K.T.: Fire performance of polymer-based composites for maritime infrastructure. Compos. Part B Eng. **155**, 31–48 (2018). https://doi.org/10.1016/j.compositesb.2018.06.037
20. Bhatnagar, A: 3.19 Lightweight fiber-reinforced composites for ballistic applications. In: Comprehensive Composite Materials II, Elsevier, pp. 527–544 (2018). https://doi.org/10.1016/B978-0-12-803581-8.09932-X
21. Marsh, G.: Composites fight for share of military applications. Reinf. Plast. **49**, 18–22 (2005). https://doi.org/10.1016/S0034-3617(05)70641-X
22. Stojanović, L. (ed.): Hrast kitnjak (Quercus patraea agg. Ehrendorfer 1967) u Srbiji, Šumarski fakultet Univerziteta u Beogradu, Udrženje šumarskih inženjera i tehničara Srbije, Beograd (2007). http://www.srpskosumarskoudruzenje.org.rs/pdf/hrast_kitnjak_u_srbiji.pdf
23. Biswal, A., Swain, S.K.: Smart composite materials for civil engineering applications. In: Polym. Nanocomposite-Based Smart Mater, pp. 197–210. Elsevier (2020). https://doi.org/10.1016/B978-0-08-103013-4.00011-X
24. Stamenović, M., Putić, S., Rakin, M., Medjo, B., Čikara, D.: Effect of alkaline and acidic solutions on the tensile properties of glass–polyester pipes. Mater. Des. **32**, 2456–2461 (2011). https://doi.org/10.1016/j.matdes.2010.11.023
25. Stamenović, M., Putić, S., Medjo, B., Rakin, M., Zrilić, M.: Effect of solution pH on crack initiation and propagation in glass-polyester pipes subjected to impact. Polym. Compos. **33**, 1321–1328 (2012). https://doi.org/10.1002/pc.22258
26. Rana, S., Subramani, P., Fangueiro, R., Correia, A.G.: A review on smart self-sensing composite materials for civil engineering applications. AIMS Mater. Sci. **3**, 357–379 (2016). https://doi.org/10.3934/matersci.2016.2.357

27. Brkic, D., Nedeljkovic, D., Putic, L., Stajic-Trosic, J., Stamenovic, M.: Gas separation properties of the dense polymer — zeolite powder composite membranes. Mater. Trans. **57**, 452–456 (2016). https://doi.org/10.2320/matertrans.M2015387
28. Bayraktar, O., Guldas, A., Goktas, M., Forming of carbon fiber reinforced composite materials by using of stretch apparatus. In: International Conference on Engineering and Natural Science Sarajevo, pp. 1866–1872 (2016). https://www.researchgate.net/publication/312525895_Forming_of_Carbon_Fiber_Reinforced_Composite_Materials_By_Using_of_Stretch_Apparatus
29. Jelic, A., Sekulic, M., Stamenovic, M., Ugrinovic, V., Putic, S.: Micromechanical analysis of fatigue and crack growth in carbon-fiber epoxy composites based on mechanical testing. Hem. Ind. **74**, 257–264 (2020). https://doi.org/10.2298/HEMIND200615022J
30. Dinesh, S., Rajasekaran, T., Dhanasekaran, M., Vigneshwaran, K., Experimental testing on mechanical properties of sandwich structured carbon fibers reinforced composites. IOP Conf. Ser. Mater. Sci. Eng. 402, 012180 (2018). https://doi.org/10.1088/1757-899X/402/1/012180
31. Xiong, J., Zhang, M., Stocchi, A., Hu, H., Ma, L., Wu, L., Zhang, Z.: Mechanical behaviors of carbon fiber composite sandwich columns with three dimensional honeycomb cores under in-plane compression. Compos. Part B Eng. **60**, 350–358 (2014). https://doi.org/10.1016/j.compositesb.2013.12.049
32. Wang, G., Yu, D., Kelkar, A.D., Zhang, L.: Electrospun nanofiber: emerging reinforcing filler in polymer matrix composite materials. Prog. Polym. Sci. **75**, 73–107 (2017). https://doi.org/10.1016/j.progpolymsci.2017.08.002
33. Roy, K., Debnath, S.C., Pongwisuthiruchte, A., Potiyaraj, P.: Up-to-date review on the development of high performance rubber composites based on halloysite nanotube. Appl. Clay Sci. **183**, 105300 (2019). https://doi.org/10.1016/j.clay.2019.105300
34. Ma, Z., Li, H., Jing, X., Liu, Y., Mi, H.-Y.: Recent advancements in self-healing composite elastomers for flexible strain sensors: materials, healing systems, and features. Sens. Actuators A: Phys. **329**, 112800 (2021). https://doi.org/10.1016/j.sna.2021.112800
35. George, J., Ishida, H.: A review on the very high nanofiller-content nanocomposites: their preparation methods and properties with high aspect ratio fillers. Prog. Polym. Sci. **86**, 1–39 (2018). https://doi.org/10.1016/j.progpolymsci.2018.07.006
36. Jelić, A., Marinković, A., Sekulić, M., Dikić, S., Ugrinović, V., Pavlović, V., Putić, S.: Design of halloysite modification for improvement of mechanical properties of the epoxy based nanocomposites. Polym. Compos. **42**, 2180–2192 (2021). https://doi.org/10.1002/pc.25967
37. Shahdin, A., Morlier, J., Gourinat, Y., Mezeix, L., Bouvet, C.: Fabrication and mechanical testing of a new sandwich structure with carbon fiber network core. J. Sandw. Struct. Mater. **12**, 569–589 (2010). https://doi.org/10.1177/1099636209106070
38. Liu, J., Zhou, Z., Wu, L., Ma, L.: A study on mechanical behavior of the carbon fiber composite sandwich panel with pyramidal truss cores at different temperatures. Sci. China Phys. Mech. Astron. **55**, 2135–2142 (2012). https://doi.org/10.1007/s11433-012-4903-6
39. George, T., Deshpande, V.S., Wadley, H.N.G.: Mechanical response of carbon fiber composite sandwich panels with pyramidal truss cores. Compos. Part A Appl. Sci. Manuf. **47**, 31–40 (2013). https://doi.org/10.1016/j.compositesa.2012.11.011
40. Xu, G., Wang, Z., Zeng, T., Cheng, S., Fang, D.: Mechanical response of carbon/epoxy composite sandwich structures with three-dimensional corrugated cores. Compos. Sci. Technol. **156**, 296–304 (2018). https://doi.org/10.1016/j.compscitech.2018.01.015
41. Kwon, D.-J., Kim, J.-H., DeVries, K.L., Park, J.-M.: Optimized epoxy foam interface of CFRP/Epoxy Foam/CFRP sandwich composites for improving compressive and impact properties. J. Mater. Res. Technol. **11**, 62–71 (2021). https://doi.org/10.1016/j.jmrt.2021.01.015
42. Zheng, C., Duan, F., Liang, S.: Manufacturing and mechanical performance of novel epoxy resin matrix carbon fiber reinforced damping composites, Compos. Struct. **256**, 113099 (2021). https://doi.org/10.1016/j.compstruct.2020.113099
43. Daniel, I.M., Gdoutos, E.E., Wang, K.-A., Abot, J.L.: Failure modes of composite sandwich beams. Int. J. Damage Mech. **11**, 309–334 (2002). https://doi.org/10.1106/105678902027247

44. Jean-St-Laurent, M., Dano, M.-L., Potvin, M.-J.: Compression after impact behavior of carbon/epoxy composite sandwich panels with Nomex honeycomb core subjected to low velocity impacts at extreme cold temperatures. Compos. Struct. **261**, 113516 (2021). https://doi.org/10.1016/j.compstruct.2020.113516
45. Kazemahvazi, S., Russell, B.P., Zenkert, D.: Impact of carbon fibre/epoxy corrugated cores. Compos. Struct. **94**, 3300–3308 (2012). https://doi.org/10.1016/j.compstruct.2012.04.034
46. Mitrovic, N., Petrovic, A., Milosevic, M., Momcilovic, N., Miskovic, Z., Maneski, T., Popovic, P.: Experimental and numerical study of globe valve housing. Hem. Ind. **71**, 251–257 (2017). https://doi.org/10.2298/HEMIND160516035M
47. Čolić, K., Sedmak, A., Legweel, K., Milošević, M., Mitrović, N., Mišković, Z.: Experimental and numerical research of mechanical behaviour of titanium alloy hip implant. Teh. Vjesn. - Tech. Gaz. 24 (2017). https://doi.org/10.17559/TV-20160219132016
48. Chen, X., Berring, P., Madsen, S.H., Branner, K., Semenov, S.: Understanding progressive failure mechanisms of a wind turbine blade trailing edge section through subcomponent tests and nonlinear FE analysis. Compos. Struct. **214**, 422–438 (2019). https://doi.org/10.1016/j.compstruct.2019.02.024
49. Sutton, M., Wolters, W., Peters, W., Ranson, W., McNeill, S.: Determination of displacements using an improved digital correlation method. Image Vis. Comput. **1**, 133–139 (1983). https://doi.org/10.1016/0262-8856(83)90064-1
50. Hild, F., Roux, S.: Digital image correlation: from displacement measurement to identification of elastic properties - a review. Strain **42**, 69–80 (2006). https://doi.org/10.1111/j.1475-1305.2006.00258.x
51. Milosevic, M., Mitrovic, N., Sedmak, A.: Digital image correlation analysis of biomaterials. In: 2011 15th IEEE International Conference on Intelligent Engineering Systems, pp. 421–425 (2011). https://doi.org/10.1109/INES.2011.5954784
52. Milošević, M., Postić, S., Mitrović, N., Milovanović, A., Travica, M., Hloch, S., Golubović, Z.: Strain measurement setup of compressively loaded mandible model with teeth and comprensations. In: N. Mitrović, M. Milošević, G. Mladenović (Eds.), International Conferences Exp. Numer. Investig. New Technol. CNN TECH 2018, Innovation Center of Faculty of Mechanical Engineering, Zlatibor, Serbia, p. 35 (2018). http://cnntechno.com/docs/2_CNN_book_of_abstracts.pdf
53. Mitrović, N., Petrović, A., Travica, M., Milošević, M., Ćudić, B.: 3D-DIC method in structural health monitoring of pressure vessels. In: Mitrovic, N., Milošević, M., Mladenović, G. (Eds.), Int. Conf. Exp. Numer. Investig. New Technol. CNN Tech 2020, Innovation Center of Faculty of Mechanical Engineering, Zlatibor, Serbia, p. 9 (2020)
54. Travica, M., Mitrović, N., Petrović, A.: Pipe ring specimen test procedure using DIC method with three cameras. In: Mitrović, N., Milošević, M., Mladenović, G. (eds.), Int. Conf. Exp. Numer. Investig. New Technol. CNN Tech 2020, Innovation Center of Faculty of Mechanical Engineering, Zlatibor, Serbia, p. 13 (2020). http://cnntechno.com/docs/4_CNN_book_of_abstracts_CIP_Final.pdf
55. Höwer, D., Lerch, B.A., Bednarcyk, B.A., Pineda, E.J., Reese, S., Simon, J.-W.: Cohesive zone modeling for mode I facesheet to core delamination of sandwich panels accounting for fiber bridging. Compos. Struct. **183**, 568–581 (2018). https://doi.org/10.1016/j.compstruct.2017.07.005
56. Shams, S.S., El-Hajjar, R.F.: Effects of scratch damage on progressive failure of laminated carbon fiber/epoxy composites. Int. J. Mech. Sci. **67**, 70–77 (2013). https://doi.org/10.1016/j.ijmecsci.2012.12.008
57. Rolfe, E., Kaboglu, C., Quinn, R., Hooper, P.A., Arora, H., Dear, J.P.: High velocity impact and blast loading of composite sandwich panels with novel carbon and glass construction. J. Dyn. Behav. Mater. **4**, 359–372 (2018). https://doi.org/10.1007/s40870-018-0163-5
58. Crump, D.A., Dulieu-Barton, J.M.: Full field strain analysis of aircraft sandwich structures. In: Proceedings of XIth International Congress and Exposition, Society for Experimental Mechanics Inc., Orlando, Florida, USA, 2008

59. Leone, F.A.J.: Full-Scale Testing and Progressive Damage Modeling of Sandwich Composite Aircraft Fuselage Structure, Drexel University, 2010. http://hdl.handle.net/1860/3381
60. Barile, C., Casavola, C., Pappalettera, G.: Digital image correlation comparison of damaged and undamaged aeronautical CFRPs during compression tests. Materials (Basel). **12**, 249 (2019). https://doi.org/10.3390/ma12020249
61. Paszkiewicz, M., Kubiak, T.: Selected problems concerning determination of the buckling load of channel section beams and columns. Thin-Walled Struct. **93**, 112–121 (2015). https://doi.org/10.1016/j.tws.2015.03.009
62. Amyot, C.-O., Pickworth, J., Tyson, J., Bucienski, J.: 3D digital image correlation: the ultimate biomechanics tool for displacements and strains testing. In: International Medical Device Conferences 2019, San Jose, CA, USA, p. 11 (2019)
63. Ikbal, M.H.: Tensile Properties of Bidirectional Glass-Carbon Hybrid Composites: Part **I**(3), 44–49 (2018)
64. Wu, W., Wang, Q., Li, W.: Comparison of tensile and compressive properties of carbon/glass interlayer and intralayer hybrid composites. Materials (Basel) **11**, 1105 (2018). https://doi.org/10.3390/ma11071105
65. Truong, G.T., Van Tran, H., Choi, K.-K.: Tensile behavior of carbon fiber-reinforced polymer composites incorporating nanomaterials after exposure to elevated temperature. J. Nanomater. **2019**, 1–14 (2019). https://doi.org/10.1155/2019/4139208
66. Kobayashi, T.: Strength and Toughness of Materials. Springer, Japan, Tokyo (2004). https://doi.org/10.1007/978-4-431-53973-5
67. J.-B. Donnet, Bansal, R.C.: Carbon Fibers, Internatio. Dekker, M. (1990)

Stress Corrosion Crack Growth Simulation by the Finite Element Method

Aleksandar Sedmak, Srdjan Tadic, Snezana Kirin, Milos Djukic, and Mohamed Al Kateb

Abstract Stress corrosion crack growth in mild steel was investigated by using the finite element simulation method. A model simulating crack growth considered an edge crack located on the metal surface, under tensile remote stress acting on a sample. Numerical analysis was performed using the Code _Aster software to simulate crack growth. Three related variables were evaluated. K, dK/da and maximum stress. Values of these variables were recorded every 2 mm of the crack growth. Results showed an increase in the values of K and maximum stress, while there was a decrease in the values of dK/da, as the crack length increased. There was a good agreement between the results obtained analytically in the literature and numerically obtained here by using finite elements. The results obtained here are consistent with what has been obtained in most of the studies that have been conducted in this regard.

Keywords Stress corrosion crack rate · Extended finite element method · Stress intensity factor

1 Introduction

Stress corrosion cracking is an environmentally assisted failure of engineering materials, characterized by gradual crack growth, and eventual final failure, as a result of simultaneous action of chemical reactions and mechanical forces at the crack tip [1]. Stress corrosion cracking is caused by three main interacting factors, as shown in Fig. 1:

(1) material susceptibility to cracking,
(2) environmental corrosive conditions,

A. Sedmak (✉) · M. Djukic · M. Al Kateb
Faculty of Mechanical Engineering, University of Belgrade, Kraljice Marije 16, 11000 Belgrade, Serbia
e-mail: asedmak@mas.bg.ac.rs

S. Tadic · S. Kirin
Innovation Center, Faculty of Mechanical Engineering, Kraljice Marije 16, 11000 Belgrade, Serbia

N. Mitrovic et al. (eds.), *Current Problems in Experimental and Computational Engineering*, Lecture Notes in Networks and Systems 323,
https://doi.org/10.1007/978-3-030-86009-7_14

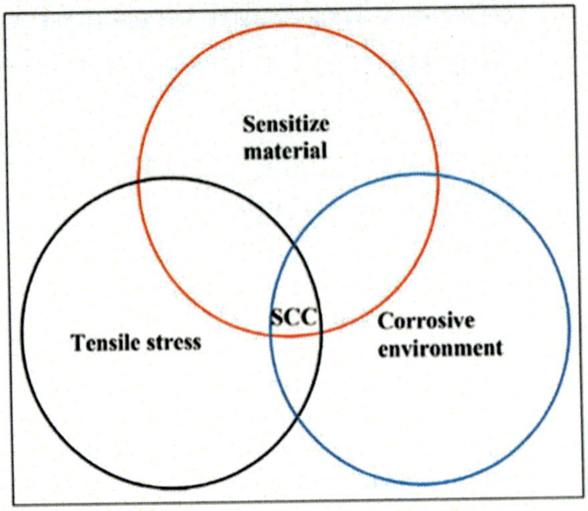

Fig. 1. Relationship between SCC factors

(3) tensile stress, applied or residual.

Also, depending on the rate of chemical reactions at the crack tip, 'hydrogen induced cracking' (HIC) is considered as a specific mechanism of stress corrosion cracking, [2–5].

During the tress corrosion crack growth, three regions are typically observed above threshold stress-intensity factor level (K_{Iscc}):

(1) low stress intensity factor K values: crack growth rate increases fast,
(2) intermediate stress intensity factor K values: crack growth rate is practically constant,
(3) stress intensity factor K values approaches its critical value, K_{Ic}: rapid crack growth appears, as well as eventual final failure [5].

There are many papers explaining chemical, electrochemical and mechanical aspects of stress corrosion crack growth, [6–10], mostly focused on actions at the crack tip. Their goal is to simulate mechanisms and modeling of stress corrosion cracking (SCC), including the finite elements analysis [8, 9].

In this paper, attention was focused on numerical simulation of stress corrosion crack growth behavior in mild steel, by using the finite element simulation method, [11, 12]. A model simulating crack growth was applied to metal surface under tensile stress to the sample. Finite element analysis of tensile stress was performed using the Code-Aster software to verify the effect of crack growth.

2 Extended Finite Element Method

Finite Element Method (FEM) has extremely important role in engineering practice, since it can deal efficiently with challenging geometric forms, different material

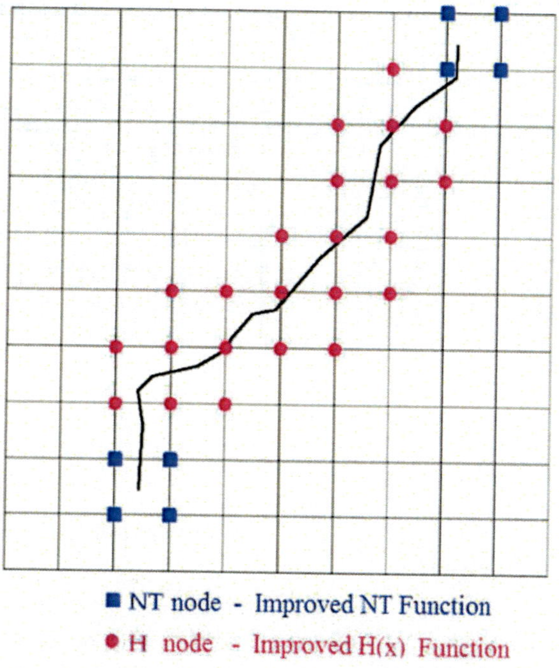

Fig. 2. Nodes NT and H(x) improved function

behavior and complex problems. It is one of very a few methods to tackle non-smooth fracture crack tip stress and strain fields, using different techniques of fracture mechanics singularity simulations, [13].

Nevertheless, if applied to the problem of crack growth, standard FEM procedure would include re-meshing at each step of crack growth. To do so, numerous techniques have been suggested, but without real success, before the extended finite element method (XFEM) has been developed, using completely different approach, [14, 15], based on additional, so-called enhancement functions (Heaviside's function – H, Near Tip functions – NT), in the nodes of elements crack cuts through, Fig. 2. The essential feature and main advantage of XFEM is the fact that mesh is independent of crack growth, so there is no need for re-meshing.

Application of XFEM to solve different engineering problems, e.g. fatigue crack growth in welded joints, has been presented in number of papers [16–20]. Here, XFEM is applied to stress corrosion crack (SCC) growth problem, by using Code-Aster FE open source software, as explained in more details in [12].

2.1 *XFEM Analysis of SCC Growth in a Tensile Specimen*

Results of testing the SCC growth rate in tensile specimen made of mild steel (YS = 450 MPa), as presented in [21], are shown here in Fig. 3, together with theoretical

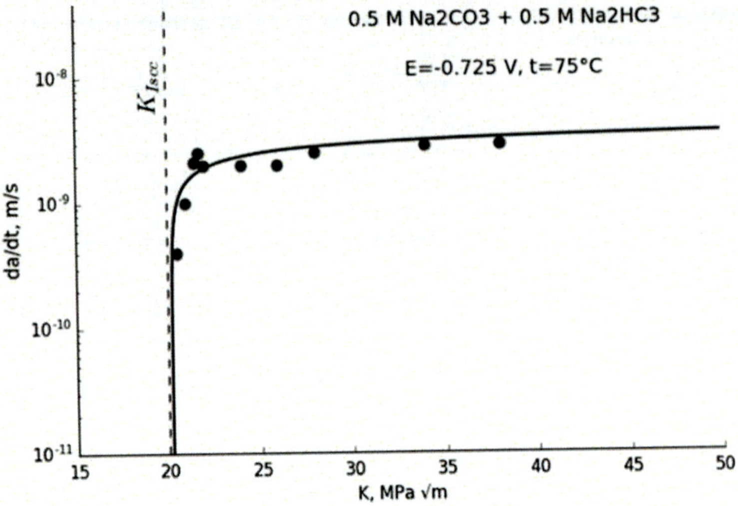

Fig. 3. SCC growth rate theoretical prediction (full line) and experimental data (black points), mild steels [21]

predictions, and used to verify results of numerical analysis, as performed by using XFEM in this research. Finite element mesh is shown in Fig. 4, representing one half of a tensile specimen, with the fine mesh in the mid-section. The crack itself was not drawn, but it was defined as a simple function in Code-Aster. Calculations were performed with crack represented as a lateral notch, $a = 2 \div 20$ mm. Applied remote stress was $\sigma = 150$ MPa. External Python procedure was written to enhance some automation in this procedure.

Results of XFEM calculation are shown in Figs. 5, 6, 7, 8, 9, 10, 11, 12, 13 and 14 for stress distribution, in Fig. 15 for stress intensity factor K vs. crack length a, in Fig. 16 for stress intensity factor rate dK/da versus crack length a, and in Table 1 for both K and dK/da versus a. One can see smooth increase of K and decrease of dK/da with growing a, as predicted by theoretical analysis, [2, 3]. Therefore, one can consider XFEM, as applied here, being verified by the experimental and theoretical results.

3 Conclusions

Based on the presented results, one can conclude that the stress intensity factor K increases with crack length increase, stress intensity crack rate dK/da decreases and maximum stress increases. This means that as crack grows, although K and maximum stress increase, the rate of K decreases, so the process decelerate, at least from that point of view, leading to the conclusion that stress corrosion cracking is a slow and stable process.

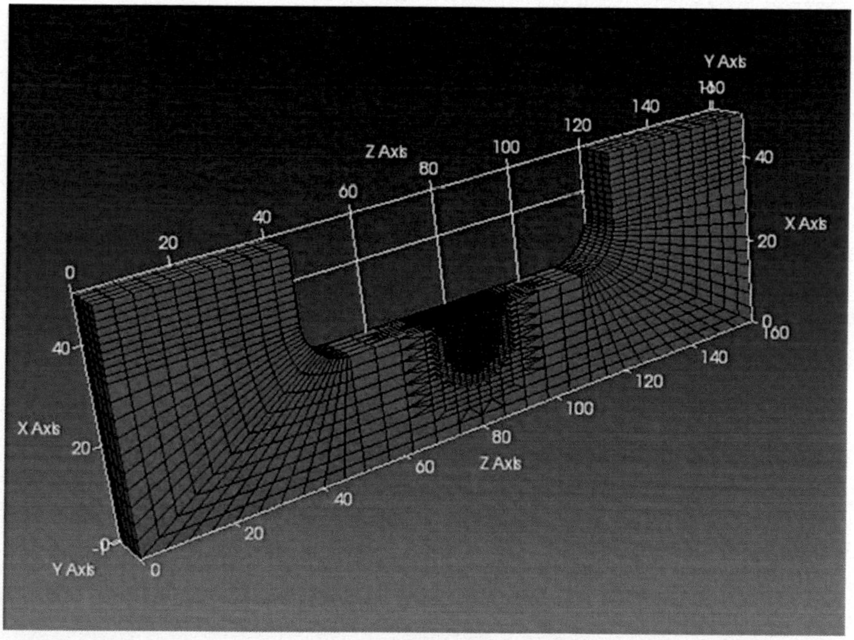

Fig. 4. Geometry of tensile test specimen. Dimensions are in mm

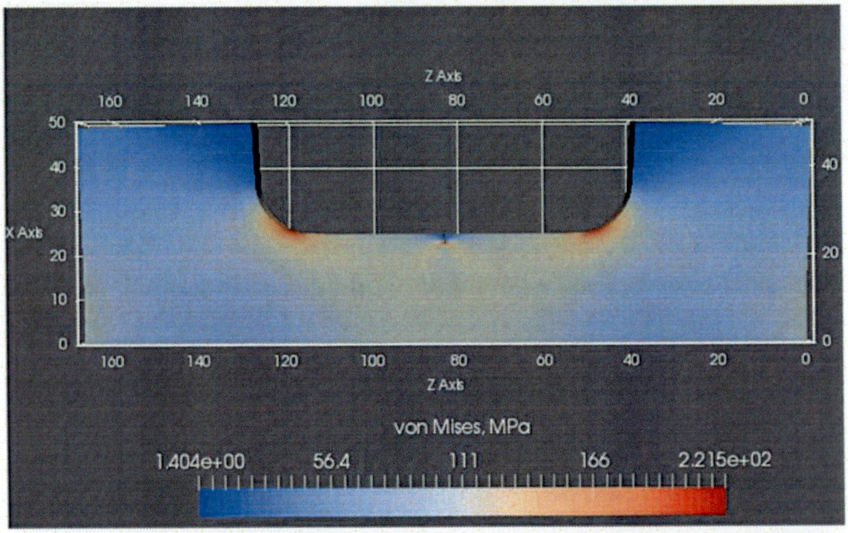

a)

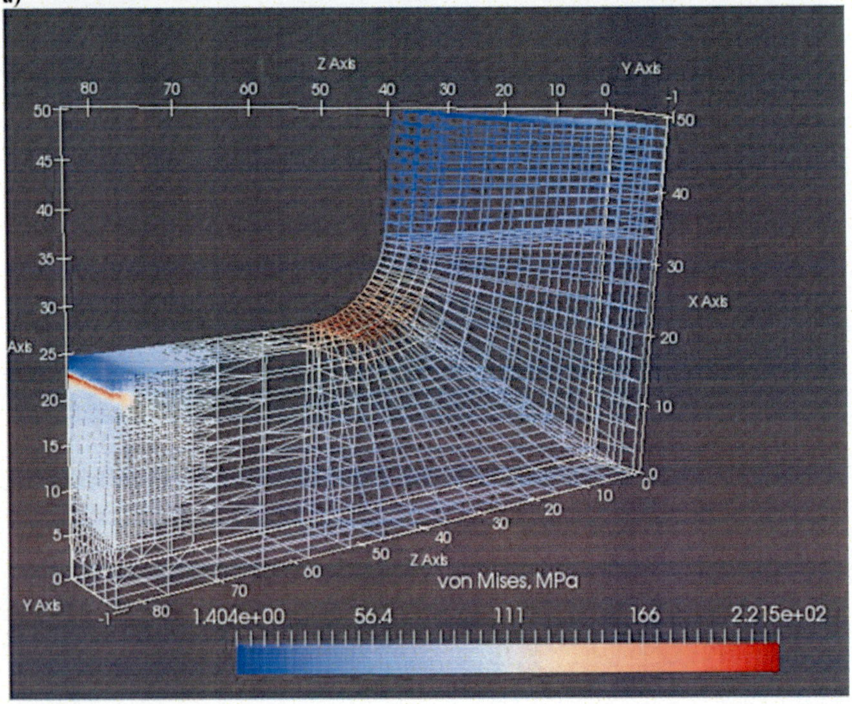

b)

Fig. 5. Stress distribution for a = 2 mm: **a** 2D view, **b** 3D view

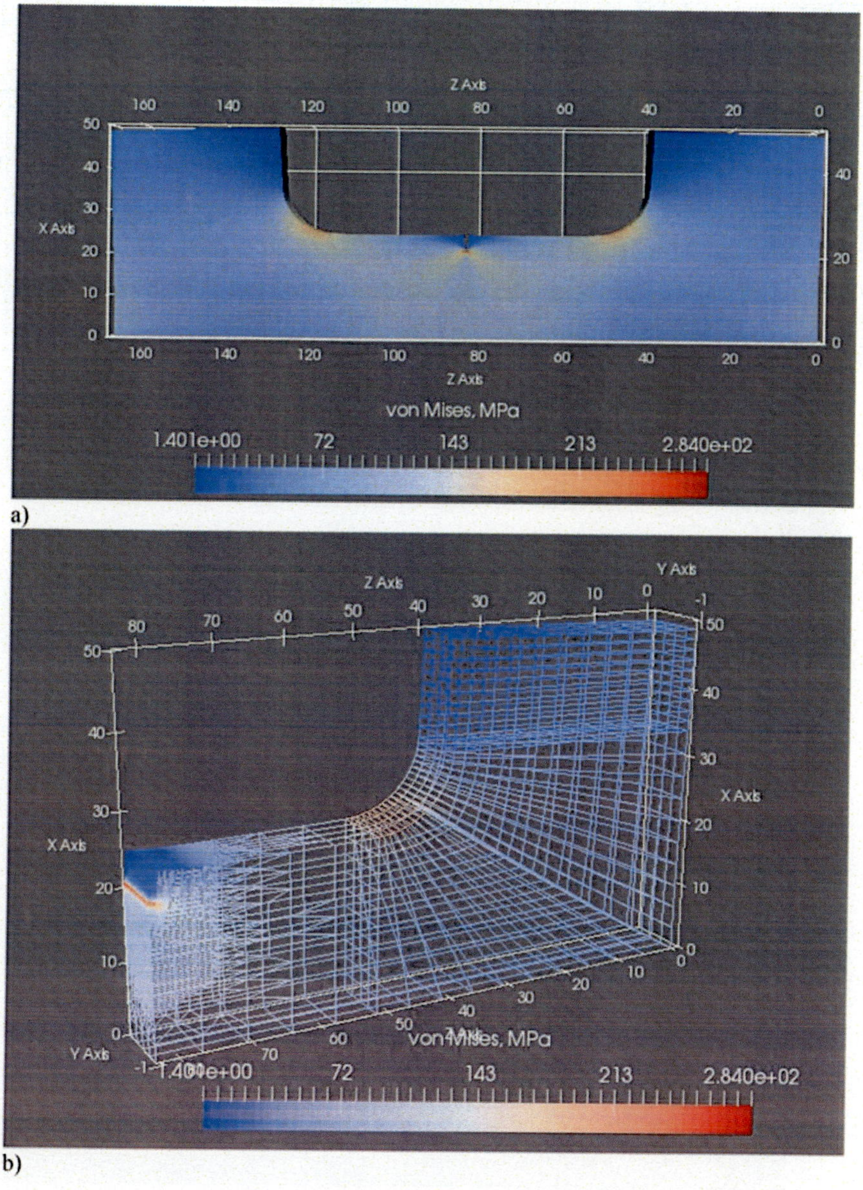

Fig. 6. Stress distribution for a = 4 mm: **a** 2D view, **b** 3D view

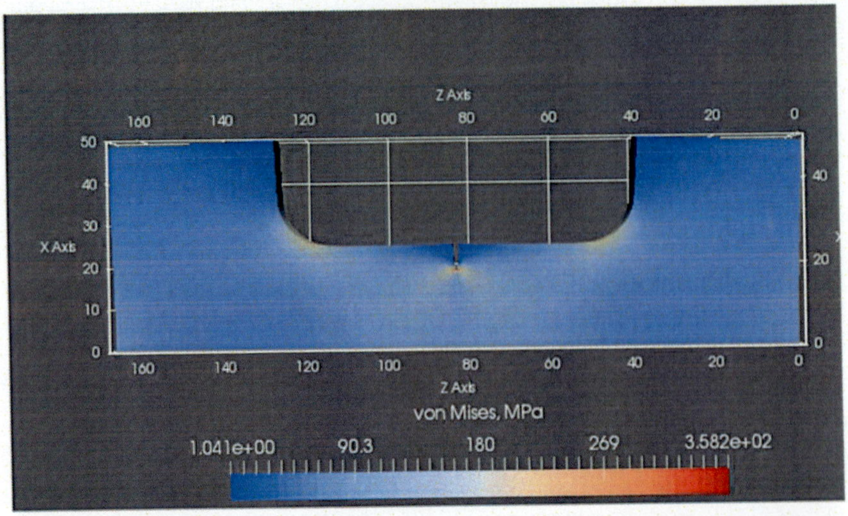

a)

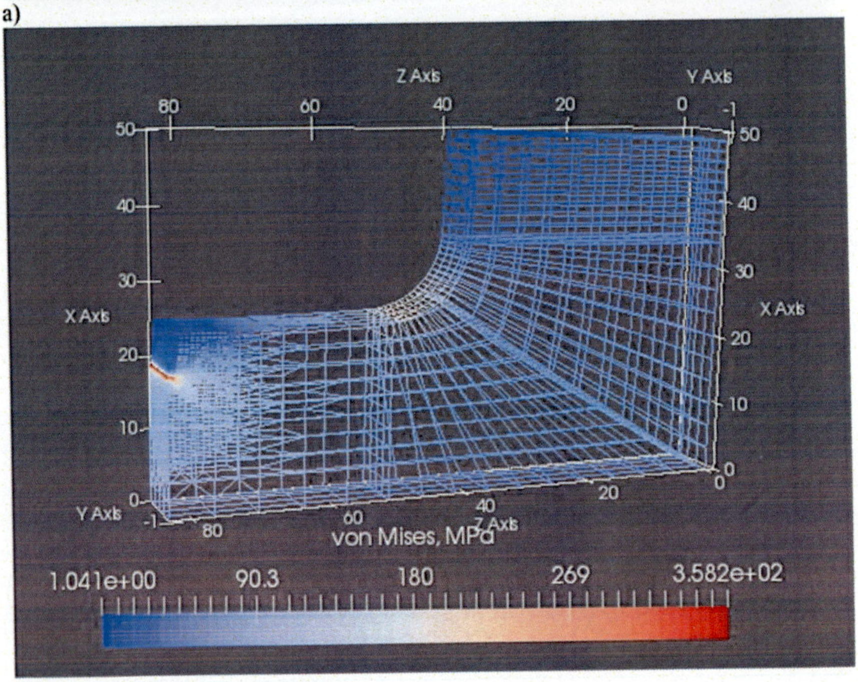

b)

Fig. 7. Stress distribution for a = 6 mm: **a** 2D view, **b** 3D view

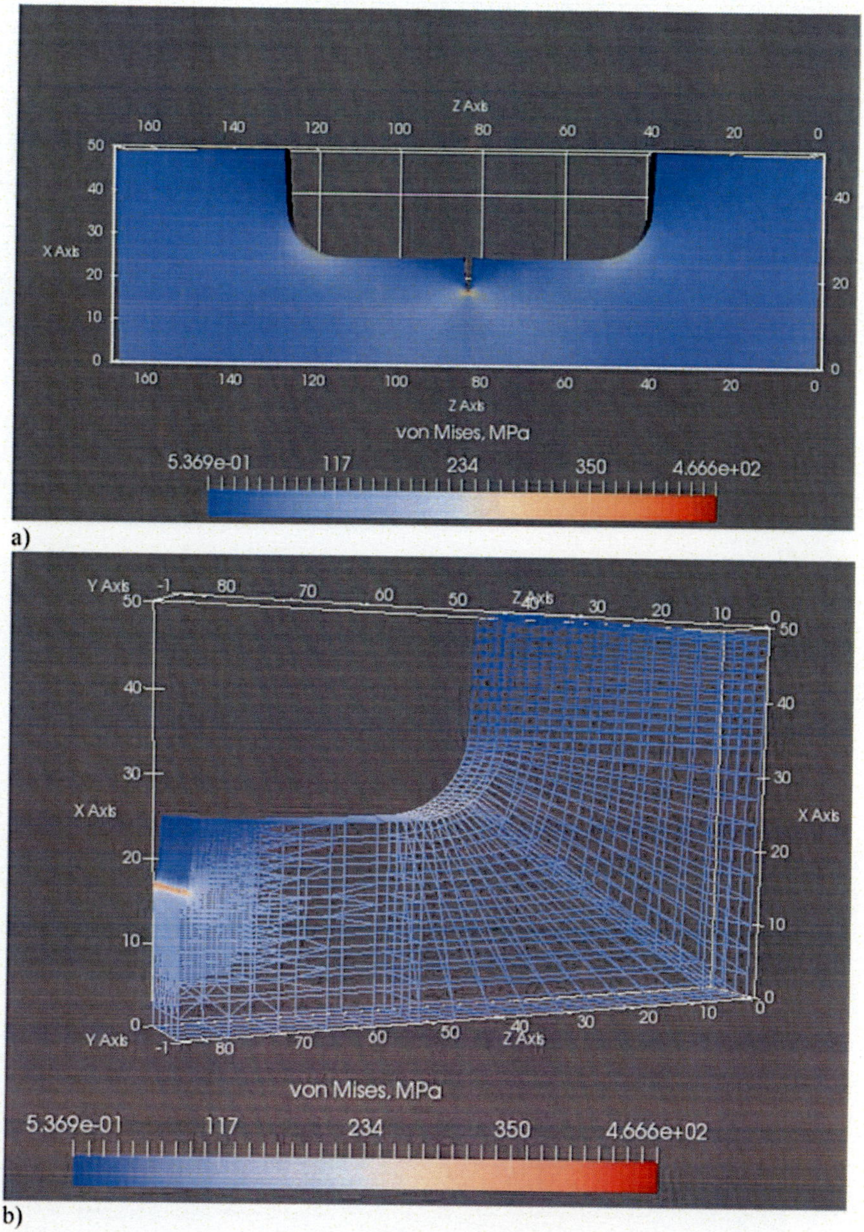

Fig. 8. Stress distribution for a = 8 mm: **a** 2D view, **b** 3D view

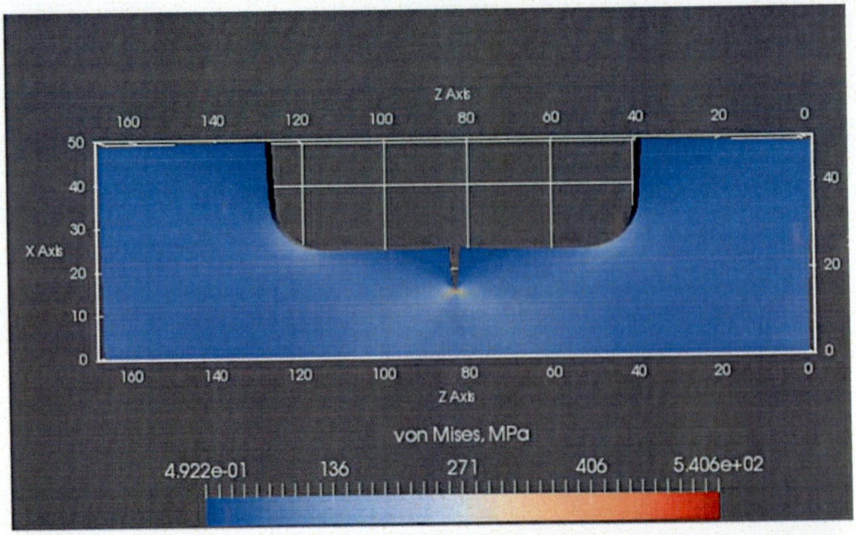

a)

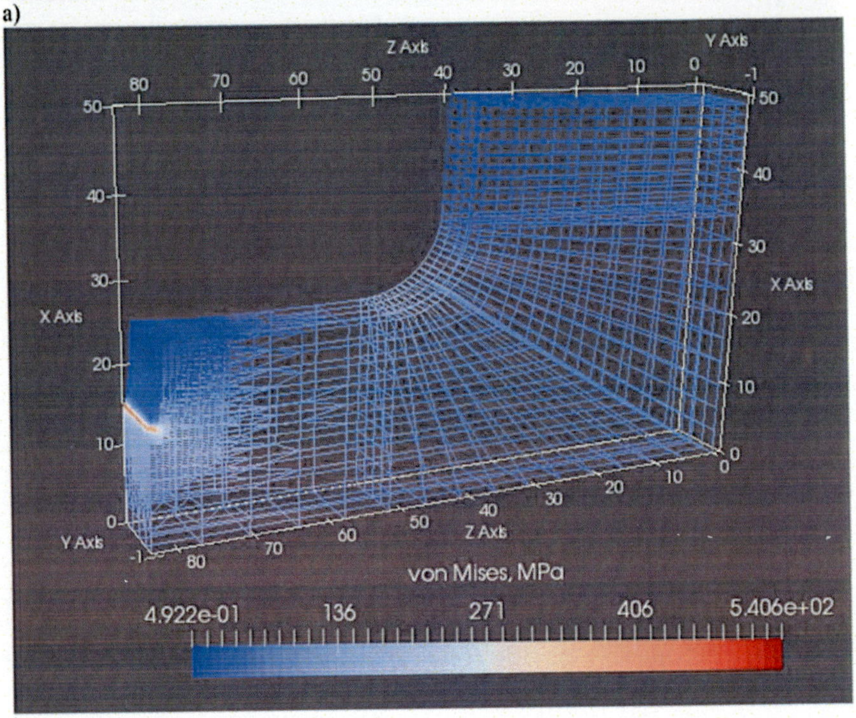

b)

Fig. 9. Stress distribution for a = 10 mm: **a** 2D view, **b** 3D view

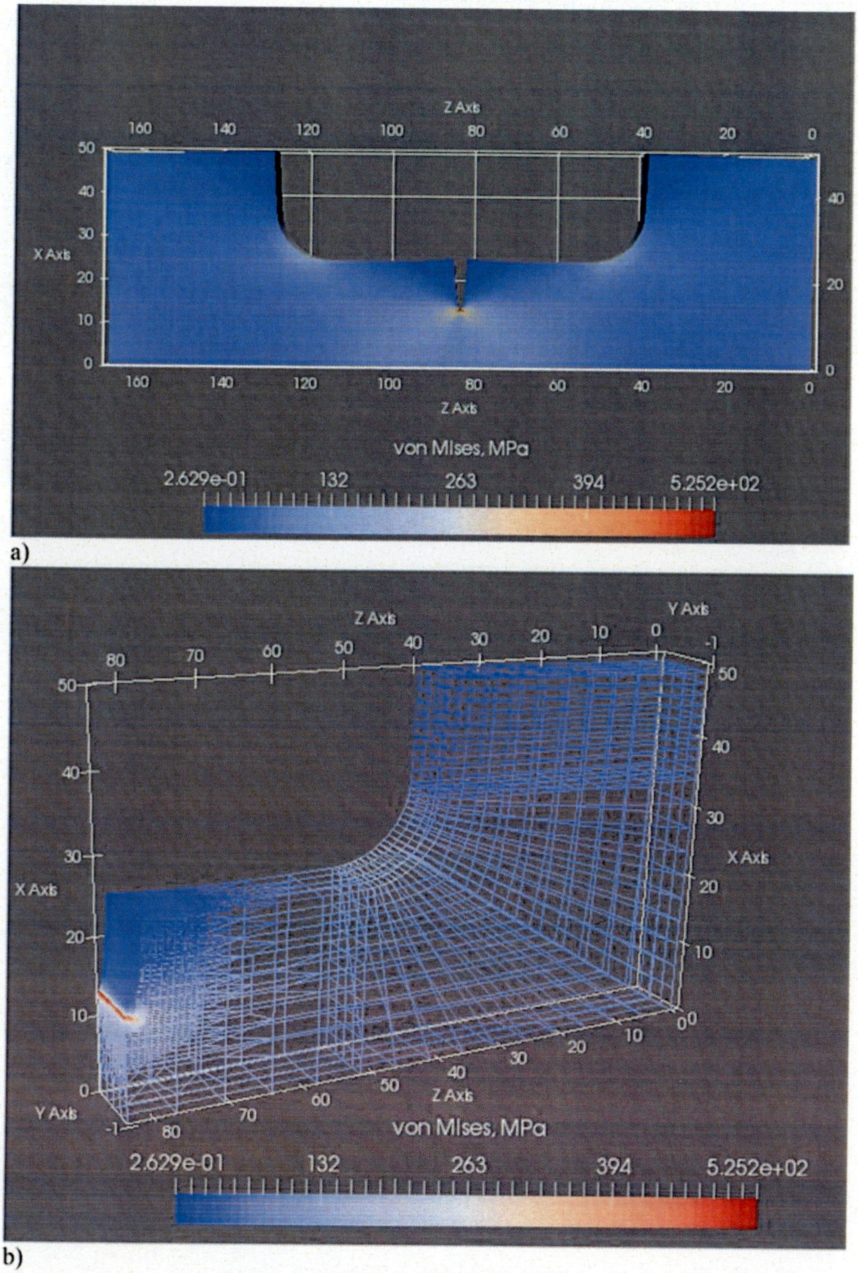

Fig. 10. Stress distribution for a = 12 mm: **a** 2D view, **b** 3D view

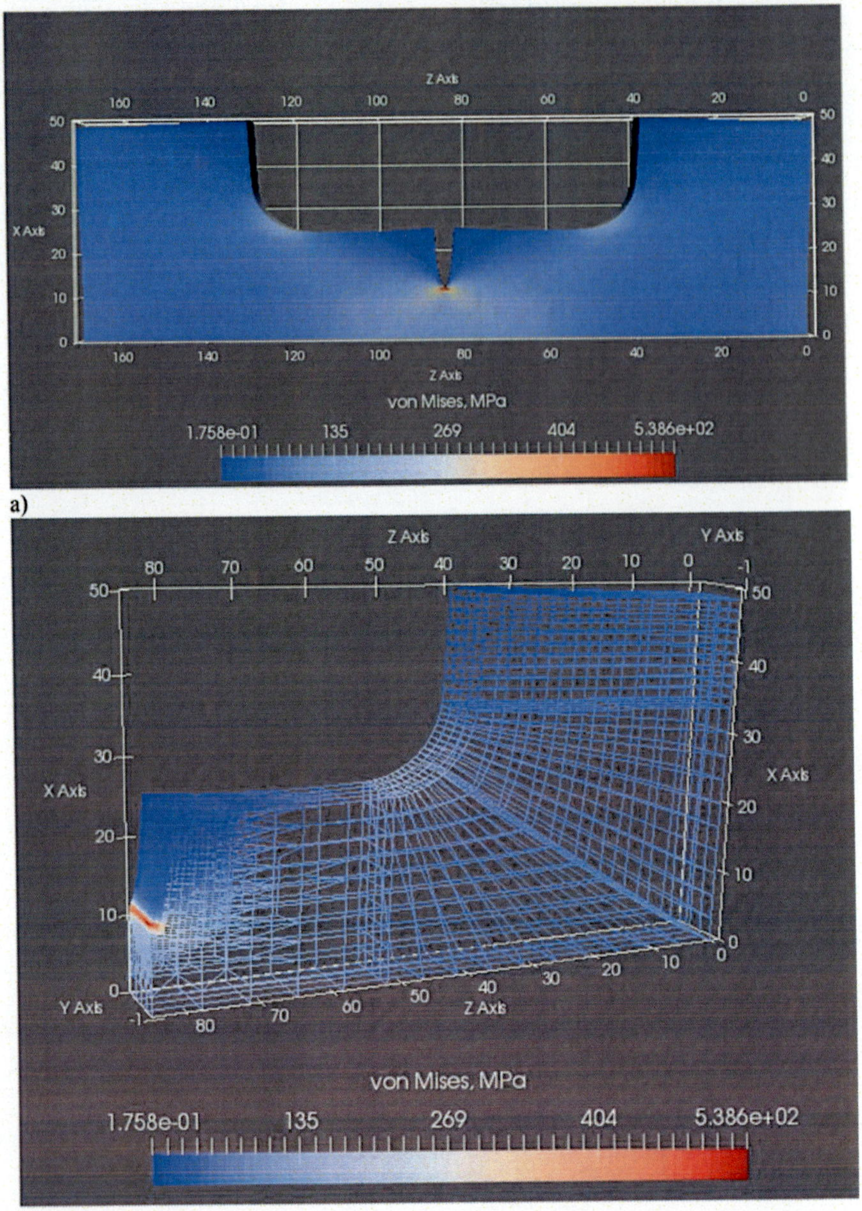

a)

b)

Fig. 11. Stress distribution for a = 14 mm: **a** 2D view, **b** 3D view

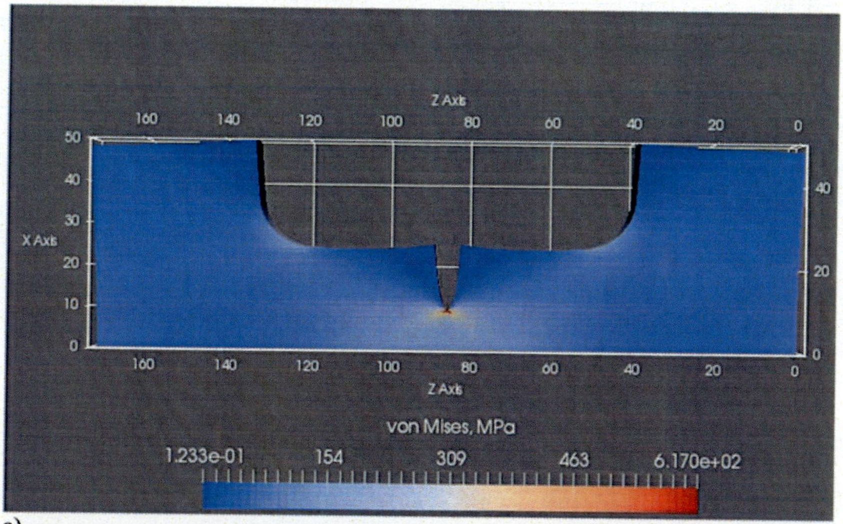

a)

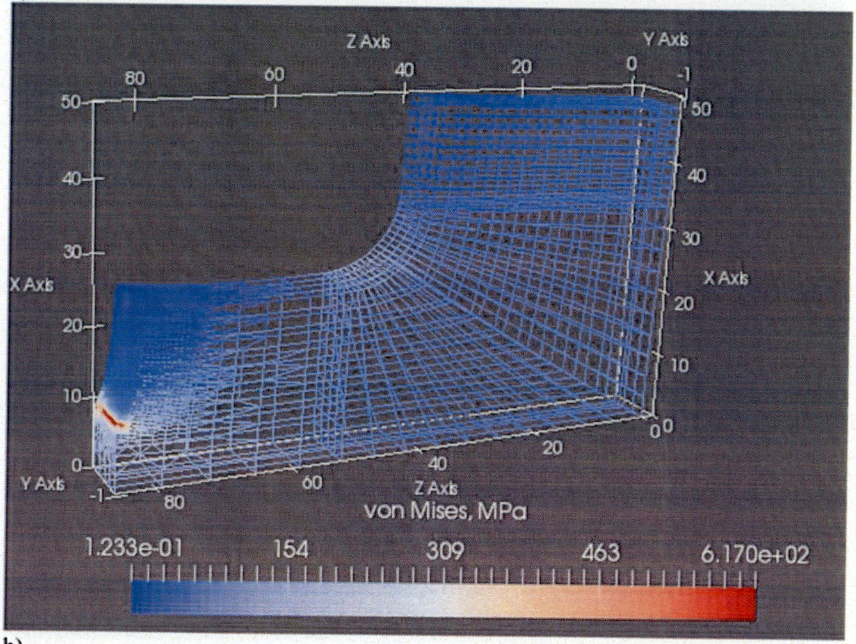

b)

Fig. 12. Stress distribution for a = 16 mm: **a** 2D view, **b** 3D view

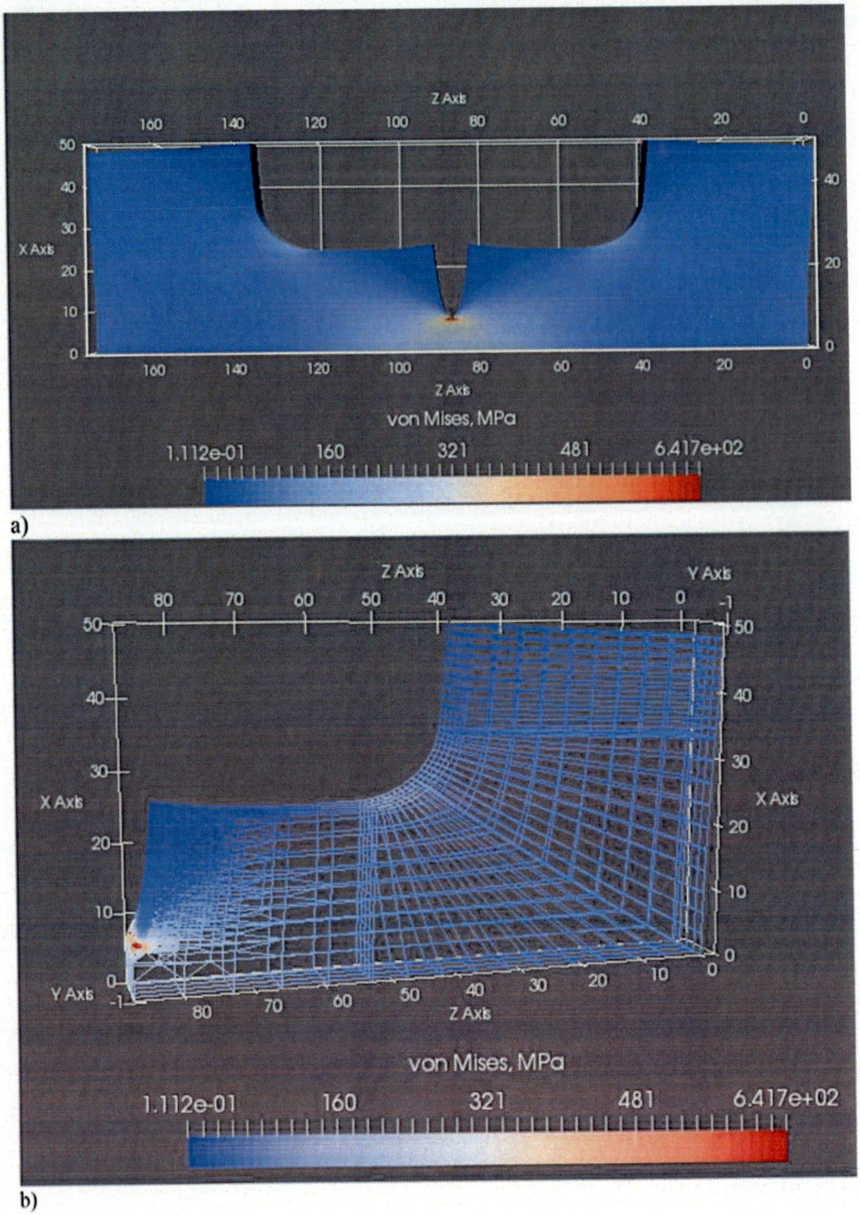

Fig. 13. Stress distribution for a = 18 mm: **a** 2D view, **b** 3D view

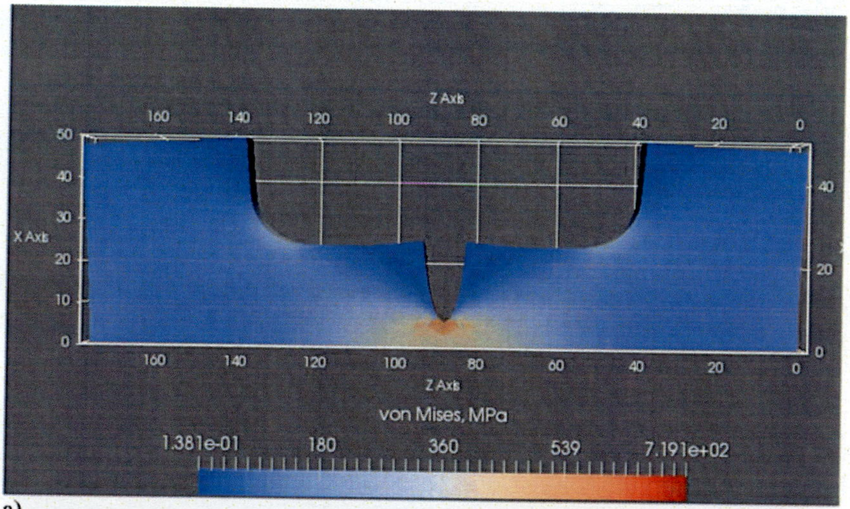

a)

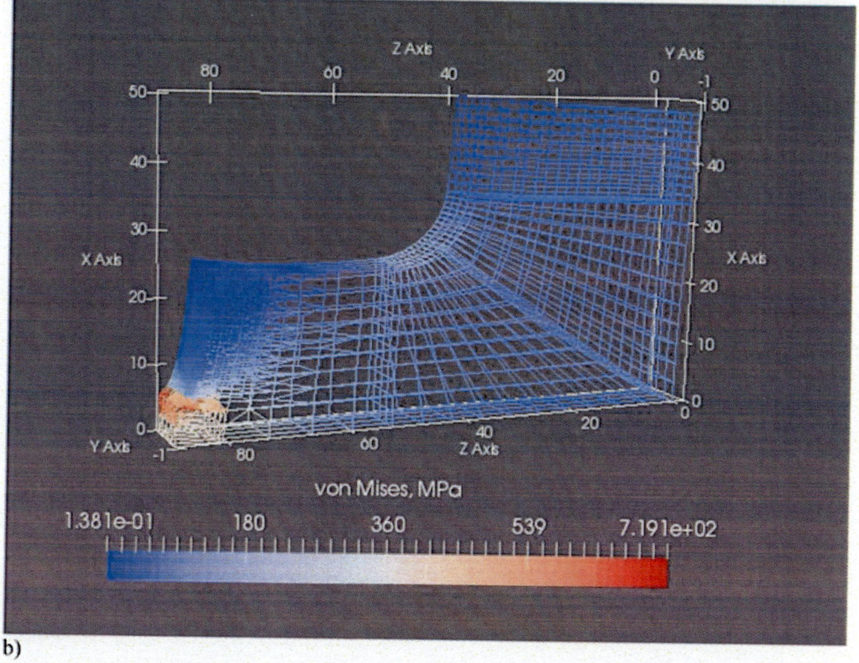

b)

Fig. 14. Stress distribution for a = 20 mm: **a** 2D view, **b** 3D view

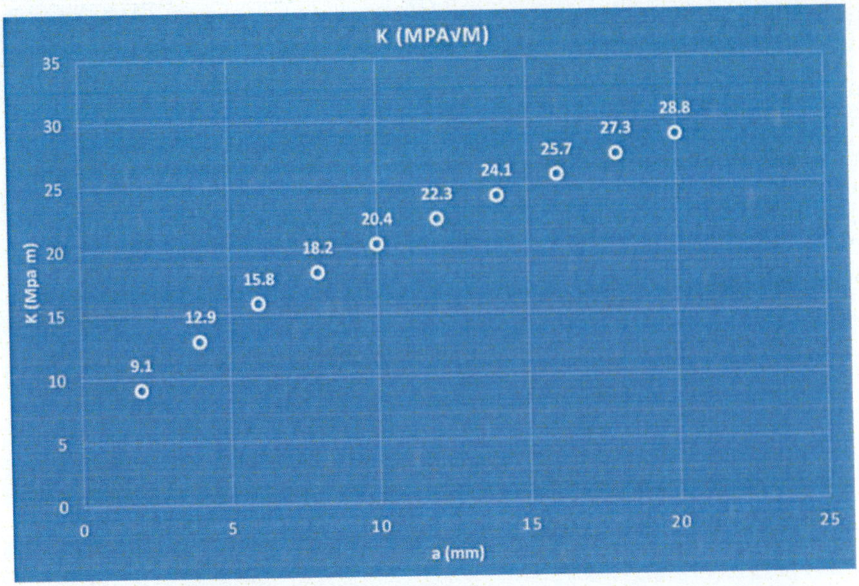

Fig. 15. Stress intensity factor K versus crack length a

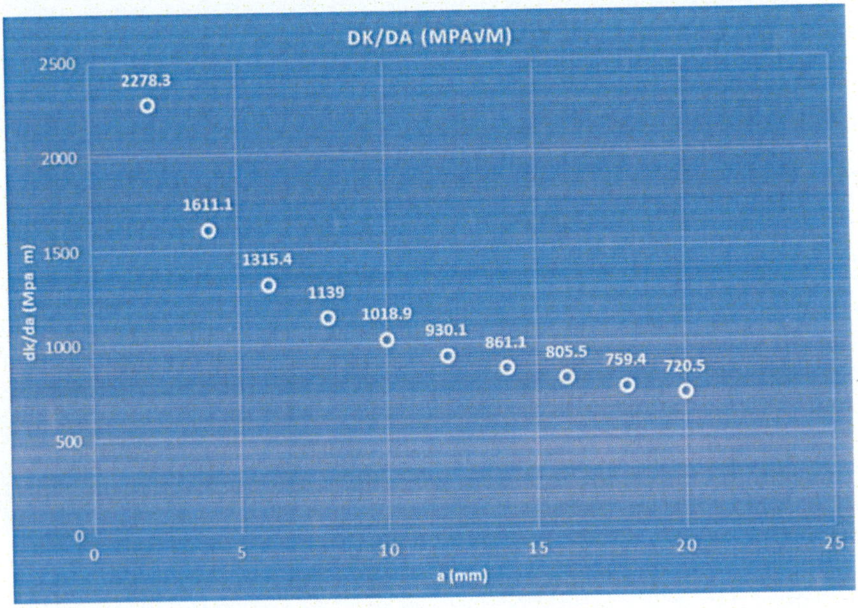

Fig. 16. Stress intensity crack rate dK/da versus crack length a

Table 1. Data for K, dK/da and max stress vs. crack length a

a (mm)	K (MPa$\sqrt{m}$)	dK/da (MPa/$\sqrt{m}$)
2	9.1	2278.3
4	12.9	1611.1
6	15.8	1315.4
8	18.2	1139.0
10	20.4	1018.9
12	22.3	930.1
14	24.1	861.1
16	25.7	805.5
18	27.3	759.4
20	28.8	720.5

References

1. Aly, O.F., Neto, M.M.: Stress Corrosion Cracking. Chapter in a book, INTECH, pp. 65–79 (2014). http://dx.doi.org/0.5772/57349
2. Djukic, M., Bakic, G., Sijacki-Zeravcic, V., Sedmak, A., Rajicic, B.: The synergistic action and interplay of hydrogen embrittlement mechanisms in steels and iron: localized plasticity and decohesion. Eng. Fract. Mech. **216**, 106528 (2019)
3. Djukic, M.B., Bakic, G.M., Zeravcic, V.S., Sedmak, A., Rajicic, B.: Hydrogen embrittlement of industrial components: prediction, prevention, and models. Corrosion **72**(7), 943–961 (2016)
4. Djukic, M.B., Bakic, G.M., Zeravcic, V.S., Rajicic, B., Sedmak, A., Mitrovic, R., Miskovic, Z.: Towards a unified and practical industrial model for prediction of hydrogen embrittlement and damage in steels. Procedia Struct. Integr. **2**, 604–611 (2016)
5. Hirose, Y., Mura, T.: Growth mechanism of stress corrosion cracking in high strength steel. Eng. Fract. Mech. **19**, 1057–1067 (1984). https://doi.org/10.1016/0013-7944(84)90151-6
6. Bland, L.G., Locke, J.S.: Chemical and electrochemical conditions within stress corrosion and corrosion fatigue cracks. npj Mater. Degrad. **1**, 12 (2017). https://doi.org/10.1038/s41529-017-0015-0
7. Turnbull, A., Ferriss, D.: Mathematical modelling of the electrochemistry in corrosion fatigue cracks in structural steel cathodically protected in sea water. Corros. Sci. **26**, 601–628 (1986). https://doi.org/10.1016/0010-938x(86)90027-2
8. Turnbull, A.: Modelling of crack chemistry in sensitized stainless steel in boiling water reactor environments. Corros. Sci. **39**, 789–805 (1997). https://doi.org/10.1016/s0010-938x(97)89342-0
9. Turnbull, A.: Modeling of the chemistry and electrochemistry in cracks a review. Corrosion **57**, 175–189 (2001). https://doi.org/10.5006/1.3290342
10. Mohanty, S., Majumdar, S., Natesan, K.: A review of stress corrosion cracking/fatigue modeling, Argonne national laboratory (2012). https://www.energy.gov/sites/prod/files/Environmental_Fatigue.pdf
11. Al Kateb, M.: Experimental and numerical investigation of corrosion crack growth in mild structural steel, doctoral thesis, University of Belgrade (2021)
12. Alkateb, M., Tadić, S., Sedmak, A., Ivanović, I., Marković, S.: Crack growth rate analysis of stress corrosion cracking. Tech. Gaz. **28**(1), 240–247 (2021)
13. Sedmak, A.: Computational fracture mechanics: an overview from early efforts to recent achievements. Fatigue Fract. Eng. Mater. Struct. **41**, 2438–2474 (2018). https://doi.org/10.1111/ffe.12912

14. Belytschko, T., Lu, Y.Y., Gu, L.: Element-free Galerkin methods. Int. J. Numer. Methods Eng. **37**(2), 229–256 (1994)
15. Jovičić, G., Živković, M., Sedmak, A., Jovičić, N., Milovanović, D.: Improvement of algorithm for numerical crack modeling. Arch. Civ. Mech. Eng. **10**(3), 19–35 (2010)
16. Zaidi, R., Sedmak, A., Kirin, S., Grbovic, A., Li, W., Lazic Vulicevic, L., Sarkocevic, Z.: Risk assessment of oil drilling rig welded pipe based on structural integrity and life estimation. Eng. Fail. Anal. **112**, 104508 (2020)
17. Milovanović, N., Sedmak, A., Arsic, M., Sedmak, S.A., Božić, Z.: Structural integrity and life assessment of rotating equipment. Eng. Fail. Anal. **113**, 104561 (2020)
18. Kraedegh, A., Li, W., Sedmak, A., Grbovic, A., Trišović, N., Kirin, S.: Simulation of fatigue crack growth in A2024–T351 "T" welded joint. Struct. Integr. Life **17**(1), 3–6 (2017)
19. Sghayer, A., Grbović, A., Sedmak, A., Dinulović, M., Doncheva, E., Petrovski, B.: Fatigue life analysis of the integral skin-stringer panel using XFEM. Struct. Integr. Life **17**(1), 7–10 (2017)
20. Durdevic, A., Zivojinovic, D., Grbovic, A., Sedmak, A., Rakin, M., Dascau, H., Kirin, S.: Numerical simulation of fatigue crack propagation in friction stir welded joint made of Al 2024–T351 alloy. Eng. Fail. Anal. **58**, 477–484 (2015)
21. Parkins, R.N., Greenwell, B.S.: The interface between corrosion fatigue and stress-corrosion cracking. Metal. Sci. **11**, 405–413 (1997). https://doi.org/10.1179/msc.1977.11.8-9.405

Start-Up Community and the Acceleration Services in the Danube Macro-Region: Cases of Austria, Bosnia and Herzegovina, Hungary and Slovenia

Bojan Ćudić, Matjaž Klemenčič, and Miloš Milošević

Abstract This paper investigates the supply (access to finance) and demand (innovation-driven SMEs and talent communities) perspectives in the Danube macro-region (DMR). The purpose of this the empirical analysis is to identify relevant challenges for all participants of the quadruple-helix model in the DMR and create recommendations for upgrading the business (support) infrastructure and venture finance in the observed region, with special focus on the acceleration services and the new available digital services. Conclusions are based on case studies of regional assessments of Austria (Styria), Bosnia and Herzegovina ((BIH) Republika Srpska (RS)), Hungary (Central Region (CR)), and Slovenia (Western Slovenia (WS) and Eastern Slovenia (ES)). The empirical analysis is based on the interviews conducted within the Accelerator project. The project consortium consisted of 15 partners representing 9 countries – 7 EU countries: Hungary, Romania, Slovenia, Bulgaria, Czech Republic, Austria and Croatia, and 2 Western Balkan countries: Serbia and BIH. The empirical analysis showed that innovative SMEs in all participating countries of the Accelerator project were hindered by access to equity, particularly in their early-stage development.

Keywords Start-ups · Access to finance · Danube macro-region

1 Introdcution

The area covered by the empirical analysis (the Danube macro-region (DMR)) stretches from Germany to Romania–Ukraine–Moldova and is home to 115 million people. It is a unique and diverse area encompassing nine EU member countries (Austria, Germany, Czech Republic, Slovakia, Slovenia, Croatia, Hungary, Romania

B. Ćudić (✉) · M. Klemenčič
University of Maribor, Koroška cesta, 2000 Maribor, Slovenia
e-mail: bojan.cudic@um.si

M. Milošević
Faculty of Mechanical Engineering, Department of Information Technologies, University of Belgrade, 11000 Belgrade, Serbia

N. Mitrovic et al. (eds.), *Current Problems in Experimental and Computational Engineering*, Lecture Notes in Networks and Systems 323,
https://doi.org/10.1007/978-3-030-86009-7_15

and Bulgaria), three potential member countries (BIH, Montenegro and Serbia), and two neighboring countries (Moldova, and South-Western and Southern regions of Ukraine). Jointly, these countries form a heterogeneous macro-region with some high and low performers in research and innovation (R&I) [8]. The economic situation in these countries is challenging, especially in those with political unrest (BIH and Ukraine, in particular). Through an improved economic situation as a result of the accelerators' work, these countries will be able to tackle the political issues more successfully. In this empirical analysis, the authors limit the regions studied to the countries of Austria, BIH, Hungary and Slovenia, or their micro-regions.

Austria is, of course, a country where stable political and economic systems prevailed from the end of World War II onwards; Hungary is a country, which, historically speaking, was a "real socialist" country with a rigid socialist/communist system until 1991. Slovenia and BIH were part of Yugoslavia, which had its own "self-management" socialist system in which the economic situation was mixed, but the country had its own history of dissolution after 1991, with a short war in Slovenia and prolonged war conflicts in BIH.

Accelerators are a rapidly growing new form of business support organizations whose aim is to support entrepreneurs through intense educational programs, including mentoring and networking for start-ups, in order to improve their ability to attract investment at the end of the program. Accelerators fill gaps to allow inexperienced entrepreneurs to start up quickly [27]. Accelerators are a unique organizational form; their structure and operational process is relatively basic and requires "lean" managerial personnel and resources [34], pp. 1–3.

Start-up accelerators support early-stage growth-driven companies for a fixed period of time, and as part of a cohort of companies. The acceleration experience is a process of intense, rapid and immersive education aimed at accelerating the life cycle of young innovative companies and entrepreneurs, compressing years' worth of learning into a couple of months [15, pp. 1–5, 26].

In general, accelerators seem to be a positive addition to start-up economic systems in specific countries and worldwide, as they tend to meaningfully improve the odds for success of the start-ups that they support [19].

In their study, Sheryl [33] suggest that accelerator-backed start-ups receive the first round of follow-up financing significantly sooner, are more likely either to be acquired or to exit, are founded by entrepreneurs from a renowned set of universities, and exhibit substantially greater founder mobility than non-accelerator-backed start-ups [33, pp. 2–25]. Moreover, authors found a strong cohort effect in accelerators that confers signaling, networking and advisory benefits to portfolio firms and influences the likelihood of entrepreneurial exit by quitting and acquisition [33, pp. 26–27].

The European start-ups economic system is still struggling to catch up with its United States (US) counterpart. Hathaway [15, pp. 1–5]. There are many reasons for this. Briefly, there is more money available in the US, more collective knowledge, and it also has more talented start-ups. Therefore, start-ups have a better chance of succeeding [1, 16, 20].

Nowadays, there are many accelerators and venture capital funds (VCFs) in the European Union (EU) that are trying to replicate the success of their peers in the US [13, 30].

Given the growing start-up environment and the birth of many start-up accelerators in the European regions as a result of EU funding, the significance of accelerators has also spread across the DMR. The main channel for the transfer of knowledge within the macro-region in this field is EU projects. Funds are available for the region because of the EU's specific interest in this region's well-being. This is mainly due to its history of conflicts. Through these funds, they wish to prevent future escalation of conflicts in the region [25].

The DMR lags behind both the US and Western Europe regarding entrepreneurship, and it is still looking for the right recipe to replicate successful accelerator models from Western Europe.

The development of the acceleration programs in most of the DMR countries is driven by the EU grants that support accelerators as an important tool for achieving goals set in the EU strategies for regional development [4].

EU regional policy and related funds are dedicated to job creation, competitiveness, economic growth, improved quality of life and sustainable development. In order to achieve these goals, the EU invested almost a third of the total EU budget in the entire policy of smart and inclusive **growth** [9].

Acceleration programs are supported through EU centralized programs, such as Horizon 2020 and the EU program for the Competitiveness of Enterprises and Small and Medium-Sized Enterprises—COSME) [8]. These programs are also available to the non-EU members of DMR countries, which are included in this empirical analysis. It would be useful for DMR countries to follow the suit of the EU and contribute substantial parts of their budgets to these kinds of accelerators.

The EU provides support programs for innovative start-ups and scale-ups through different schemes, but the most comprehensive one is the European Innovation Council (EIC) [5]. The EIC's enhanced pilot stage funding supports fast company growth and market-creating innovation. It also facilitates the scaling-up of innovative companies by providing business acceleration services [5].

Given that accelerators are oriented towards innovative start-ups and scale-ups with international ambitions, and that this model of business support infrastructure has proved to be one of the best tools for achieving the goals of the EU regional development policy, DMR countries should invest in this type of program to keep promising and talented entrepreneurs in their regions [23]. But also, the whole process of upgrading the business support services in the region has to be based on a systematic approach [17]. Due to these assumptions, this empirical analysis investigates the supply (access to finance) and demand (innovation-driven SMEs and talent communities) aspects in the DMR.

The purpose of this empirical analysis is to identify relevant challenges for all participants of the quadruple-helix model in the DMR and create recommendations for upgrading the business (support) infrastructure and venture finance in the observed region, with special focus on the acceleration services and the new available digital services (e.g., digital innovation hubs (DIHs)).

2 Methodology

The empirical analysis is based on the interviews conducted within the Accelerator project. The project consortium consisted of 15 partners representing 9 countries—7 EU countries: Hungary, Romania, Slovenia, Bulgaria, Czech Republic, Austria and Croatia, and 2 Western Balkan countries: Serbia and BIH.

The Accelerator project and the empirical analysis are supported by the EU Danube Transnational Programme (DTP). This is a financing instrument of the European Territorial Cooperation (ETC), better known as Interreg. ETC is one of the goals of the EU cohesion policy and provides a framework for the implementation of joint actions and policy exchanges between national, regional and local actors from different member states. DTP promotes economic, social, and territorial cohesion in the Danube Region through policy integration in selected fields. The project partners conducted the empirical analysis in their countries.

The empirical analysis showed that innovative SMEs in all participating countries of the Accelerator project were hindered by access to equity, particularly in their early-stage development. Thus, Accelerator engaged in exploring and piloting the innovative path of acceleration programmes (a new type of investment readiness programme) towards improved business support in the Danube region. Accelerator's main objective was to enhance access to innovation finance through improving the institutional framework conditions and related policy instruments by developing the practical solution of acceleration services and influencing the concerned strategic framework at partner regions and programme level. The primary target groups were SMEs in need of capital, but which lack the skills to acquire it. Secondly, project partners target business support organisations with whom new and improved acceleration services are bedded and these organisations are linked through a Danube-region transnational network.

Their approach had a strong transnational dimension: beyond the transnational network, regions with well-performing innovation systems and partners with successful acceleration schemes assist weaker regions or partners with limited experience. Accelerator results in introduced acceleration programmes with less experienced partners and improved accelerator programmes with experienced partners. The partnership actively integrated public bodies to channel joint policy recommendations on facilitating the spread of acceleration services and worked out a joint strategy on the promotion of acceleration services and on their integration in EU measures in the Danube region.

3 Profiles of the DMR and the Observed Countries/Regions

An overview of the target countries is presented in Table 1, which compares economic country data and other relevant data for economic growth. The authors suggest that the annual percentage of the population growth is one of the indicators of the overall

Table 1. Profiles of the observed countries in 2018

	Austria	BIH	BIH	
World view				
Population growth (annual %)	0.6	−0.8	−0.2	0
Surface area (sq. km) (thousands)	83.9	51.2	93	20.7
Population density (people per sq. km of land area)	107.2	64.9	107.9	102.6
Economy				
GDP (current US$) (billions)	455.74	19.78	155.7	54.24
GDP growth (annual %)	2.7	3.1	4.9	4.5
GDP per capita (current US$)	51,496	5,958	15,937	26,203
Inflation, GDP deflator (annual %)	1.6	1.4	4.5	2.3
Agriculture, forestry, and fishing, value added (% of GDP)	1	6	4	2
Industry (including construction), value added (% of GDP)	25	24	26	29
Exports of goods and services (% of GDP)	55	41	87	85
Imports of goods and services (% of GDP)	51	57	82	76
States and markets				
Time required to start a business (in days)	21	81	7	8
Domestic credit provided by the financial sector (% of GDP)	123.9	62	55.4	64.2
Tax revenue (% of GDP)	*25.4*	*20.3*	*23.2*	*18.4*
Military expenditure (% of GDP)	0.7	1.1	1.1	1
Mobile cellular subscriptions (per 100 people)	*170.8*	*98.1*	*113.5*	*117.5*
Individuals using the Internet (% of population)	*87.9*	*69.5*	*76.8*	*78.9*
Net migration (thousands)	*100*	*−3*	*30*	*6*
Foreign direct investment, net inflows (BoP, current US$) (millions)	11,246	485	−75,179	1,514

Source World Development Indicators database, 2018
Figures in italics refer to periods other than those specified

economic situation in a specific country of the DMR.[1] It is composed of two segments: the percentage of the growth of residents and the number of net migrations. If we focus on the observed countries in 2018, we can see that Austria had an annual population growth of 0.6%, Slovenia around 0%, while Hungary and BIH had negative rates. Today, the decline in population numbers is one of the biggest challenges among

[1] There are other examples in other countries. The economies of Japan and Germany went into recovery around the time their populations began to decline (2003–2006). However, in 2015 Germany accepted over a million refugees. Both the total and per capita GDP in both countries grew more rapidly after 2005 than before. But both countries fell into the global recession of 2008–2009.

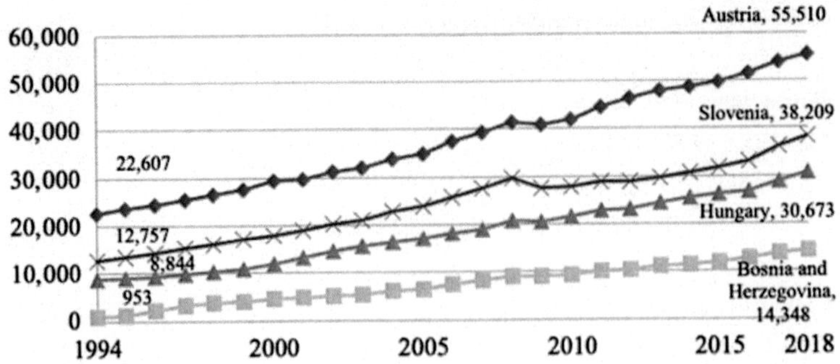

Fig. 1 GDP per capita in observed countries from 1994 to 2018 (PPP (current international $))
Source World Bank, 2019

many countries of Central-Eastern and South-Eastern Europe.[2] The decline is caused by both the negative natural population growth (more deaths than births) and the negative net migration rate.

The authors suggest that the best way to compare economic situations in the observed countries is to present the data about GDP per capita.[3] Bearing in mind that Slovenia, BIH and Hungary were socialist countries, they were more or less unprepared for the modern competitive market and the free market economy. On the other hand, Austrian companies were adjusted to the competitive economy, and they provided a higher level of value-added and better standard of living for their workforce. With a GDP per capita of $22,607 in 1994, Austria had a significantly higher standard of living than Slovenia and Hungary ($12,757 Slovenia and $8,844 Hungary). However, in 1994 BIH had an incredibly low level of GDP per capita of just $953 due to the armed conflicts between 1992 and 1995 [3, pp. 23–25].

After 24 years, in 2018, the ranking of the countries remains the same, even though BIH increased its GDP per capita by 15 times, Hungary by 3.4 times, Slovenia by 3 times and Austria by 2.5 times. These figures show that all the presented countries made a great effort in their economic development in the last two and half decades. Figure 1 represents GDP per capita disparities among the observed countries. GDP per capita is based on purchasing power parity (PPP) [32].

The EU has had a strong interest in the issue of regional disparities since its establishment. One of the most important aims of the EU is to ensure economic and social cohesion among member states and within them. The EU also takes care of the candidate countries, which have access to many EU funds under the same conditions as the EU-28 states. The availability of Structural Funds and the Cohesion Fund aimed at achieving the convergence goal has created new impetus for regional policy.

---

[2] On the other hand, official policies of the observed countries fight against the immigration of refugees.

[3] EUROSTAT measures disparity by the sum of absolute differences between regional and national GDP per capita, weighted by the share of population and expressed in percent of national GDP.

One of the best available tools for comparing business conditions between regions in the EU is the Regional Innovation Scoreboard (RIS). This is a regional extension of the European innovation scoreboard, assessing the innovation performance of European regions on a limited number of indicators. The RIS 2019 covers 238 regions across most of the EU countries, Norway, Serbia and Switzerland.[4]

Table 2 shows the RIS in 2019 compared to target regions. Presented data cover regions that are slightly different to those defined in this research, but they are certainly indicative.

The data for Austria are presented as a model for the peer regions. Also, they show a difference in many aspects of innovation across the regions of the country.

In the contexts of target regions, Table 2 shows the dominant position of South Austria for most of the indicators, but also the relative strengths of other observed regions. The WS and Budapest region show relative strengths in, for example, Population with tertiary education, while presenting weaknesses, for example, in SMEs innovating in-house and Product or process innovators. The RS also shows differences in the specific aspects of innovation, where the well-educated population is one of the strengths, but also there are many weaknesses in this region, for example in the R&D expenditure of the business sector and Patent Cooperation Treaty (PCT) patent applications.

More detailed information about the observed countries/regions is presented in the following text.

From 1867 until 1918 the Habsburg Empire was a multinational state. It was one of Europe's major powers at the time, the second-largest country in Europe and the third most populous one. This was a period of economic boom in the area of the northern part of today's Austria, which is still felt today. The Empire dissolved, partly united with other regions, into: the Republic of Austria; the Kingdom of Serb, Croats and Slovenes; Hungary; Romania; Czechoslovakia; and Poland.

Today, **Austria** is a Central European country comprised of nine federated provinces, covering an area of 83,879 km2, and with a population of nearly nine million people.

The country has a high standard of living and in 2018 was ranked as 20th in the world for its Human Development Index. In the same period, Slovenia was 25th, Hungary held 45th position, and BIH performed poorly – it was ranked 77th [31, pp. 22–25].

In 2018, *Forbes* magazine placed Austria as the 22nd country in the world for doing business.[5] Austria has the strongest financial sector among the observed countries. Styria (Austrian term: *Steiermark*) is the second largest of the nine provinces (a.i. *Bundesländer*) in Austria and located in the southeast of the country. In 2017, according to Eurostat, its total population was 1.24 million.

[4] The RIS 2019 covers: Austria, Belgium, Bulgaria, Croatia, Czechia, Denmark, Finland, France, Germany, Greece, Hungary, Ireland, Italy, Lithuania, Netherlands, Norway, Poland, Portugal, Romania, Serbia, Slovakia, Slovenia, Spain, Sweden, Switzerland and the United Kingdom.

[5] In the same research, conducted by *Forbes*, Slovenia was placed as the 31st country, Hungary 40th, and BIH was ranked as the 98th country of the 161 countries included.

Table 2. Regional innovation scoreboard 2019

	Population with tertiary education	Scientific co-publications	R&D expenditure business sector	Product or process innovators	SMEs innovating in-house	Innovative SMEs collaborating with others	PCT patent applications
East Austria	0.551	0.775	0.663	0.630	0.594	0.681	0.426
West Austria	0.422	0.575	0.760	0.697	0.663	0.856	0.626
South Austria	0.404	0.629	1	0.58	0.528	0.711	0.545
Republika Srpska e	0.210 e	0.323 e	0.147 e	0.340 e	0.347 e	0.280 e	0.200 e
Budapest region	0.666	0.569	0.595	0.297	0.272	0.277	0.288
East Slovenia	0.51	0.317	0.607	0.34	0.329	0.361	0.548
West Slovenia	0.648	0.826	0.654	0.394	0.374	0.459	0.206

Note e—estimation
Source European Commission and authors' estimation for RS [6, 7, 11]

Styria is one of the strongest Austrian regions in terms of R&D, as it launches the most innovative products and services compared to its observed peers. This is mostly due to Styria's high-powered educational system, research community, business environment and its residents. Moreover, the transfer of knowledge and technology is the region's core competence, as most of Austria's competence centers are located in Styria, where university spin-offs are quite common and independent research institutions cooperate regularly with businesses [9], [10]. Also, Austria has deep regional disparities, as the Austrian north is much more developed than the Austrian south.

BIH is a South-Eastern European country, located within the Balkan Peninsula. The country has had a rich history, and especially recently was affected by wars, which had a strong negative impact on its economy.

The country went from *Banovina* of Bosnia in the twelfth century to the Kingdom of Bosnia in the fourteenth, then succumbed to the rule of the Ottoman Empire, from the mid-fifteenth to the nineteenth century, followed by the annexation by the Austro-Hungarian Monarchy in 1908. Between the two World Wars, BIH was part of the Kingdom of Yugoslavia, and after World War II it was one of the federal republics of the newly formed Yugoslavia. It proclaimed its independence in 1992, followed by the armed conflicts that lasted until the end of 1995. After the war, the country was divided into the Federation of Bosnia and Herzegovina ((FBIH) consisting of 10 cantons), RS, and the *Brčko* District, with a very complex political and administrative system.

During the wars of the 1990s, the economy suffered enormous material damage, followed by the dual challenge of rebuilding a war-devastated country and introducing transitional market reforms to its formerly mixed economy.

Today, BIH is a transitional economy with limited market reforms. The economy is dominated by the processing industry and agriculture sectors, followed by the trade and service sectors, although nowadays more focus is being placed on the innovative start-ups and scale-ups. It relies heavily on the export of metals, energy, textiles and furniture as well as on remittances from the diaspora and foreign aid.[6]

The biggest problem that the country is facing today is the emigration of its inhabitants and inflow of refugees.

RS is a political and territorial entity within BIH. It covers an area of 24,858 km2 (49% of BIH territory) and has 1.4 million inhabitants.

---

[6] Remittances from the diaspora have played an important role in the BIH economy. The official estimates show that about 2 million people from BIH and their children live in approximately 50 countries around the world. Those members of the diaspora are often perceived as "money senders," given that remittances account for 14% of the country's GDP. According to data from the Central Bank BIH, the diaspora sent $1.42 billion to the BIH in 2017, and in the period from 1998 annually it sent an average of slightly more than $1.5 billion to BIH. Although the potential of the diaspora is high, and could have a significant role in the development, poverty reduction and economic growth in BIH, the interaction of the BIH authorities with the diaspora is sporadic and unstructured, and they are rarely included or consulted in the design of policies or decision-making processes. The money sent to relatives is mostly used for personal consumption.

BIH's constitution stipulates that all functions and authorizations that are not explicitly designated to the institutions of BIH must belong to the entities, and this also includes directions of market economy. Bearing in mind the above regulations, and the fact that SMEs account for more than 99% of all business entities in the RS, authorities regard the SMEs' development as relying solely on the entities within BIH. SME development within the RS is designated to the authority of the Ministry of Economy and Entrepreneurship of RS. RS has adopted all the necessary laws and strategies, established institutions and secured financial support for the regional economic growth. But still, there are many obstacles in the system of support for SMEs, where RS lags significantly behind its observed peers.

Hungary is an East-Central European country, covering 93,030 km2, with about 10 million inhabitants.

By the twelfth century, historical Hungary became a regional power, which included most of today's neighbors from south and east, reaching its cultural and political height in the fifteenth century. As part of the Austro–Hungarian Empire, it became a major European power. After World War I, the Hungarian part was limited to its major ethnic territory. Many Hungarians were left to live as national minorities in neighboring countries. From 1949 until 1989 the country was a socialist republic, with a socialist model of economy.

In the 1990s, Hungary transitioned from a centrally planned socialist economy to a market-driven one, with a per capita income approximately two thirds of the EU-28 average. In recent years, the government has implemented economic policies to boost household consumption and has used EU-funded development projects to generate growth.

Hungary is a high-income economy and has the world's 58th largest economy by PPP. Similarly to BIH and other Balkan countries, systemic economic challenges include corruption, labor shortages driven by demographic declines and emigration, widespread poverty in rural areas, vulnerabilities to changes in demand for exports, and a heavy reliance on Russian energy imports.

The Central Region (CR) (Hungarian term: *Közép-Magyarország*) is one of the seven statistical regions in Hungary (NUTS-1 and NUTS-2). The CR has an area of 6,919 km2 and a population of 2,993,948 inhabitants. It includes Budapest (the capital of the nation) and the Pest County. Budapest is the economic, commercial, financial, administrative and cultural center of Hungary. Its economic, social, institutional, educational and R&D-related performance indicators are far above the national average. With a high concentration of research capacities in the capital, its innovation performance is outstanding among all Hungarian regions, albeit meager in comparison to other capital regions in Europe.

According to the Small Business Act Fact Sheet 2018, Hungary's performance remains below the EU average in skills and innovation, and its attempts to catch up have brought only moderate results.

Slovenia is an East-Central European country; it covers 20,273 km2 and has a population of two million.

Historically, Slovene ethnic territory was divided among many territorial units, and also among Austria, Hungary and Italy after 1918. At that time, the largest part of Slovene ethnic territory became part of the Kingdom of Serbs, Croats and Slovenes. The Yugoslav Slovenia was enlarged by one third after World War II, and it remained part of Yugoslavia until 1991. After 1945, Yugoslavia had a socialist economy, which later developed into an "indirectly controlled" market economy, with a "self-management" approach at its core. This system at the beginning achieved impressive results in rates of economic growth, where (from 1948 until 1979) the average annual GDP growth rate was 6.2% and growth of the industrial production index was at a high level (slightly below 8%). It was followed by more than ten years of decay from the mid 1970s onwards.

In the period of the transition from socialism to a free enterprise society from 1990 onwards, Slovenia accepted the model of "shock therapy," based on complete liberalization, stabilization and privatization. Results of implementing this model were similar to those in other socialist economies—not so positive.

In 1991, after the introduction of a multi-party representative democracy, Slovenia became the first republic to split from Yugoslavia, thus becoming an independent state. Expectations related to the perfection of market self-regulation that followed were disappointing: the transition from socialism to a free-market economy has led to the redistribution of social wealth and income without development. Although in the latter years the free enterprise society in Slovenia led to industrial and social growth, this was, especially in the beginning, a difficult period for all the former socialist countries.[7]

In 2004, Slovenia entered the EU and NATO. In 2007 it became the first former communist country to join the Eurozone, and in 2010 Slovenia joined the Organisation for Economic Co-operation and Development (OECD). Slovene economy benefits from a well-educated workforce, a well-developed infrastructure, and its location at the crossroads of major trade routes. The strategic importance of Slovenia in Central and Eastern Europe (CEE) is also recognized today by many global leaders. Thus, Chinese Premier Li Keqiang said in April of 2019 that "*China is willing to better align the Belt and Road Initiative with the development strategy of Slovenia*" [12]. Other global forces are also interested in this former Yugoslav state.

Slovenia's main industries are motor vehicles, electric and electronic equipment, machinery, pharmaceuticals and fuels. There is a big difference in prosperity among the country's macro-regions.

WS (Slovenian word: *Zahodna Slovenija*) is the base of a strong R&D sector, which is located mostly in Ljubljana. The regional groups are Central Slovenia, Upper Carniola, Gorica, and Coastal–Karst statistical regions. The economically wealthiest regions are Central Slovenia (which includes the capital Ljubljana) and

[7] The "self-management" system, which was present in former Yugoslavia, in its true sense, nevertheless provided a number of advantages over a modest Marxist–Leninist–Stalinist-type of economy (which Hungary had). The system allowed competition between similar enterprises, and it allowed equally for worker and manager innovation and the practice of free-market-type relations. The self-management system allowed the republics of former Yugoslavia to move more easily to the market model of business. A key problem after the initial decades of intense development is that socialism has fallen into the trap of economic inefficiency of the self-governing enterprise and the tendency of employees to favor personal consumption over accumulation and investment in development.

western Slovenian regions (Goriška and Coastal–Karst), while the least developed regions are Mura, Central Sava and Littoral–Inner Carniola.

ES (Slovenian word: *Vzhodna Slovenija*) is the least developed of the two observed regions of Slovenia and its GDP per capita is slightly above two thirds of the EU-28 average. It comprises the Drava, Koroška, Savinja, Central Sava, Lower Sava, Southeast Slovenia and Inner Carniola–Karst statistical regions.

4 Innovation-Driven SMEs and Talent Communities in the DMR (Demand Side)

The economies of the DMR rely heavily on SMEs. However, they are still not able to fully exploit the SMEs' potential due to structural insufficiencies, lack of internationalization, and internal disparities related to the research and innovation (R&I) field and the intensity of the transfer of knowledge from the scientific to the business community (industry).

According to the Global Innovation Index (GII) for 2019, Austria (as the best among the observed countries) scored 50.94 points (21st position), followed by Slovenia with 45.25 (31st position), Hungary with 44.51 (33rd position) and, lagging behind, was BIH with a score of 31.41 (76th position) [2, pp. 35–38].

There are also big differences among these countries in terms of indicators for starting a business, where interestingly the best ranked is Slovenia at 42nd with a score of 91.42, making it the most start-up-oriented community among the observed countries (unfortunately, the rate of closure of start-ups is very high), followed by Hungary in 62nd place with a score of 87.27 and Austria in 85th with 83.72. However, BIH is seriously lagging behind most countries in the world, as it is 122nd (with a score of 65.09) out of 127 countries (Cornell University et al. 2019, pp. 218–346).

If we compare numbers of researchers per million of population in the target region in 2017, convincingly the best score goes to Austria (59.96), placing it at 11th in the world, Slovenia was ranked 24th with a score of 46.20, followed by Hungary with 31.01 (33rd place), while BIH has 3.84 and ranked 69th.

This also corresponds to the ranking of the countries based on their gross expenditure on R&D. In 2017, Austria allocated 3.10% of its GDP to R&D, while Slovenia allocated 2.21%, Hungary 1.39% and BIH only 0.22%.

The same seems to apply to the ICT use index (in 2018), where we can see that Austria ranked 29th with a score of 74.7, Slovenia 43rd with 65.7, Hungary 48th with 63.6 and BIH 74th with 48.1.

An indicator of the vibrancy of the innovation-driven SMEs and talent community would be the number of international patent applications filed by residents at the PCT (per billion PPP US $ GDP).[8] In general, this indicator has been very low across the DMR, where the countries of the region, on average, only managed to secure 2.08 patents (in 2017), which is more than 40% less than the EU average of 3.70 patents,

[8] Patent application per billion GDP measured in terms of purchasing power.

again with a significant range between Austria with 3.42 patents, Slovenia with 1.04 patents, Hungary with 0.67, and finally BIH with just 0.26.

In research conducted in 2018, companies from the DMR identified not only access to finance as relevant but also many other challenges (e.g., availability of skilled staff or experienced managers, finding customers, regulation, etc.).[9] According to the same survey, founders of the companies in the EU have mostly relied on private capital (family, friends or similar) with a share of more than 83%; a similar situation exists in the non-EU DMR countries, with a share of 80%.

5 Access to Finance and Start-Up Support in the DMR (Supply Side)

Globally, access to finance is constantly one of the main challenges in SME development, with a lack of both adequate supportive organizations and knowledge or awareness of available sources for financing for SMEs and entrepreneurs. The situation is the same in the DMR, where the geographical position of the specific macro-region plays an important role in the availability of funds. Financial resources tend to be more accessible in cities, while other (more rural) regions often lack adequate support, usually due to the absence of supporting organizations and investors. Furthermore, start-up companies are preferably looking to base their premises or key services in better developed areas, leaving underdeveloped regions deprived of potential growth [28, pp. 15–35, 29] unless there are good transporting connections within regions.

According to the empirical analysis conducted by the University of Applied Sciences in Graz through the Accelerator project of the business support economic system of Austria (Styria), there are many tools in this region that support start-ups and scale-ups (FH JOANNEUM 2017, pp. 2–18).

Austria and Styria as a region offer a comprehensive system of public funding and private programs. The most important funding sources in Austria are the Austrian Research Promotion Agency (FFG) and the Austrian Wirtschafts service GmbH (AWS). They offer grants, guarantees or subsidized loans, from pre-seed and seed funding to business angels' support.

There are many public initiatives for start-ups launched in order to increase the number and quality of start-ups in Austria. Thanks to these initiatives, Austria has emerged as a start-up hub, especially in the fields of information technology, media and life sciences as well as creative industries. *Forbes* selected Austria as one of the seven start-up hotspots in Europe.

---

[9] In 2008, the European Central Bank and DG Enterprises and Industry of the European Commission established the Survey on the Access to Finance of Enterprises (SAFE). These surveys, conducted across the EU member states and several additional countries (including Western Balkan countries), were held in June–July 2009, in August–October 2011, in August–October 2013, and in September–October in every year in the period 2014–2018. The most recent wave covers 36 countries: the EU-28 member states as well as Iceland, Turkey, Montenegro, Albania, Serbia, North Macedonia and BIH.

However, private domestic venture capital companies such as Speed invest and business angels assist start-ups by providing financing and investing millions of euros in capital. But the most important difficulties that came to light in the process are, on the one hand, a too small risk capital scene in Austria and, on the other, the low interest of international equity funds for Austrian companies.

Because of the problem on the supply side in Austria, including in Styria, the Styrian government established an investors' service desk. The objective is to bring headquarters and centers of competence (technology and R&D centers from international corporations) to Styria. In addition, the team deals with company and R&D headquarters that are already based in Styria and supports existing companies in developing headquarters or centers of competence.

One of the activities is innovation prizes on a national level, called "*Innovation awards*," and there are regional awards too—for example, in Styria the Fast Forward Award.

Also, there are intellectual property (IP) rights, awareness-raising and coaching activities for start-ups and SMEs where the services are built around the IP hub located at the patent office, which provides comprehensive support services on IP-related issues.[10]

Elsewhere, the Federal Ministry of Digital and Economic Affairs set up a blockchain strategy, along with a virtual test village called "*Kettenbruck*," which was put in place to test blockchain technology applications for public services.

Therefore, with the existing system of education, quality workforce, business infrastructure, available sources of funding, and legal framework, Styria is one of the leaders in providing a favorable business environment for the development of start-ups and scale-ups both in the DMR and in the EU.

5.1 Case Study: Accelerator Programs University Spin-Off Centers

The Science Park Graz (SPG) is the incubator center of the University of Graz and the Technical University of Graz. The Center for Applied Techniques (ZAT) is the incubator of the University of Leoben.

Backed by a network of experts, both incubators support university graduates from all fields (pre- and postgraduates as well as research assistants) by providing professional counseling and coaching, infrastructure and financing during the pre-start-up period.

The SPG and ZAT mentoring programs gather the expertise of Styrian university institutions and successful entrepreneurs in order to equip innovative start-up

[10] The AWS is an important stakeholder in IP issues for all of Austria and has a range of measures to help companies analyze and ascertain their IP. It works closely with the Austrian patent office in this regard.

founders with valuable practical experience. The SPG and ZAT provide a wide-ranging offer both to start-up founders as well as to experienced managers wishing to share their knowledge with start-up teams. Furthermore, investors also have the opportunity to benefit from growth-oriented companies and their innovative products and services.

SMEs in **BIH** are primarily supported by development agencies at the regional and local level, but also chambers of commerce, incubators, clusters, etc., financed by different donors from local authorities, EU organizations, bilateral programs, etc. Based on reports, the number of programs for young entrepreneurs is expanding in BIH, from university programs through nongovernmental organizations and government initiatives to professional incubators and accelerators and EU programs (EC 2018b). The geographical scope is limited to larger towns.

Given that the country is divided into two entities—economic, and political and administrative entities—support to SMEs has regional character. The institution that coordinates support to SMEs in the RS entity is the Republic Agency for the Development of SMEs (RARS). RARS is authorized to provide professional services of support for the establishment, management and development of start-ups and scale-ups. It also offers professional services in order to encourage investments in SMEs, support the establishment of entrepreneurial infrastructure, innovator activity, the creation of new products and the introduction of new technologies. In the other entity, the Federation of BIH, RARS' work is done by several development agencies, which target the cantonal level.

In all strategic documents, the RS government has particularly emphasized the strong need for horizontal and vertical communication within the existing development network, with RARS acting as a communication channel among businesses and information, providing experiences and best practices' exchange, and encouraging a proactive approach in the implementation of strategies and development plans [21, 22].

Though some companies in BIH complain about access to finance, many have reasonably good access, at least to bank loans, compared with the rest of the countries in the region. In financing, their working capital, and in fixed asset purchases, companies rely predominantly on bank borrowing and credit purchases.

The legal framework for access to finance in the RS has been developed and funds for SMEs are available through regular commercial banks' credit lines, the RS Investment Development Bank credit lines, credit and guarantee lines of the RS Guarantee Fund and micro-credit and leasing organizations.

Currently, equity financing is not developed in BIH, as there are no registered venture capital or seed capital funds due to the lack of adequate laws that regulate the activities of VCFs, tax regulation that does not recognize these kinds of investments, limited exit opportunities, the market being too small, etc.

Individual talented start-ups and scale-ups are directed to the VCFs from other countries and cross-border investments.

The most successful examples of the equity financing concept in BIH were implemented through cooperation with organizations from more developed countries.

5.2 *Case Study: Cross-Border Cooperation: The Challenge to Change Program*

The Challenge to Change (C2C) program in BIH has been established thanks to the help of the Swedish International Development Cooperation Agency (SIDA) and the Embassy of Sweden in BIH. It was implemented by local business support organizations. An additional project partner was the Region Östergötland from Sweden. The main goal of the project was to strengthen economic development in BIH, and cooperation between Sweden and BIH, where the focus was on the innovative projects of the local start-ups and scale-ups.

As there are many innovative business ideas in BIH that have not been implemented due to a lack of risk-sharing capital, SIDA has identified the need to develop new instruments supporting innovation and business growth and therefore support the establishment of the C2C Fund in BIH. The total budget of the C2C was €4 million. The C2C was open to SMEs from all over BIH (RS and FBIH) and from Sweden, as well as start-ups that have innovative business ideas, products or services that may lead to increased employment, increased competitiveness and sustainable socioeconomic development in BIH. Companies that met the established criteria were offered a co-funding grant of up to a maximum €30,000 (and 50% of the investment).

This case shows that cross-border cooperation in the sense of investments, as well as the transfer of knowledge, is very important for the local start-ups and scale-ups in the regions with a lack of equity financing, although the first such attempt was done with very low funding.

Velmainex Ltd. is an example of a company that received a grant from this program. Over the course of a full three decades of development, the small family company from the north part of BIH has developed a business model that is based on high-quality branded products.

Being aware that business models of textile industries in most of the EU, US and Japan companies (countries with the highest level of value added in this sector of business) are based on the R&I, the company follows these trends.

The concept of the company's development is based on the identification of new technologies in the area of textile processing, and the added value concept. This way of doing business enables the company's management to offer better working conditions for their employees—first of all financially—and in return the employees produce higher quality work and are more committed to the company.

The aim of the approved project was to produce an innovative quilt and pillow of top standardized quality based on the use of nanotechnologies. The product is intended for export (markets of neighboring EU countries, Austria, Switzerland, Italy and Germany).

5.3 *Case Study: The Innovation Center Banja Luka*

The Innovation Center Banja Luka (ICBL), established in 2010, is the first combined modern equipped center for support and development of entrepreneurship in the RS. Its purpose is to support the development of private companies with knowledge-based products, services, business processes and applications of innovative and advanced technologies. Today, ICBL is quite well accepted by the private sector and as a result new private ventures/ initiative are being developed.

ICBL was founded by the Ministry of Science and Technology of RS, the City of Banja Luka, RARS, the University of Banja Luka and the University of East Sarajevo, with support from the Ministry of Foreign Affairs of the Kingdom of Norway.

ICBL offers all the necessary support elements for the development of early-stage business ideas or growing businesses through incubation, coupled with a wide range of professional consulting, training and educational services. The only resource that ICBL needs to become an accelerator is equity financing for its tenants, for which ICBL might use EU funds.

Similarly, as in other EU countries, **Hungarian** SMEs were financed by private investments via a network of friends and family, but nowadays recognize increasingly the role of acceleration and equity financing.

Although still in development, Hungary is one of the leading countries in VCF finance in the EU due to recent very large investments provided by the state. It has a long-standing Venture Capital Association, established in the 1990s, although it has been active only in the past 10 years. VC and equity transactions appeared in the 2000s, driven by the support of the European Regional Development Fund (ERDF) in 2007–2013. In the beginning, services and support were only concentrated in Budapest.

The rise of accelerator programs in Hungary is partly due to national measures and partly due to market players recognizing the need for them. The aim of the central programs is to prepare the companies for equity financing. The Hungarian equity financing environment is supported by both public and private investors. Although earlier there were regional and mainly state-financed agencies to deal with innovation-driven SMEs, today there is a shift towards encouraging market players to carry out acceleration tasks by providing them with financial support.

Hungary has an extensive list of accelerator programs, typically offering successful services to support the start-up economic system. It is important to note that the majority of these programs are relatively new and strongly dependent on EU funds. In terms of other supportive services (mentorship and initiatives), these would need further improvements to offer adequate expert advice and support to SMEs. The most prominent accelerators are: Kitchen Budapest, Colabs and Telenor Accelerate.

Key players in the Hungarian innovation economic system are the Hungarian Venture Capital Association and the National Trade House, which play a significant role in the start-up community. Hungary's case suggests the value of learning from different types of accelerators: private—with strong finances, competences and network; public—more locally oriented, and which often do not ask for shares (equity) but just (repayment) of the initial investment (as a loan); and university—which is strong in expertise and provides options such as royalty fees of sales.

The country implemented several new initiatives and programs in the past years, where early-stage companies can access funding as well as additional services, training, coaching and networking events, and can learn more about different opportunities [24].

Through the 'Supplier Development' program, larger companies can help their SME suppliers develop and produce high value-added products while adopting Industry 4.0 technologies.

Boosting the digitalization of the competitive companies' program is designed to help SMEs use new ICT technologies by providing them with high-level digital support for product development, design, manufacturing and assembly, construction, trade and service processes.

5.4 Best Cases: ICatapult

iCatapult is a unique accelerator and business development company, which focuses on taking Central European technologies to the global market. The main idea of this accelerator is that many non-US start-ups have the potential of going global from day one, they just need the boost to avoid cultural, language and financial barriers and the lack of a cohesive network.

This accelerator works with the early-stage start-ups with a strong technical background. The focus is on global web, mobile, or Internet of Things (IoT) technologies that have achieved some form of validation—preferably by a working prototype and traction. The main request for founders is that they must be open-minded and willing to pivot according to new discoveries from the market. They invest between $10 K to $25 K, with an ownership participation of 10% to 15% of equity.

Since 2008, **Slovenia** has been on an upward track on skills and innovation, having implemented policy measures that address most of the Small Business Act recommendations in this area. These include R&D infrastructures like university incubators and technology parks, as well as grants, innovation vouchers, tax incentives and supportive coaching services.

Between 2010 and 2015, competence centers for human resources development were established in 19 industry sectors, contributing to the improvement of workforce skills. Up to 2017 another 11 competence centers were established where 240 companies employing over 35,800 employees are now involved. Additionally, nine strategic development and innovation partnerships were established (in 2016) by 400 companies and 100 knowledge institutions.

Innovation processes in SMEs have benefited from a new strategic approach set out in the smart specialization strategy, followed by action plans (in 2017) for the implementation of common strategic goals in the fields of innovation, human resources development and internationalization [14].

With regard to access to finance, Slovenia has reached the EU-28 average, as it made significant progress due to actions by the Slovene government to enable better access to public financial support, including a guarantee for credits and making banks more willing to provide business loans.

Slovenia currently lacks legislation for new forms of equity financing, particularly crowdfunding. According to data collected by the Crowdfunding community, Slovenians collected nearly €1.8 m in 2016.

The Business Angels Network of Slovenia was established in 2010. In 2019, this association connected the most ambitious entrepreneurs with qualified investors in Slovenia, but also in the CEE region and beyond. There are many successful examples.

In the field of equity financing, Slovenia made an important step forward by adopting a law on venture capital funds (VCFs) in 2007. Furthermore, in 2010 it prepared a tender where it acted as a co-investor and invested 49% together with newly founded VCFs. From the original eight funds, four are still operational.

Young start-up companies can also get seed capital. The Slovenian entrepreneurship fund prepared two instruments that together with the financial investment also offer the expert help that start-ups need in their beginnings. Thus, the Slovene Enterprise Fund created an initiative called "Twin" (*dvojček*), which together with the financial investment also offers consulting support that includes: ensuring the active participation of at least ten private investors; connection with mentors; a three-month content program; expert support of entrepreneurial consultants; and the preparation of multimedia reports. Most of the companies that received consulting support through "Twin" were overwhelmingly satisfied. Companies also wrote their opinions about the incentive, and from the comments it is possible to see the wish of participating companies to form more homogeneous communities where they can share knowledge and experiences.

5.5 *Case Study: Accelerators in Slovenia*

There are three acceleration programs for start-ups in Slovenia, one private (ABC accelerator) and two public ones from the Slovene Enterprise Fund and Initiative Startup Slovenia (SK75 and SK200).

ABC Accelerator was established in April 2015 with the purpose of connecting innovative start-up companies with the market of the whole of South-Eastern Europe. In exchange for 8% equity, the start-up gets access to the infrastructure of BTC City and other partners, €15,000 cash and an intense 3-month educational program. The conditions for entering the accelerator are a good team and a product that is over 90% developed and will be, after the finishing touches, prepared for the market. Along with the ABC Accelerator they opened an extension called ABC HUB, which represents the connection with the mentor and investor network of the ABC Accelerator, as well as flexible and fixed co-working spaces with conference rooms and a lecture hall. They also opened a unit in Germany—a springboard for the German and other

European markets—and a unit in Silicon Valley in San Jose, CA that represents an entry point to the North-American market.

Startup GeekHouse Acceleration Program is designed for innovative start-up companies with a potential for global growth that wish to find their product-market fit as soon as possible. It ensures €75,000 of capital in the form of a convertible loan, an intense educational program, start-up mentors and individual consultants. Within the program, teams can benefit from free co-working facilities in seven different locations across Slovenia, specialized for start-up companies. The accelerator program also helps start-ups to achieve media visibility and promotion at home as well as abroad.

SK200—Go-Global Acceleration Program is designed for start-ups that have found product-market fit and are ready for fast growth. ***Start-ups can get €200,000 of capital in*** the form of an equity investment, which allows companies to start expanding to foreign markets, further develop products and overcome the need for financing until the next investment round of VCFs. Beneficiaries receive ***the investment preparation program***, where they learn the rules of equity financing, prepare an application form and meet potential investors; a ***personal start-up mentor*** with experience in expanding a company to foreign markets to help the team with strategic decisions, international contacts and expansion to individual markets; ***comprehensive administrative support*** to deal with equity investment-related administrative tasks; a***ccess to a network of investors*** as well as VCFs and accelerators at home and abroad; and support for ***promotion*** in the form of media visibility, networking, marketing and copywriting.

6 Future Outlook, Conclusions and Recommendations

The countries in this empirical analysis represent examples of different success levels and development stages of business economic systems, where Austria represents an overall best business economic system practice model, Slovenia a model for successful transition from a socialist to a free market system, Hungary an EU member country struggling to catch up with its Central European peers, and BIH a country standing at the crossroads of the past and future, facing the challenge of high emigration. There are also significant disparities among the different macro- and micro-regions within the observed countries.

This empirical analysis is focused on start-ups and scale-ups on one (demand) side and the available equity funds on the other (supply) side in the DMR. In this regard, according to the analysis conducted within the Accelerator project, the DMR countries and regions show a wide range of the level of development of their SME economic systems.

As the innovation-driven enterprises have the potential to grow significantly larger than ordinary SMEs and have a substantial job creation potential, the main challenge on the demand side seems to be the identification of the front-runners within the large group of SMEs seeking support.

On the supply side, while some regions have good availability of financial resources and well-established access to public and private investors' support (mainly Austria and WS), others are lagging behind their peers, as they have only recently started establishing accelerators, equity financial models and cross-border investment by the VCFs. Limited access to venture capital finance, across the DMR, is mainly caused by the overall economic conditions (bureaucracy, tax regulation, small market size, lack of attractiveness for international investors), lack of interactions between entrepreneurs and investors, and a low level of awareness of challenges on both sides.

Investors are often faced with lack of investment readiness coupled with a lack of managerial and sales skills of entrepreneurs. This is particularly prevalent in the less developed regions (mainly in RS, but also in ES), where investments are perceived as being significantly more risky, making investors less willing to provide seed and venture capital to support innovative ideas.

Currently, SMEs from the region prefer debt financing to equity financing (7%), and most of them (61%) prefer bank loans. SMEs in the observed region are more confident in their ability to secure loans from banks (70%) than securing funds from equity investors and VCs (19%), therefore it is highly likely that SMEs in the DMR will keep using bank loans as the dominant source of funding. Interestingly, 36% of the surveyed SMEs do not expect significant changes in the business environment and access to finance in the upcoming period. However, it is promising that alternative sources of financing are recognized by SMEs as an important driver for enhanced access to finance in the region (e.g., crowdfunding, business angels, EU grants, etc.) [18].

On the other hand, SMEs expressed the need for more technical support, ranging from administration, legal and human resources, training, mentoring, coaching and investment-readiness programs, and networking events where they could meet with potential investors. The link between investors and entrepreneurs surfaced as one of the key challenges across the DMR, leading to the conclusion that accelerators might be the most appropriate tools for addressing this issue.

The leading regions in the DMR (Styria and WS) invest a lot in the modern business support infrastructure, where digital innovation hubs (DIHs) are recognized as upgraded versions of accelerators that provide support to start-ups independently of their geographical location, with technical universities or research organizations at their core. DIHs act as one-stop shops where companies, especially SMEs, start-ups and mid-caps, can get access to technology-testing, financing, advice, market intelligence and networking opportunities.

Among the many elements that have an impact on the country/region's development (e.g., the public funds (state or regional), R&D&I, available business infrastructure, the effort that is put into internationalization, the effort that is put into linking large companies with start-ups and SMEs, the effort that is made to facilitate the start-up or spin-off of businesses), the authors suggest that the number of scientific and technical journal articles and results/findings (per billion PPP US$ GDP) and their level of application in the industry are significant indicators to predict the future growth and development of the country/region alike.

Presented data suggest that countries and regions that invest in R&D will be economic leaders in the future. Therefore, Austria and Styria will remain some of the best places to live in Europe. Also, it is expected that WS will improve its position, because of its significant investments in R&D. It is also expected that (CR) Hungary and ES regions will keep their current positions, or improve in many segments (e.g., business infrastructure, cooperation with VCFs, education, etc.), while BIH will most likely remain the most challenged country in the region. Of particular concern for BIH is the low level of investment in education, infrastructure and R&D&I as well as the rapid emigration of young talented professionals. Consequently, it is necessary to direct attention towards investments in education, R&D&I capacity and support for establishing and developing companies with a global perspective (including direct financial support, investment-readiness and financial literacy programs) and following examples from more developed neighboring countries.

However, all the countries of the DMR have the problem of a low level of application of scholarly and technical articles and results/findings in the industry. As a result, there is a low level of awareness about the need for cooperation between academia, industry, government and civil society, and the quadruple-helix model of innovation. Based on good practice, examples from the leaders in the DMR (e.g., Austria, Slovenia) and the overall assessment of the region, lead to the following **recommendations**. The region needs to:

establish more accelerators as hubs (especially the DIHs) for continuous collaboration of all actors involved in the innovation economic system;
establish and strengthen the support specifically intended for early-stage innovative start-up companies and provide services that would help entrepreneurs to create new projects and further increase the confidence of investors;
facilitate access to global markets and networks, in order to attract more international investors and help SMEs expand internationally;
enhance the cooperation of the diverse stakeholders (policymakers, SMEs, startups, universities and research institutes and civil society) through co-creative activities following the quadruple-helix approach.

To promote cooperation and networking, it is necessary to strengthen the functioning of the joint initiatives, such as virtual platforms, organizing joint workshops, exchange events and preparation of the comprehensive reviews of the startup economic system of the whole Danube area. These actions would support and contribute to the activation of the entrepreneurial and innovative potential of the DMR. Improvements will also need to be made in the availability of alternative equity financing for SMEs, especially crowdfunding and business angels. Moreover, greater collaboration among the DMR countries, with knowledge and best practice exchange, could benefit all the regions in further developing services that could contribute to the development of the start-up economic system and establish a supportive environment for the faster growth of SMEs.

Acknowledgements The article is created within the programme group P6-0372, and it is supported by Slovenian Research Agency (ARRS) – No. 5442-1/2018/89.

References

1. Aslund, A., Djankov, S.: *Europe's growth challenge.*Oxford University Press, Oxford (2017).
2. Cornell University, INSEAD, WIPO: *The Global Innovation Index 2019: energizing the world with innovation.* Ithaca, Fontainebleau, and Geneva (2019)
3. Cutts, M.: The humanitarian operation in Bosnia, 1992–95: dilemmas of negotiating humanitarian access. UNHCR, pp 23–25. https://www.unhcr.org/research/working/3ae6a0c58/humanitarian-operation-bosnia-1992-95-dilemmas-negotiating-humanitarian.html Accessed 4 August 2019 (1999).
4. Dileo, I.: The Eu macro-regions: approaches and patterns. Olsztyn Econ. J. **10**(3), 209–221 (2015)
5. EASME – Executive Agency for Small and Medium-Sized Enterprises: Business acceleration support and coaching services (2019). https://ec.europa.eu/easme/en/eic-sme-instrument/business-acceleration-support-and-coaching-services Accessed 5 Sept 2019
6. EC – European Commission: EU regional and urban development (2019). https://ec.europa.eu/regional_policy/en/policy/what/investment-policy Accessed 1 Oct 2019
7. EC – European Commission: Regional innovation monitor plus (2019). https://ec.europa.eu/growth/tools-databases/regional-innovation-monitor/base-profile/styria Accessed 27 Aug 2019
8. EC – European Commission: Smart specialisation platform: Danube region (2018). https://s3platform.jrc.ec.europa.eu/danube-macroregion-activities Accessed 16 August 2019
9. EC – European Commission: Ecosystem portfolio start-up: Central and Eastern Europe Network. MY-GATEWAY project (2018). https://mygatewayproject.eu/ Accessed 10 Aug 2019
10. EC – European Commission: COSME: Europe's programme for small and medium-sized enterprises (2019). https://ec.europa.eu/growth/smes/cosme_en Accessed 8 Aug 2019
11. FH JOANNEUM – University of Applied Sciences: Analysis of the relevant policy framework of Austria Styria; demand and supply analysis of Austria – regional assessment of Styria. Report. Graz, pp 2–18 (2017)
12. Global Times: China ready to better align BRI with Slovenia's development strategy: Li. Source: Xinhua, Published: 2019/4/12. (2019). http://www.globaltimes.cn/content/1145653.shtml Accessed 27 Sept 2019
13. Grilli, L., Mrkajic, B., Latifi, G.: Venture capital in Europe: social capital, formal institutions and mediation effects. Small Bus. Econ. 393–410 (2018). https://doi.org/10.1007/s11187-018-0007-7
14. Haarich, S.: Mapping of EU Member States / Regions: Research and Innovation Plans & Strategies for Smart Specialization RIS3. European Commission: Case Study Report Danube region (2017).
15. Hathaway, I.: What start-ups accelerators really do. Harvard Business Review (2016). https://hbr.org/2016/03/what-start-up-accelerators-really-do Accessed 1 Oct 2019
16. Henrekson, M., Sanandaji, T.: Stock option taxation: a missing piece in European innovation policy? Small Bus. Econ. **51**(2), 411–424 (2018). https://doi.org/10.1007/s11187-018-0008-6
17. Jadwiga, G.: The impact of business support organizations on the innovative activity of Lubusz industrial enterprises. Management **22**(1), 85–100 (2013)
18. Kwaak, T. et al.: Survey on the access to finance of enterprises SAFE. European Commission (2018)
19. Madaleno, M., Nathan, M., Overman, H., Waights, S.: Incubators, Accelerators and Regional Economic Development. London School of Economics and Political Science, London (2018)
20. Novikova, J.: From research to market: what the EU can learn from the USA? Institute of European Studies, Bruxelles (2019)
21. RARS – Republic Agency for the Development of Small and Medium Enterprises of Republic of Srpska: Accelerator, analysis of the relevant policy framework of the Republic of Srpska. Banja Luka, pp. 6–10 (2017)
22. RARS – Republic Agency for the Development of Small and Medium Enterprises of Republic of Srpska: Accelerator, demand and supply analysis: regional assessment of the Republic of Srpska. Report. Banja Luka, pp. 4–19 (2017)

23. Roig-Tierno, N., Alcázar, J., Ribeiro-Navarrete, S.: Use of infrastructures to support innovative entrepreneurship and business growth. J. Bus. Res. **68**(11), 2290–2294 (2015). https://doi.org/10.1016/j.jbusres.2015.06.013
24. SZTA – Széchenyi Venture Capital Fund Management Ltd.: Accelerator pilot programme scheme. Report. Budapest, pp. 3–15 (2018).
25. Serwatka, A.: Accelerators for start-ups in Europe. Copernic. J. Financ. Account. **7**(1), 67–81 (2018)
26. Shu, Y., Romi, K., Thomas, S.: Where do accelerators fit in the venture creation pipeline? Different values brought by different types of accelerators. Entrep. Res. J. **8**(4), 1–13 (2018)
27. Stayton, J., Mangematin, V.: Seed accelerators and the speed of new venture creation. J. Technol. Transf. **4**, 1163–1187 (2019). https://doi.org/10.1007/s10961-017-9646-0
28. TPLJ – Technology Park Ljubljana: Accelerator, analysis of the relevant policy framework of Slovenia: regional assessment of East Slovenia. Report. Ljubljana, pp. 10–15 (2017)
29. TPLJ – Technology Park Ljubljana: Accelerator, demand and supply analysis of Danubian region: regional assessment of East Slovenia. Report. Ljubljana, pp 15–35 (2017)
30. Tykvová, T.: Venture capital and private equity financing: an overview of recent literature and an agenda for future research. J. Bus. Econ. **88**(3), 325–362 (2018). https://doi.org/10.1007/s11573-017-0874-4
31. UNDP – United Nations Development Programme: Human development indices and indicators. United Nations Development Programme, New York (2018). http://hdr.undp.org/sites/default/files/2018_human_development_statistical_update.pdf Accessed 10 Aug 2019
32. WB – World Bank BIH fact sheet, innovation and entrepreneurship assessment. World Bank Report 2019, pp 15–34 (2019)
33. Winston, Smith, S., Hannigan, TJ., Gasiorowski, L.: Accelerators and crowd-funding: complementarity, competition, or convergence in the earliest stages of financing new ventures? Boulder: University of Colorado-Kauffman Foundation, pp 1–20 (2013)
34. Wright, M., Drori, I.: Accelerators: successful venture creation and growth. Edward Elgar Pub, Cheltenham pp 1–3 (2018)

Fatigue Life Evaluation of the Damaged Passenger Boarding Bridge Supports

Martina Balac, Aleksandar Grbovic, Gordana Kastratovic, Aleksandar Petrovic, and Lajos Sarvas

Abstract Bridge constructions are developing rapidly as they are necessity in the transport network. The Passenger Boarding Bridge (PBB) provides passengers with means to embark and disembark a ship through weather protected tunnels. It is obvious that the endurance and safety of such structures are of paramount importance. It has been shown that forces and moments at three-tunnel boarding bridge supports were high in the critical cases, as well as stresses in pins and lugs. The aim of this research is to analyze crack propagation in the damaged supporting part of the PBB. After determination of critical components to demonstrate how dangerous the appearance of the crack could be and to evaluate the fatigue life of the damaged supporting part, numerical simulations based on the improved finite element method (FEM) were carried out. To do this, the determination of actual loads was conducted with special consideration. The predicted number of cycles to complete failure was low, indicating the fact that special attention has to be paid on the design of PBB's critical supporting components.

Keywords Passenger Boarding Bridge · Fatigue life · Finite element method

1 Introduction

The PBB or Passenger Loading Bridge (PLB) provides passengers with means to embark and disembark a ship through weather protected tunnels. The PLB presents unique challenge in terms of design and construction. The main parts of the PLB can be seen in Fig. 1.

M. Balac (✉) · A. Grbovic
Faculty of Mechanical Engineering, University of Belgrade, 11000 Belgrade, Serbia
e-mail: mbalac@mas.bg.ac.rs

G. Kastratovic
Faculty of Transport and Traffic Engineering, University of Belgrade, 11000 Belgrade, Serbia

A. Petrovic · L. Sarvas
JT2 Batajnicki drum 261a, 11000 Belgrade, Serbia

N. Mitrovic et al. (eds.), *Current Problems in Experimental and Computational Engineering*, Lecture Notes in Networks and Systems 323,
https://doi.org/10.1007/978-3-030-86009-7_16

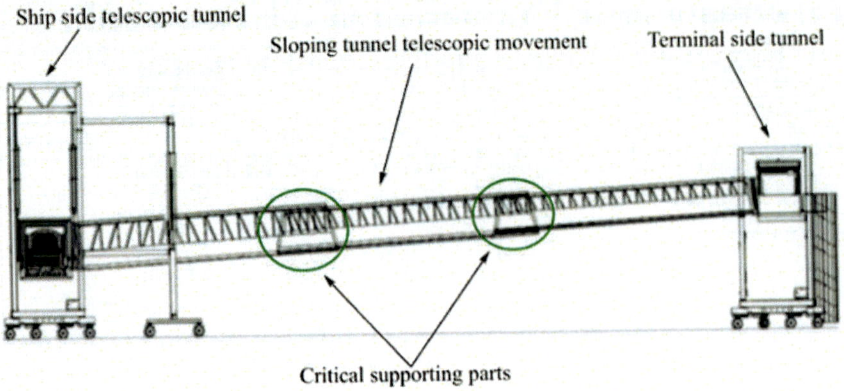

Fig. 1. The Seawing PLB.

2 Determination of Loads and Structural Analysis

The PBB shall support the following static loads:

a. Dead load (that is inertial load of every tunnel, T1 = 14,400 kg, T2 = 16,400 kg, T1 = 17,500 kg).
b. Live floor load of 100 psf (equivalent to 488.24 kg/m^2).
c. Roof load of 20 psf (equivalent to 97.65 kg/m^2).

Live floor load is an extremely high load per square meter, since the maximum number of standing people per square meter is five (see Fig. 2), while the "normal" number of people is two to three (see Fig. 3). To better visualize the standing density in people per square meter, Fig. 4 shows the 3D crowd visualizer of 2 people per square meter (source: https://www.gkstill.com/Support/crowd-density/CrowdDensity-1.html). If we take 5 people per square meter the load of 488.24 kg/m^2 suggests that people are standing and not moving. Usually, the load is calculated by taking the average mass of 80 kg (+ 20 kg of luggage), assuming that 3 persons per square meter is maximum walking density (that is 300 kg/m^2). Anyhow, the load of 488.24 kg/m^2 was used in the analysis. The similar can be said for the roof load.

3 2D Finite Element Analysis of the PBB

Figure 5 shows the overall dimensions of the PBB and was used for the evaluation of distributed load, as shown in the equations below. Two support (points A and F) represent pinned supports (translation fixed, rotation free), while at points B, C, D, and E horizontal rollers (vertical translation fixed) are assumed. This is a very rough approximation of reality because points B, C, D, and E are not connected to the ground (but, at the same time, cannot move freely because tunnels serve as supports),

Fig. 2. Five people per meter square.

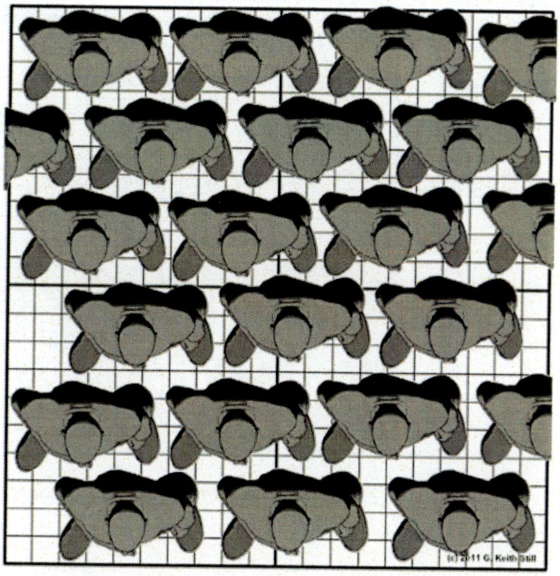

Fig. 3. Three people per meter square.

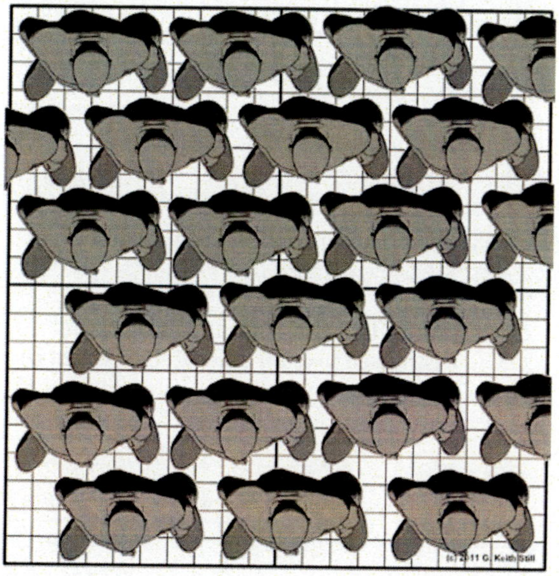

and only in the 3D model of the beam (as shown in Fig. 6) mutual influence of tunnels 1, 2 and 3 can be considered. B-C and D-E represent overlaps.

The following areas of the floors and roofs are given:

Tunnel T1 – floor 40.4 m^2, roof 43.5 m^2

Tunnel T2 – floor 30.8 m^2, roof 48.6 m^2

Fig. 4. The 3D crowd visualizer of 2 people per square meter.

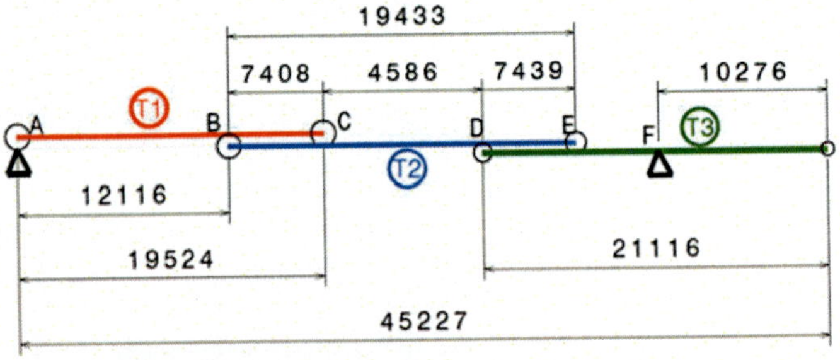

Fig. 5. The overall dimensions of the PBB in [mm].

Tunnel T3 – floor 43.5 m^2, roof 59.4 m^2
Distributed load for each beam segment is:

$$beam_{AB} = \left(\frac{14400kg \times 9.81m/s^2}{19.524m}\right)_{\text{T1 wght}} + \left(\frac{40.4m^2 \times 488.24 \times 9.81N/m^2}{19.524\text{m}}\right)_{\text{floor load}} + \left(\frac{43.5m^2 \times 97.65 \times 9.81N/m^2}{19.524m}\right)_{\text{roof load}} = 19,280.7\ \text{N/m} \tag{1}$$

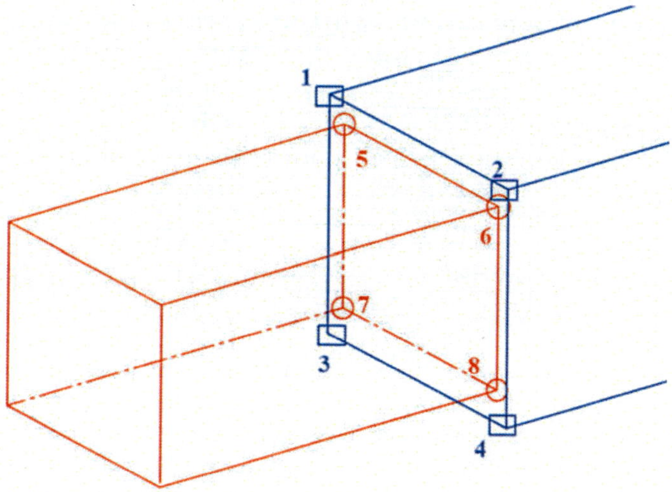

Fig. 6. Red beam serves as support for blue at 1 and 2, while blue supports red at 7 and 8.

$$beam_{BC} = \left(\frac{14400kg \times 9.81m/s^2}{19.524m}\right)_{T1\ wght} + \left(\frac{16400kg \times 9.81m/s^2}{19.433m}\right)_{T2\ wght} + \left(\frac{30.8m^2 \times 488.24 \times 9.81N/m^2}{19.433m}\right)_{floor\ load} + \left(\frac{48.6m^2 \times 97.65 \times 9.81N/m^2}{19.433m}\right)_{roof\ load} = 25501.32\ N/m \quad (2)$$

$$beam_{CD} = \left(\frac{16400kg \times 9.81m/s^2}{19.433m}\right)_{T2\ wght} + \left(\frac{30.8m^2 \times 488.24 \times 9.81N/m^2}{19.433m}\right)_{floor\ load} + \left(\frac{48.6m^2 \times 97.65 \times 9.81N/m^2}{19.433m}\right)_{roof\ load} = 18,265.9\ N/m \quad (3)$$

$$beam_{DE} = \left(\frac{16400kg \times 9.81m/s^2}{19.433m}\right)_{T2\ wght} + \left(\frac{17500kg \times 9.81m/s^2}{21.116m}\right)_{T3\ wght} + \left(\frac{43.5m^2 \times 488.24 \times 9.81N/m^2}{21.116m}\right)_{floor\ load}$$

$$+\left(\frac{59.4m^2 \times 97.65 \times 9.81N/m^2}{21.116m}\right)_{roof\ load} = 28,970.6\ N/m \quad (4)$$

$$\boldsymbol{beam_{EFG}} = \left(\frac{17500kg \times 9.81m/s^2}{21.116m}\right)_{T3\ wght} + \left(\frac{43.5m^2 \times 488.24 \times 9.81N/m^2}{21.116m}\right)_{floor\ load} + \left(\frac{59.4m^2 \times 97.65 \times 9.81N/m^2}{21.116m}\right)_{roof\ load} = 20,691.7\ N/m \quad (5)$$

The model with applied forces and constraints is shown below (Fig. 7).

Table 1 gives all of the spans in the model. In order to carry out FEM analysis arbitrary material and cross-section of the 2D beam have been defined since the goal was to estimate bending moments and shear forces at supports (Tables 2, 3).

Based on presented data, the Free Body Diagram (FBD) and deformed mesh are obtained and shown below (Fig. 8).

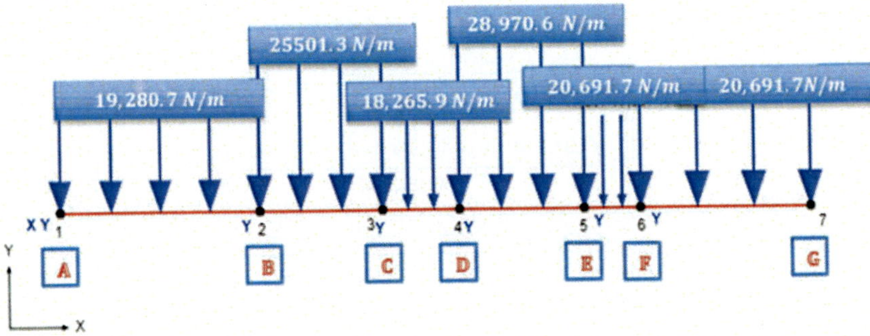

Fig. 7. 2D beam with distributed load and restricted displacements

Table 1. Spans in the model

ID	Point 1	Point 2	Material	Cross Section
1	2:(12.1163 m, 0 m)	3: (19.5293 m, 0 in)	ASTM A572, Grade 50 (Plate, <= 4.0 in)	I-Beam, 0.786 m x 0.384 m
2	1: (0 m, 0 m)	2: (12.1163 m, 0 m)	ASTM A572, Grade 50 (Plate, <= 4.0 in)	I-Beam, 0.786 m x 0.384 m
3	3: (19.5293 m, 0 m)	4: (24.1106 m, 0 in)	ASTM A572, Grade 50 (Plate, <= 4.0 in)	I-Beam, 0.786 m x 0.384 m
4	4: (24.1106 m, 0 m)	5: (31.5546 m, 0 m)	ASTM A572, Grade 50 (Plate, <= 4.0 in)	I-Beam, 0.786 m x 0.384 m
5	5: (31.5546 m, 0 m)	6: (34.9509 m, 0 m)	ASTM A572, Grade 50 (Plate, <= 4.0 in)	I-Beam, 0.786 m x 0.384 m
6	6: (34.9509 m, 0 m)	7: (45.22748 m, 0 m)	ASTM A572, Grade 50 (Plate, <= 4.0 in)	I-Beam, 0.786 m x 0.384 m

Table 2. The properties of each material in the model

Material Name	Yield Strength [MPa]	Ultimate Strength [MPa]	Elastic Modulus [GPa]	Poisson's Ratio
ASTM A572, Grade 50 (Plate, <= 4.0 in)	344.7	448.2	206.8	0.3000

Table 3. The properties of cross material in the model

Description	Area [cm^2]	Moment of Inertia [cm^4]
I-Beam, 0.786 m x 0.384 m	397.8	424,031

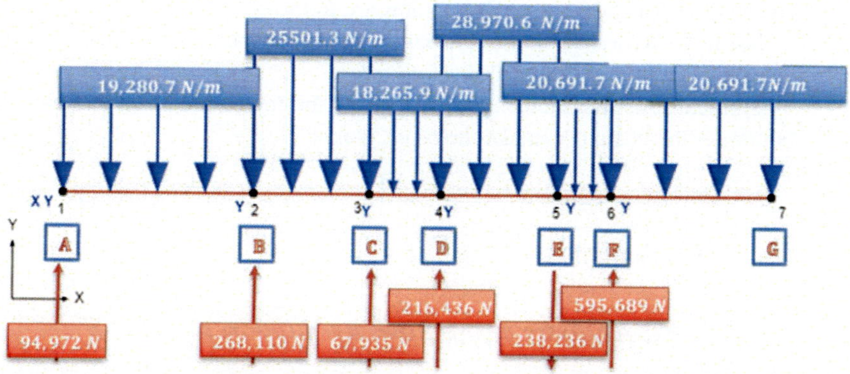

Fig. 8. Free body diagram. The sum of reaction forces is $R = 1,004,906N$ (or 102,436.9 kg).

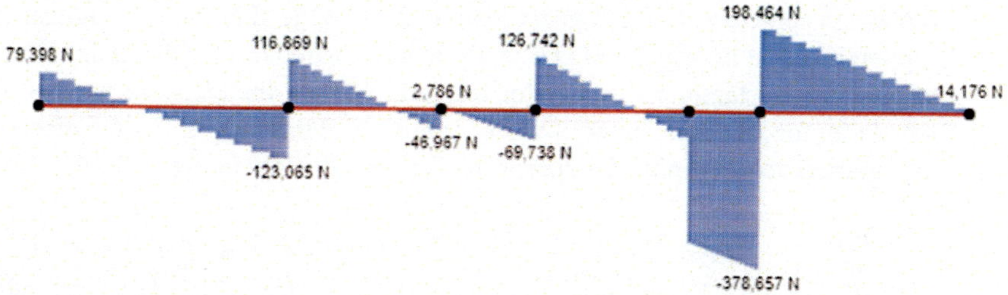

Fig. 9. Shear diagram.

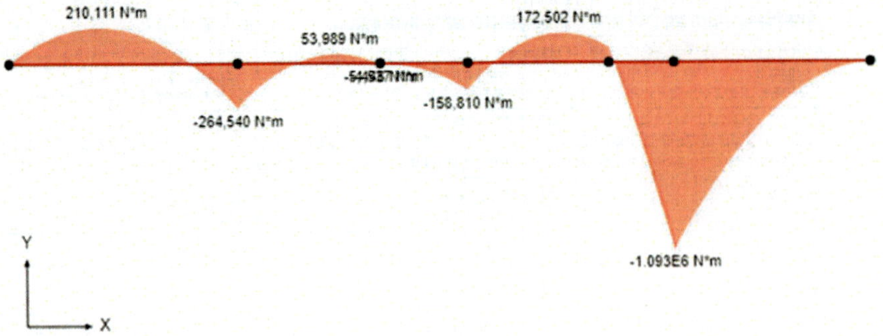

Fig. 10. Moment diagram.

The shear and moment diagrams are shown in Figs. 9 and 10. The standard sign conventions for shear-moment diagrams are followed:

Shear: Positive shear causes clockwise rotation of the beam, negative shear causes counter-clockwise rotation.

Moment: Positive moment will compress the top of the beam and elongate the bottom of the beam (i.e. make the beam "smile").

4 3D FE Analysis

4.1 3D FE Analysis of the Pin-Lug Support at Point F

The loads (reactions at supports) to which supporting parts (lugs and pin) at point F were subjected are taken from Fig. 8, where it can be noted that point F as pinned support is subjected to the highest load, comparing to the point A, another pinned support and that is why the 3D FE analysis was carried out for this point. The tables below give the factors of safety for the joint and for the clamped parts. It can be noted that for all cases factor of safety is greater than 1. It is up to the discretion of the engineer to determine the appropriate factor of safety to use in design (Tables 4, 5, 6).

In Figs. 11, 12, 13, 14 and 15, a 3D model of the pin-lug support at point F, finite element mesh, frictional contacts, and boundary conditions used in FE analysis are shown. The purpose of the 3D analysis is evaluation of the values of safety factors.

Table 4. Strength results of the male lug

Total Applied Force	P_app	297,844 N
Lug Strength for Oblique Loading	P_ult	1,270,329 N
Factor of Safety	FS	**4.27**

Table 5. Strength results of the female lug

Total Applied Force	P_app	148,922 N
Lug Strength for Oblique Loading	P_ult	755,025 N
Factor of Safety	FS	**5.07**

Table 6. Overall strength of the joint, accounting for the male lug, the two female lugs, and the pin

Total Applied Force	P_app	297,844 N
Lug Strength for Oblique Loading	P_ult	1,270,329 N
Factor of Safety	FS	**4.27**

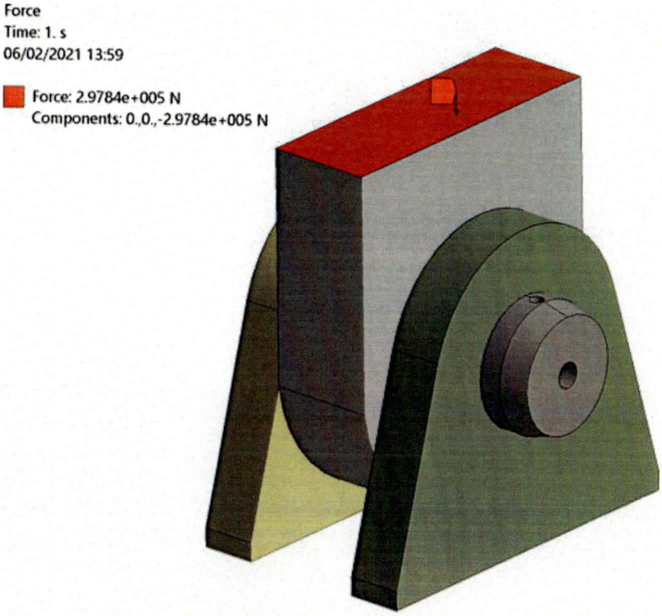

Fig. 11. Load on 3D model (F = 297, 844.5N).

Figure 16 shows that the maximum value of equivalent stress in male lug is 199 MPa, but this value represents stress singularity that occurs at a single point (see "Max" red arrow in Fig. 16) and is a result of discontinuity in geometry (sharp edges). Several randomly selected stress values in the most stressed area show that average stress is between 95 and 136 MPa. The value 164 MPa shown in Fig. 16 is another point stress and several other points like this can be seen in the same Figure.

Figure 17 shows that the maximum value of equivalent stress in male lug is 206 MPa, but this value represents stress singularity that occurs at a single point (see Max in Fig. 17) and is a result of discontinuity in geometry (sharp edge). Several

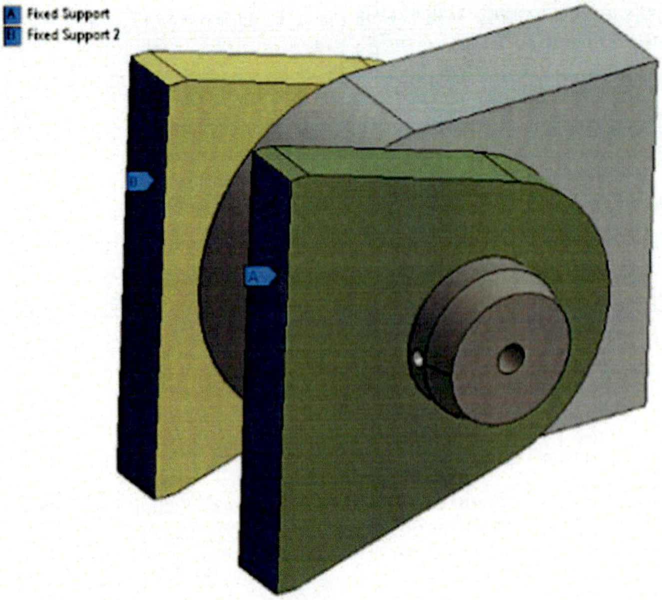

Fig. 12. Fixed supports used on female lugs.

Fig. 13. FE mesh used in calculations.

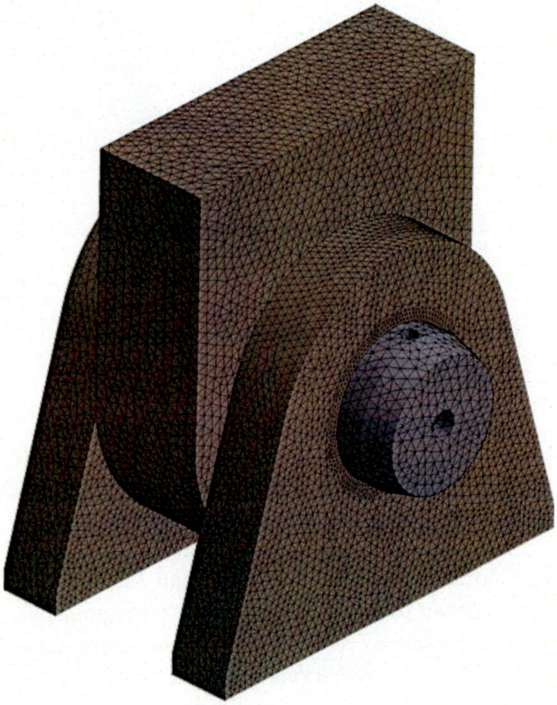

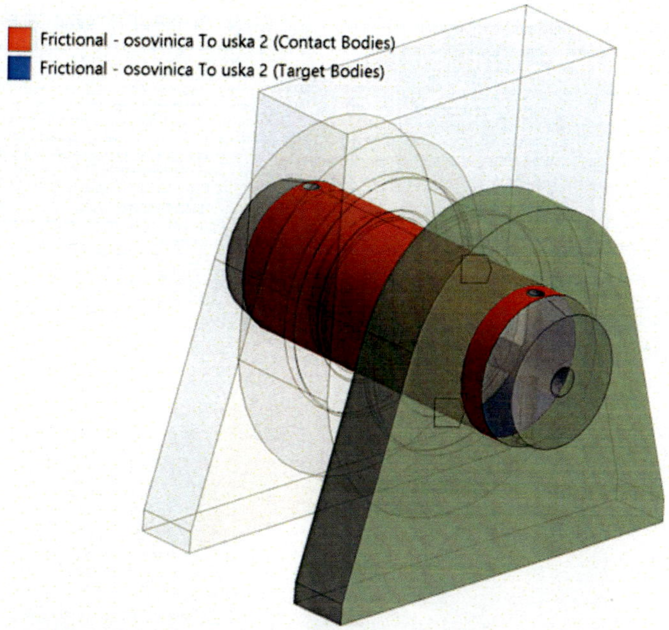

Fig. 14. Frictional contact between pin and lugs.

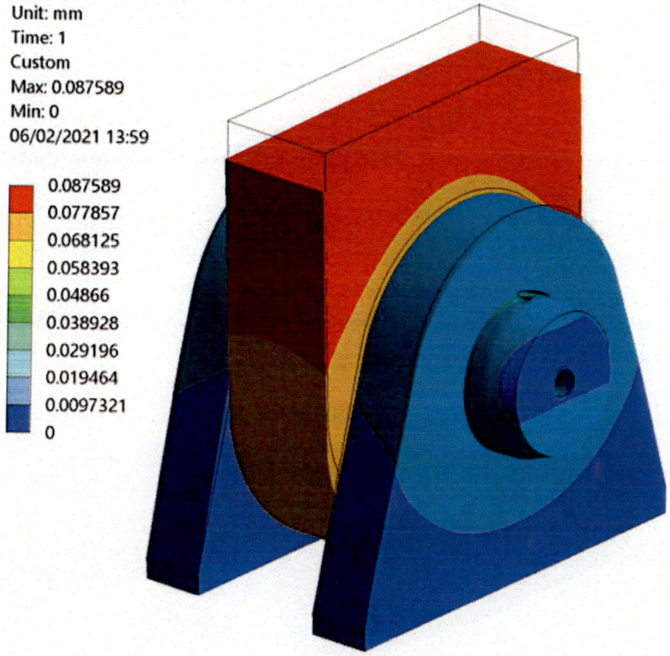

Fig. 15. Maximum deformation obtained in FE simulation is 0.088 mm.

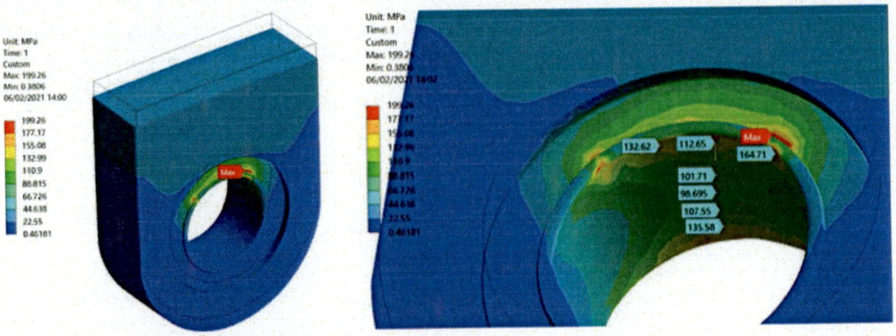

Fig. 16. Stress distribution in male lug.

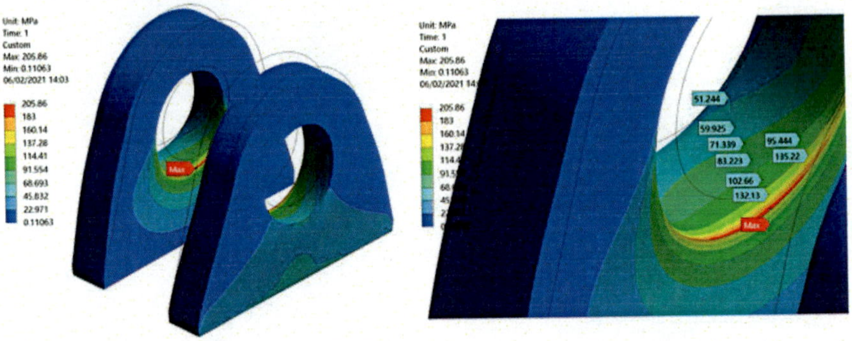

Fig.17 Stress distribution in female lugs.

randomly selected stress values in the most stressed area show that average stress is between 50 and 135 MPa.

Figure 18 shows that the maximum value of equivalent stress in pin is 133 MPa, and several randomly selected stress values in the most stressed area show that average stress is between 60 and 115 MPa (Tables 7, 8, 9).

The factor of safety values obtained by FEM are slightly lower but still significantly higher than 1.

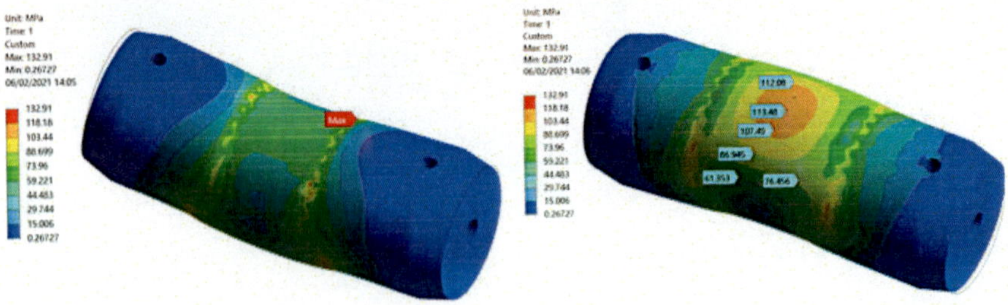

Fig. 18. Stress distribution in pin.

Table 7. Strength results of the male lug

Maximum applied stress	S_app	136 MPa
Ultimate Strength of Material	S_ult	448 MPa
Factor of Safety	FS	**3.29**

Table 8. Strength results of the female lug

Maximum applied stress	S_app	135 MPa
Ultimate Strength of Material	S_ult	448 MPa
Factor of Safety	FS	**3.32**

Table 9. Strength results of the female lug

Maximum applied stress	S_app	115 MPa
Ultimate Strength of Material	S_ult	448 MPa
Factor of Safety	FS	**3.90**

4.2 3D FE Analysis of the Support Assembly at Point B

Supporting parts (wheels, plate and pin) at point B are given in Fig. 19. In fact, at all 4 points (B, C, D, and E horizontal rollers in Fig. 5) where one tunnel is in a contact with another tunnel, the same supporting elements are used, and what differs at these points are loads that have to be carried. The loads (reactions at supports) are taken from Fig. 8, where it can be noted that point B is subjected to the highest load, comparing to the points C, D and E. This is the reason why 3D FE analysis

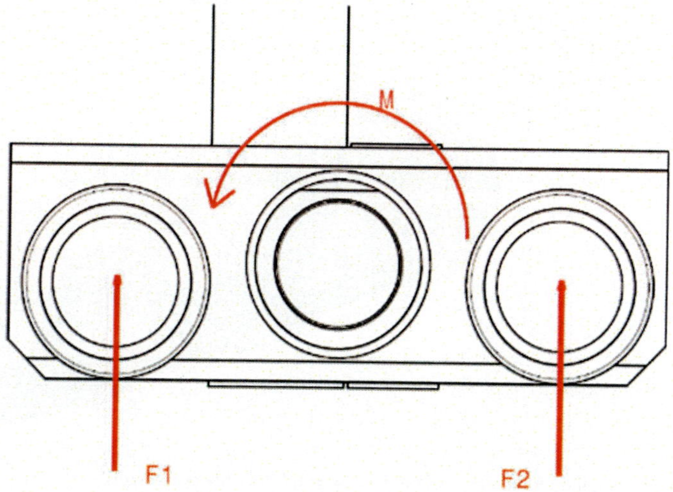

Fig. 19. Loads on load-carrying assembly at point B (also at C, D, and E).

of the support assembly at this point is presented here. Since it is expected that pins will be the load-carrying elements with the highest stress, the safety factors will be evaluated according to the stress calculated on pins. The Fig. 19 shows that reactions forces will be applied on both wheels equally; that is, any reaction force in Fig. 8 first must be divided by two (since there is pair of supports at all points), and then obtained value must be divided again by two (since there are two wheels on each support). The moment on the pin (shown in Fig. 19) equals zero when F1 and F2 are equal (this is an ideal case); otherwise, the twisting moment (torque) occurs. Possible values of twisting moments are also analyzed and presented later in this report.

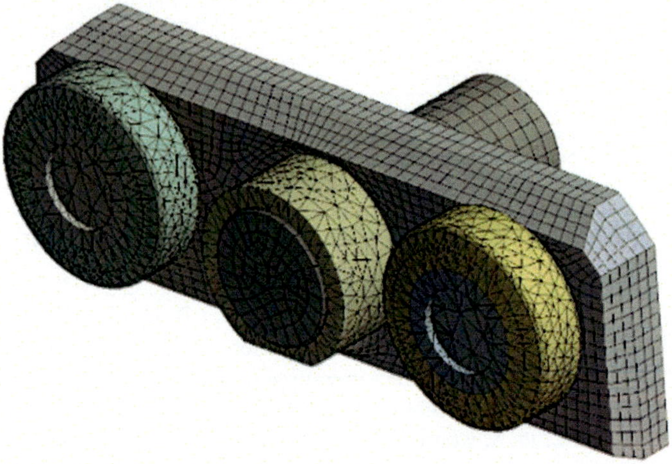

Fig. 20. FE mesh used in calculations.

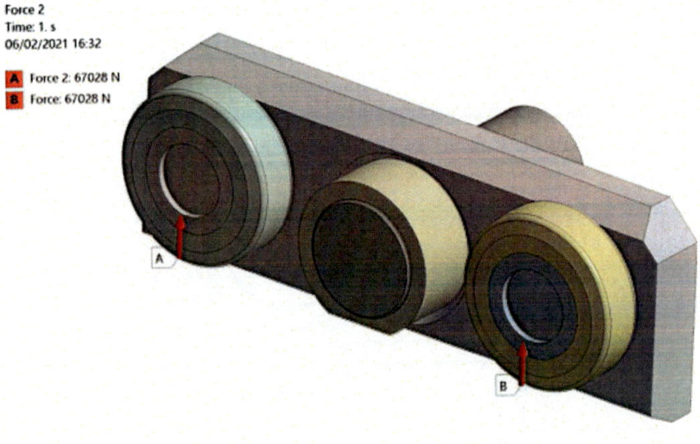

Fig. 21. Applied loads ($F_B = 268, 110/4 = 67, 027.5N$ per wheel).

In Figs. 20, 21 FE mesh, loads, and boundary conditions used in the calculations based on FEM are shown.

Results of FE analysis of the support assembly at point B can be seen in following Figs. 22, 23, 24 and Table 10.

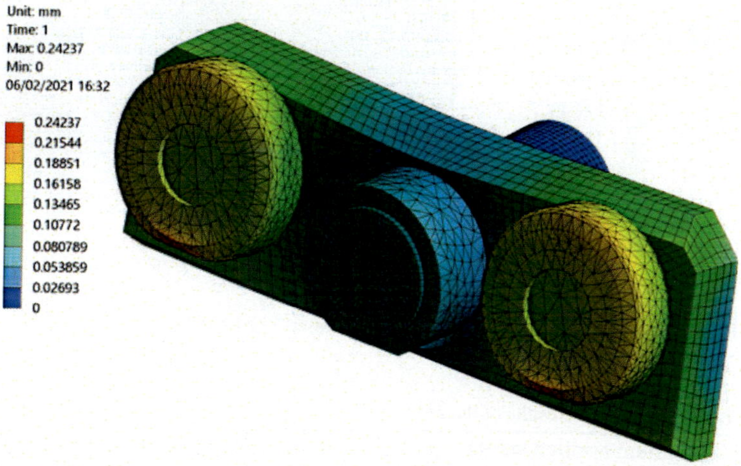

Fig. 22. Maximum deformation of FE model is 0.24 mm.

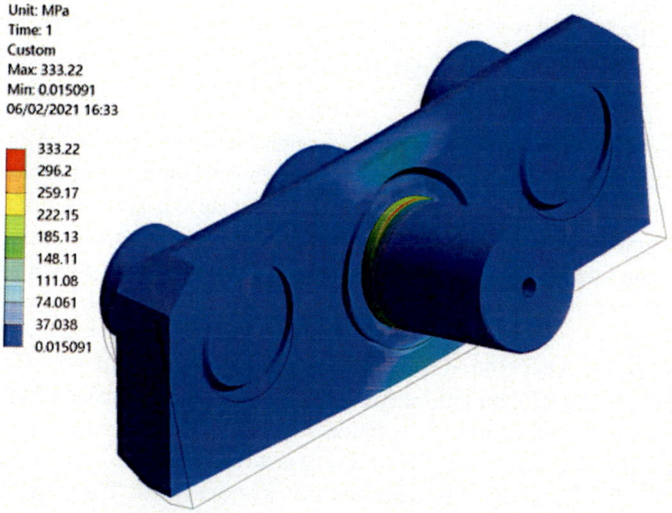

Fig. 23. Stress distribution in FE model.

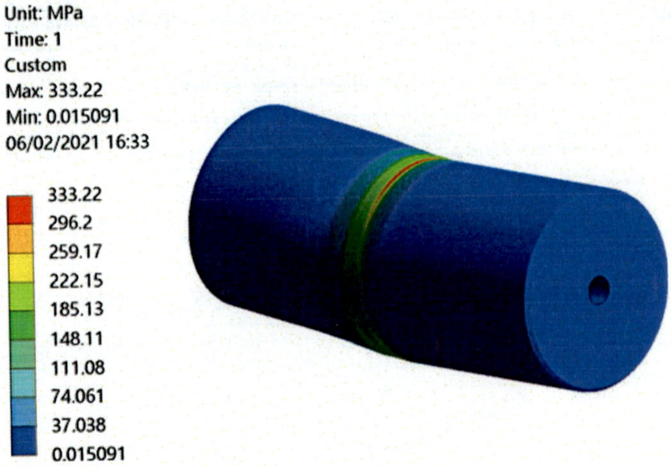

Fig. 24. Stress distribution in pin (maximum value 333.22 MPa).

Table 10 Strength results of the pin

Maximum applied stress	S_app	333 MPa
Ultimate Strength of Material	S_ult	700 MPa
Factor of Safety	FS	**2.10**

4.3 Fatigue Crack Growth Analysis

It has been shown in previous analyses that the highest stresses during the operation of PLB are expected to occur in the pin of the horizontal roller support at point B, so potential damage in that place could represent serious problem if it starts spreading. To demonstrate how dangerous the appearance of the crack could be and to evaluate the fatigue life of the damaged pin, 3D fatigue crack growth simulations were performed in order to assess the number of cycles to complete failure. These simulations were carried out by the usage of the improved FEM implemented thru Ansys Workbench. Ansys introduced Separating Morphing and Adaptive Re-meshing Technology (SMART), an innovative feature which automatically updates the mesh but only near the crack at each solution step and in that manner delivers solutions for stress intensity factors (SIFs) for every newly defined position of the crack front.

For this purpose, only damaged pin was analyzed. The part of the pin that is connected to the wheels assembly was fixed and the applied loads were in accordance with previous analysis. Firstly, the initial crack was placed in the fixed part of the pin where maximum stress values occur (Fig. 25). Initial crack size was 0.3 mm.

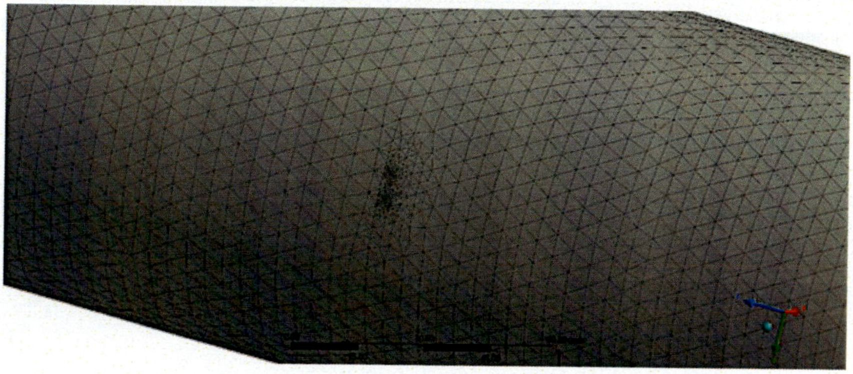

Fig. 25. The initial crack in the fixed part of the pin.

After initiation, crack propagated in 20 steps and reached size of 4.3786 mm. The stress distribution after 20 steps of crack growth is shown in (Fig. 26).

The stress intensity factors were calculated after each step (Fig. 27). It is interesting to see that the front of the crack propagated outside the plane of the initial crack.

Finally, number of cycles for the crack size accumulation of 4.0786 mm was obtained (Fig. 28). This number is 360. The simulation was stopped after twentieth step because in the next step the computed number of cycles was less than 1, indicating occurrence of sudden crack growth, i.e. brittle fracture and complete failure of the pin.

Secondly, the initial crack was placed outside of the fixed part of the pin (Fig. 29). Initial crack size was also 0.3 mm.

The stress distribution after crack opening is shown in (Figs. 30 and 31).

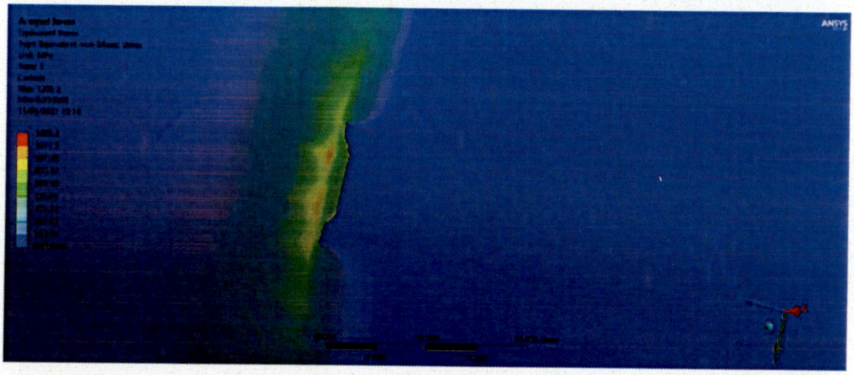

Fig. 26. The crack in the fixed part of the pin (last step 20).

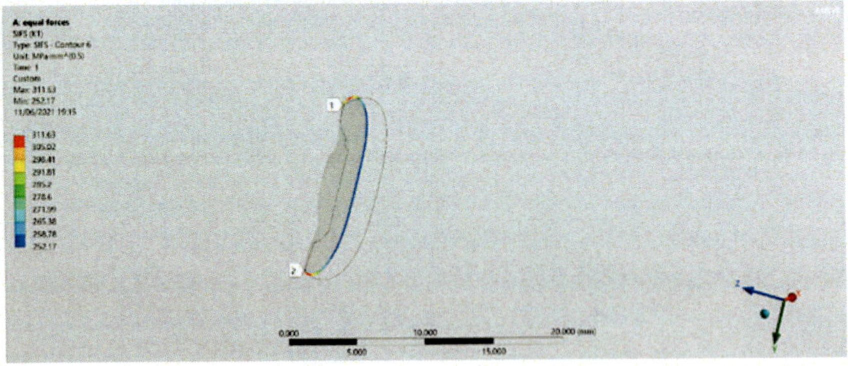

Fig. 27. The SIFs for the crack in the fixed part of the pin (step 20).

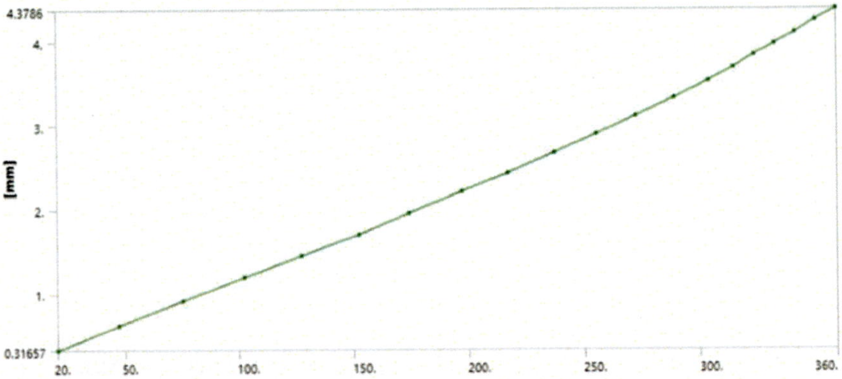

Fig. 28. Number of cycles vs crack length.

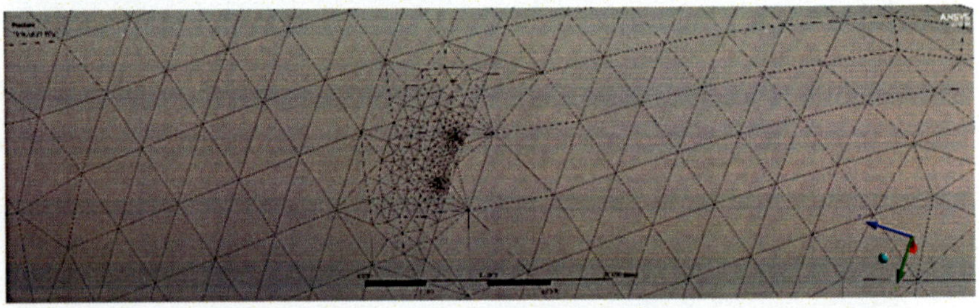

Fig. 29. The initial crack outside the fixed part of the pin.

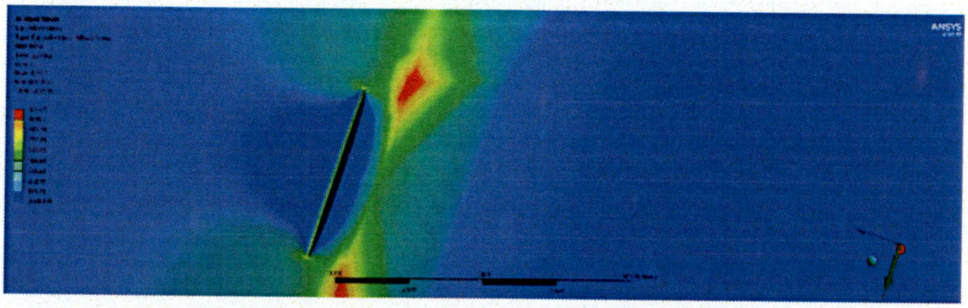

Fig. 30. The crack outside the fixed part of the pin.

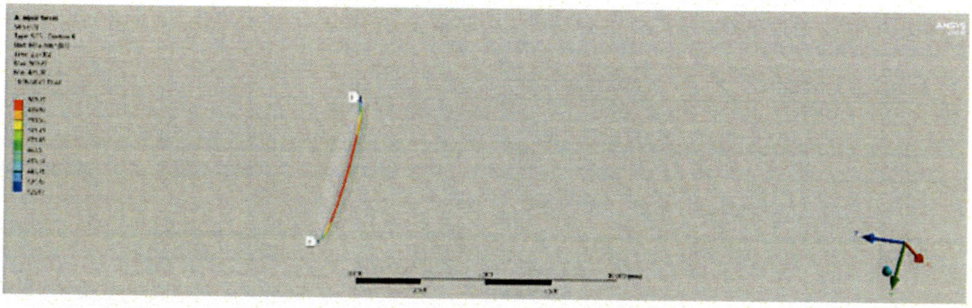

Fig. 31. The SIFs for the crack outside the fixed part of the pin.

5 Conclusion

The Cruise industry is in expansion nowadays and is facing harder and harder challenges every day. Cruise ships are increasing in size and passenger volumes are continuously rising; thus, attention must be paid to fast turnaround times, cost-effective operations, and safe and secure passenger embarkation and de-embarkation. The seaport passenger boarding bridge presented in this paper should provide passengers with a superior boarding experience, but—at the same time—it must be highly operational, flexible, lightweight, and failure-free. It has been shown in this paper that forces and moments at three-tunnel boarding bridge supports were high in the critical cases, as well as stresses in pins and lugs. Crack occurrence in the highest stress areas could lead to catastrophic consequences since it was shown that even the smallest crack (up to 3 mm in length) causes static fracture of the pin. On the other hand, if a crack occurs in a lower stress area, it grows toward a high-stress area under dynamic load and after a certain number of cycles (and relatively stable crack growth), the pin breaks suddenly. This is the reason why special care must be paid to crack occurrence and growth prevention and, as a result, regular maintenance and a visual examination schedule must be strictly defined. Also, it is recommendable to use fatigue resistive steels for pins and lugs since these are the most critical parts of passenger boarding bridge supports. It will, without any doubt, increase the costs of boarding bridge production, but the safety of passengers has no price and is the ultimate goal of any design.

References

1. Wenhui, W., Yan, Z.: Design and Finite Element Analysis of Corrugated Plate Boarding Bridge. Advanced Materials Research **97–101**, 3727–3730 (2010)
2. Na Won, H.: Koo Hwan Jun and Bin Soo Yeol: Study on the Effect of the Lift Column Layout Design on Structural Strength of the Passenger Boarding Bridge. Trans. Korean Soc. Mech. Eng. C **3**(4), 307–312 (2015)
3. Triantafyllidis, G.K., Kazantzis, A.V., Drambi, E.K, et al.: Fracture Characteristics of Torsion-Bending Fatigue and Impact Fatigue Failure of Two Steel Pins in a Crawler Excavator. J Fail. Anal. and Preven. **9**, 23–27 (2009)
4. Hong-Feng, M., Zheng-Wen, Y., Le, G.: The research on fatigue test of the pin for the folded arm lorry crane, Proceedings of the 3rd International Conference on Material, Mechanical and Manufacturing Engineering, pp.1874–1877 (2015)
5. Papadopoulou, S., Pressas, I., Vazdirvanidis, A.: Pantazopoulos, G: Fatigue failure analysis of roll steel pins from a chain assembly: Fracture mechanism and numerical modeling. Eng. Fail. Anal. **101**, 320–328 (2019)
6. Boljanovic, S., Maksimovic, S.: Fatigue crack growth modeling of attachment lugs. Int. J. Fatigue **58**, 66–74 (2014)
7. Sumanth, M.H., Ayyappa, T.: Comparative analysis of aircraft wing fuselage lug attachment bracket. Int. J. Technol. Res. Eng. **5**(11), 4422–4429 (2018)
8. Solob, A., Grbovic, A., Bozic, Z., Sedmak, S.: XFEM based analysis of fatigue crack growth in damaged wing-fuselage attachment lug. Engineering Failure Analysis 112(4):104516, (2020).
9. Babuska, I., Melenk, J.M.: The partition of unity method. Int. J. Numer. Meth. Eng. **40**, 727–758 (1998)
10. Jovicic, G., Zivkovic, M., Sedmak, A., Jovicic, N., Milovanovic, D.: Improvement of algorithm for numerical crack modelling. Archives of Civil and Mechanical Engineering **10**(3), 19–35 (2010)
11. Balac, M., Grbovic, A., Petrovic, A., Popovic, V.: FEM analysis of pressure vessel with an investigation of crack growth on cylindrical surface. Eksploatacja i Niezawodnosc - Maintenance and Reliability **20**(3), 378–386 (2018)
12. Fatigue Crack Paths in Light Aircraft Wing Spars: Grbovic, A., Kastratovic, G., Sedmak, A Popovic, M. Int. J. Fatigue **123**, 96–104 (2019)
13. Grbovic, A., Kastratovic, G., Sedmak, A., Eldweib, K., Kirin, S.: Determination of optimum wing spar cross section for maximum fatigue life. Int. J. Fatigue **127**, 305–311 (2019)
14. Aleksic, B.V., Grbovic, A., Hemer, A., Milovic, Lj., Aleksic, V.: Evaluation of Stress Intensity Factors (SIFs) Using Extended Finite Element Method (XFEM). International Conference on New Trends in Fatigue and Fracture. pp.355–369 (2018).
15. Kastratovic, G., Aldarwish, M., Grbovic, A., Vidanovic, N., Eldwaib, K.: Stress intensity factor for multiple cracks on curved panels. Procedia Structural Integrity **13**, 469–474 (2018)
16. Kastratovic, G., Vidanovic, N., Grbovic, A. Mirkov, N., Rasuo, B.: Numerical Simulation of Crack Propagation in Seven-Wire Strand. Computational and Experimental Approaches in Materials Science and Engineering, Proceedings of the International Conference of Experimental and Numerical Investigations and New Technologies, CNNTech 2019, LNNS 90, pp. 76–91, (2019).
17. Aleksic,B., Grbovic, A., Milovic, Lj., Hemer, A., Aleksic, V.: Numerical simulation of fatigue crack propagation: A case study of defected steam pipeline, Engineering Failure Analysis 106,104165, (2019).
18. Djukić, D., Grbovic, A., Kastratovic, G., Vidanovic, N., Sedmak, A.: Stress intensity factors numerical calculations for two penny shaped cracks in the elastic solid, Engineering Failure Analysis 112, 104507, (2020).
19. Elayeb, A.A., Grbovicć, A., Kastratovic, G., et al.: Fatigue crack growth in a structure exposed to high temperature, Engineering Failure Analysis 127,105493, (2021).

Effects of Flue Gas Recirculation on NOx Emissions from Gas-Fired Utility Boilers

Nikola Tanasić, Mirjana Stamenić, and Vladimir Tanasić

Abstract Gas-fired utility boilers are most common in the district heating system in the city of Belgrade. Thus, they are an important stationary source of NOx emissions, which are particularly harmful for people and environment. In this paper the effectiveness of the flue gas recirculation (FGR) system that is applied to reduce NOx emissions of the utility hot water and steam boilers in district heating plant is experimentally investigated. Two types of systems are analysed; the first with the application of flue gas recirculation fan at hot water boilers; and the second without the application of flue gas recirculation fan at steam boilers. In both cases the rate of flue gas recirculation was 10%. The results of measurements showed that applied FGR systems efficiently reduced NOx emissions from utility boilers by 20 to 50% and below the reference values defined by legislative regulations. Emissions are higher at steam boiler which is expected because of the higher flue gas temperature at boiler furnace. Since the measured values for hot water boilers are close to the reference values, it was concluded that there is a potential for further improvements by increasing the flue gas recirculation rate.

Keywords Utility gas-fired boilers · Reduction of NOx emissions · Flue gas recirculation (FGR)

N. Tanasić (✉)
Department of Transportation, Mechanical and Safety Engineering, The Academy of Applied Technical Studies Belgrade, 11000 Belgrade-Zemun, Serbia
e-mail: ntanasic@tehnikum.edu.rs

M. Stamenić
Faculty of Mechanical Engineering, Department of Process and Environmental Engineering, University of Belgrade, 11000 Belgrade, Serbia

V. Tanasić
PUC, Belgrade District Heating Company, 11000 Belgrade, Serbia

N. Mitrovic et al. (eds.), *Current Problems in Experimental and Computational Engineering*, Lecture Notes in Networks and Systems 323,
https://doi.org/10.1007/978-3-030-86009-7_17

1 Introduction

1.1 EU Regulations on Energy Efficiency and Environmental Protection

The late 1990s and the turn of the decade starting in the year 2000 was a crucial period for Europe's energy and climate policy. After adopting the Kyoto Protocol in December 1997, the European Union intensified its work to implement the agreed greenhouse gas reduction targets. It was soon agreed that after a joint EU approach towards adopting the Kyoto Protocol, cooperation was also required to implement a crucial implementing regulation. In September 2001, the Directive 2001/77/EC [1] was adopted on the promotion of electricity produced from renewable energy sources in the internal electricity market, including indicative national targets for each EU Member State on the share of electricity produced from renewable energy sources ('RES-E') in their total electricity consumption for the year 2010. One month later, this was followed by a proposal for a framework Directive for greenhouse gas emissions trading, and in 2003 Directive 2003/87/EC [2] established the EU Emissions Trading System ("the EU ETS"). A series of other directives followed to support further the growth of renewable energy on the European energy markets and implement greenhouse gas emission reduction targets. These, among others, include Directive 2003/30/EC [3] on promoting renewable energy in transport, the 2009 renewable energy Directive [4], and the 2009 Effort sharing decision that had agreed on national greenhouse gasses (GHG) reduction targets for non-ETS sectors.

While it was commonly agreed that the EU energy and climate change strategy required targets and regulations for GHG emission reductions, renewable energy, and energy efficiency, reaching agreements on binding targets for the latter remained the most challenging element in this strategy. In 2007 EU leaders agreed on the so-called 2020 strategy, including the 20% target to reduce GHG emissions in 2020 from 1990 levels, a 20% share of energy from renewable energy sources, and a 20% improvement in energy efficiency compared to the projected use of energy in 2020. After several years of intense discussions between the European Commission, the European Council, and the European parliament and strong leadership from temporary EU council chair Denmark the Energy Efficiency Directive (EED) [5] was finally adopted in October 2012.

The EED repealed two existing Directives that were the main elements to steer energy efficiency in Europe to that date: Directive 2006/32/EC [6] on energy end-use efficiency and energy services and Directive 2004/8/EC [7] on the promotion of cogeneration. The Directive also included updates to two earlier adopted Directives on eco-design (Directive 2009/125/EC [8]) and labelling of energy-related products (Directive 2010/30/EU [9]). The EED complements and includes links to Directive 2010/31/EU [10] on the energy performance of buildings ("EPBD") that was adopted in May 2010. A clear link, for example, is that the EPBD defined quality criteria for building renovations while the EED sets ambition levels for such renovations. The EED also strengthened concepts that already existed in earlier regulations but had not resulted in sufficient energy efficiency improvements. This, for example, includes the provisions for individual metering and informative billing that were already included in Directive 2006/32/EC and to which the EED added obligations at the Member state level and more strict formulations and requirements of implementing conditions.

The version of the Directive that was adopted was significantly weakened compared to an earlier draft version published in 2011. To counteract this lowering of ambitions, two important evaluations were agreed upon. First, by 30 June 2016, the Commission would have to evaluate the implementation of Article 7 on the energy efficiency obligations or equivalent measures adopted at the Member State level and, where necessary, provide a legislative proposal to strengthen this Article. Second, by June 2018, the Commission would have to evaluate the progress made by the Member States in removing the barriers to implementation of energy efficiency measures as referred to in Article 19(1) and, where appropriate, specify for further measures. Both evaluations were conducted, and an updated EED—Directive 2018/2002 [11]—was adopted in December 2018 as part of the so-called "Clean Energy for all Europeans package" that also included updates to the Energy Performance of Buildings Directive (Directive 2018/844 [12]) and a recast of the Renewable Energy Directive (Directive 2018/2001 [13]). The update of, among others, the EED was combined with prolongation and strengthening of the overall EU-wide energy efficiency target set at least 32.5% energy efficiency improvement by 2030 compared to the 2007 modelling estimate for 2030. The amended Directive also includes a possible upward revision in the target in 2023 in case of substantial cost reductions due to economic or technological developments.

1.2 *Scope of the Work*

The increasing emissions of flue gases from district heating plants have great impact on the environment and people's health in large urban areas. Among other gasses, nitrogen oxides (NOx) are particularly harmful. NOx is considered to be the most significant gaseous pollutant emitted from natural gas combustion [14]. When NO and NO_2 are discharged into the atmosphere, nitric acid fog and nitrate are generated through a series of physicochemical reactions. Under sunlight, nitrate combines with hydrocarbons and ozone to produce photochemical smog, which has a great impact on the atmospheric environment and human health [14]. Gas-fired utility boilers are most common in the district heating system in the city of Belgrade. Thus, they are an important stationary source of NOx emissions.

Many researchers have done research on reducing the NOx emissions from industrial and utility boilers. Moreover, boiler burners have been modified in many aspects to meet increasingly strict emissions regulations, resulting with the fact that almost all new boilers are equipped with burners capable to improve the combustion efficiency and reduce the NOx emissions. In fact, a lot of studies have been carried out to prevent NOx generation, and numerous combustion technologies have been developed, such as flue gas recirculation (FGR), fuel/air staging, lean-burn premixed combustion, divided flame, and enhanced mixing of air and fuel [15]. Among them, FGR is one of the most widely used methods of reducing NOx emission from combustion [16]. FGR usually includes two types, external and internal recirculation. In the external recirculation, which is the case in this study, flue gas is recirculated outside of the furnace and added to the combustion air.

In this paper flue gas recirculation (FGR) methodology that is applied to reduce NOx emissions of the utility hot water and steam boilers in district heating plant is experimentally investigated.

1.3 Research Review

In paper [15] a novel type of low NOx burner for gas boiler with FGR using a non-premixed combustion method was tested in an industrial boiler and numerically in order to find the optimum operating parameters of the burner, such as excess air ratio and FGR rate. The results of industrial experiments indicated that adjusting these parameters within a reasonable range can achieve a low NOx emission performance of the burner (<30 mg/m^3). Flame temperature decreases with the FGR rate and result in the decreasing of the NOx emission rates. Additionally, the region of high NOx concentration roughly coincides with the high temperature region. To achieve the low NOx emission, the authors proposed optimum excess air coefficient of 1.15 due to the better temperature uniformity, and the value of FGR rate in the range of 28 to 40%. The flue gas circulation rate of traditional FGR system cannot be too high (10 to 20%). As the flue gas circulation rate increases, it will cause flame fluctuations and unstable combustion. Compared with the traditional, the new type of FGR burner described in paper [15] allows higher FGR rates and thus has lower NOx emissions.

Other author [17] also concluded, based on experiments, that FGR is an effective measure to reduce NOx emissions significantly. In paper [18] numerical simulation is applied to study the influence of excess air ratio and air preheating temperature on the NOx emissions of a gas boiler. In their study, increasing the excess air ratio resulted in a maximum value of thermal NOx concentration at the boiler stack when the excess air ratio equalled to 1.2. With the increase of combustion air temperature, temperature inside boiler furnace increased and the thermal NOx concentration increased rapidly. Other authors [19] also used numerical approach in order to optimize combustion parameters for fuel-staged low NOx burner with internal and non-internal FGR systems. In paper [20] a decoupled numerical modelling method was developed to simulate the combustion process of biomass in a 40 t/h grate firing boiler. The objective of this study was to explore the NOx reduction mechanism as well as to investigate the potential of internal FGR on the combustion process and emissions formation. In was concluded that as the flue gas recirculation ratio increased, final NOx emission showed a decreasing trend. The addition of biomass and co-firing with other fuels generally tends to reduce NOx emissions in utility boilers and gas engines. This is because of lower overall fuel-bound nitrogen content in biomass [21].

In paper [22] the flue gas composition and the flue gas temperature at outlet of the economizer were measured. The influence of FGR system on the efficiency of the incinerator-waste heat boiler and NOx emission in a waste incineration power plant with a waste disposal capacity of 500 t/day was explored experimentally. The results showed that by increasing the ratio of the recirculating flue gas from 0 to 16.43% the NOx concentration at the outlet of the economizer decreased by 29.8%.

In paper [23] results of theoretical numerical research dealing with CO and NOx emission performed in the process of optimization of the performance of low-power atmospheric burners are presented. In paper [24] flame geometry and location was investigated using CH* chemiluminescence imaging technique. It was concluded that the presence of CO_2 in fuel to air mixture decreased heat of combustion, flame temperature and flame speed.

Recently legislative regulations for NOx emission are increasingly more stringent. While in Serbia reference emission levels for NOx are set at 100 to 150 mg/m^3, in EU and other developed countries reference emission levels are significantly lower. For example, according to the revised Emission Standard of Air Pollutants for Boilers in Beijing [25], the NOx emission of the new built boilers must be lower than 30 mg/m^3 since April 1, 2017 and for the retrofitted boilers, the emission standard of NOx has dropped from 150 mg/m^3 to 80 mg/m^3 [14].

2 Technical System Description

2.1 Flue Gas Recirculation (FGR)

Flue gas recirculation (FGR) or exhaust gas recirculation (EGR) implies recirculation of part of the flue gas to the combustion chamber to replace part of the fresh combustion air, having the double effect of cooling the flame temperature and limiting the O_2 content for nitrogen oxidation, thus limiting the NOx emissions [26].

The recirculation of flue gas results in the reduction of available oxygen in the combustion zone and, as it directly cools the flame, in the decrease of the flame temperature: therefore, both fuel bound nitrogen conversion and thermal NOx formation are reduced. The recirculation of flue gas into the combustion air has proven to be a successful method for NOx abatement in high temperature combustion systems, such as wet-bottom boilers and oil- or gas-fired installations.

As shown in Fig. 1, part of the flue gas (10 to 30% at temperatures of about 350 to 400°C) is withdrawn from the main flue gas flow downstream of the air preheater, usually before any particulates have been removed, and then recycled to the boiler. The recirculated flue gas can be mixed with combustion air upstream of the burner or with staging air. Special burners designed to operate on recirculated flue gas are required with high flows to be recycled [26]. A similar technique also applies to some

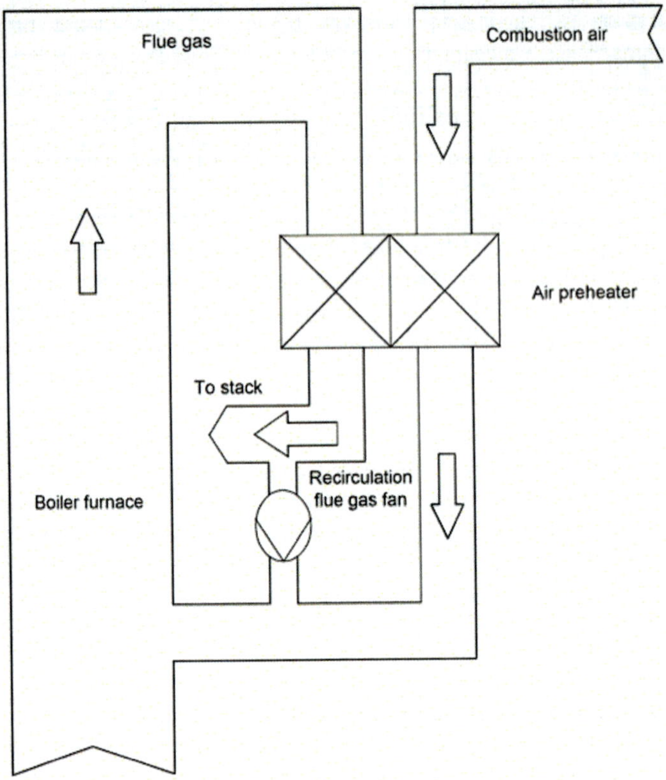

Fig. 1. A schematic view of FGR systems in water-tube boilers [26].

engines: EGR can be a method for reducing NOx emissions from large-bore two-stroke engines, but it requires solving certain problems associated with the cooling and cleaning of the exhaust gas before it is recirculated in the engine.

General NOx reduction rate is from 30 to 50% for gas fired boilers combined with over fire air [26].

Retrofitting an existing boiler with flue gas recirculation presents some adaptation difficulties, mostly due to efficiency losses of both the boiler and the burners, except when recirculating very small amounts of flue gas. This NOx abatement measure can be used alone or combined with air staging for retrofitting. Recirculation of flue gas results in additional energy consumption due to the recirculation fan. FGR addition may slightly modify the heat exchange, with a slight increase in boiler flue gas temperatures (e.g. from 222 °C to 228 °C in an example of gas-fired plant [26]), resulting in a slight decrease in energy efficiency (e.g. 0.3 percentage points in the example plant case [26]).

The retrofitting requires the installation of ducting and controls and may be constrained by plant configuration. Component upgrades may be required to accommodate the larger volume and higher temperature incoming air/FGR mixture (e.g.

fans). Flame instability may limit the amount of FGR possible (and NOx reduction) unless burner changes are made as well. Additional operating risks due to FGR failures and their impact on burner combustion need to be managed. If there is an excessive amount of recirculated flue gas, this can lead to some operational limitations, e.g. corrosion problems when burning a fuel containing sulphur; efficiency losses due to a temperature increase at the stack; or increased energy consumption for the ventilators. Therefore, operators usually aim to limit the amount of recirculated flue gas (to approximately 30%) and compensate for the higher NOx emissions by using advanced low-NOx burners [26].

2.2 *Low NOx Burners*

Combustion is a complex phenomenon that involves numerous parallel and consecutive chemical reactions and multiple mechanisms of heat and mass transfer. The combustion process can occur under certain conditions: the reactant concentration must be within particular limits, and there must be a heat source that will ensure the start of chemical reactions (combustion). Diagram (Fig. 2) shows the combustion domain depending on fuel heating value, air–fuel ratio ("excess air"), and the temperature of combustion air. The diagram indicates that the domain where combustion can occur increases with higher temperatures and higher fuel heating values. Higher temperatures lead to higher NOx values in flue gas since the combustion temperature increase with combustion air preheating temperature. However, lowering the oxygen concentration in combustion air (by recirculating the flue gas and mixing the part of the fuel gas with combustion air) could keep the concentration of NOx in flue gas at an acceptable level (below the reference limiting values defined by law).

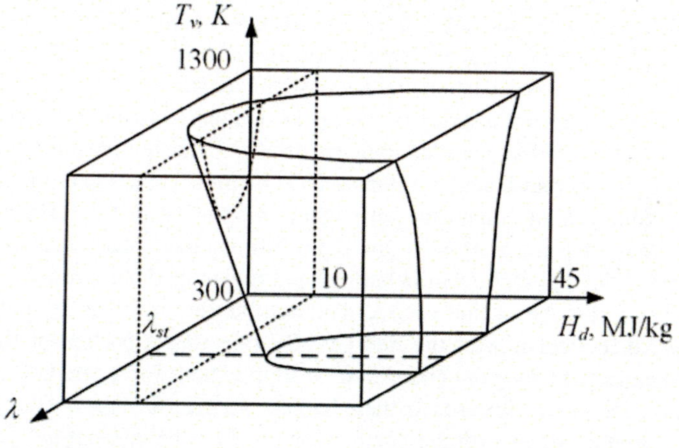

Fig. 2. Combustion domain depending on fuel heating value, air–fuel ratio ("excess air"), and the temperature of combustion air [27]

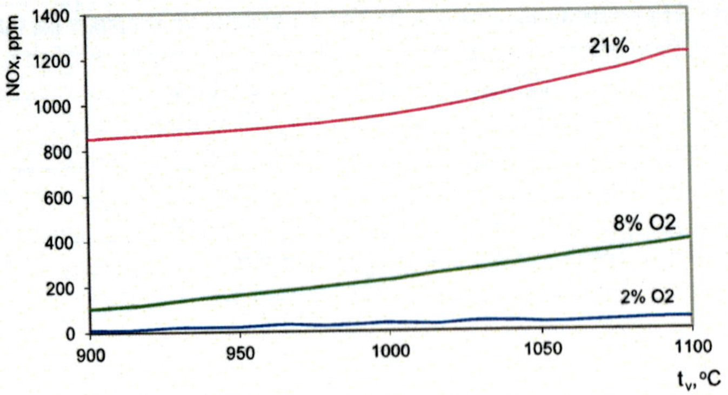

Fig. 3. NOx emission related to the oxygen concentration and preheating temperature of combustion air [28]

Three main factors affect the kinetics in forming "thermal NOx":

1. High maximum temperature in some points within the combustion chamber;
2. The retention time of reactants (O_2 and N_2) in spots of maximum temperatures;
3. Oxygen concentration within the occupied space by flame.

While there is a need to be as efficient as possible, on the other hand, there are constraints regarding environmental protection and reducing the emission of pollutants. For example, efficiency implies using highly preheated air to reduce fuel consumption. There are two often-used systems of efficient burners: recuperative and regenerative. An advanced, environmentally friendly solution to decrease NOx concentration in the flue gases uses the dilution of the combustion air by mixing the air with flue gas. Figure 3 shows the influence of the level of dilution of combustion air with flue gas on NOx emission in regard to the air preheating temperature (t_v).

The technique, including ultra- or advanced- low-NOx burners, is based on the principles of reducing peak flame temperatures; boiler burners are designed to delay but improve the combustion and increase the length of the flames, allowing the heat radiated during combustion to reduce peak temperatures. The air/fuel mixing reduces the availability of oxygen and reduces the peak flame temperature, thus retarding the conversion of fuel-bound nitrogen to NOx and the formation of thermal NOx, while maintaining high combustion efficiency. According to the different principles to reduce the formation of NOx, low-NOx burners have been developed as air-staged burners, flue gas recirculation burners and fuel-staged burners [26].

The technique of flue gas recirculation burners involves separated flames with internal recirculation of the flue gas. By injecting a portion of the flue gas into the combustion zone or combustion air supply, both flame temperatures and oxygen concentrations are lowered. In these burners, 15 to 25% of the hot flue gas is internally recirculated, along with the combustion air, into the burnout air. The flue gas acts as a diluent, reducing the flame temperature and partial pressure of oxygen, thus reducing NOx formation up to 20% [26].

2.3 Case Study Plant

Belgrade district heating company (PUCBE) is public utility company that operates production and distribution of thermal energy within the district heating system in the city of Belgrade, Serbia. The district heating system consists of numerous large and small heating plants and district heating network with substations. Each heating source consists of one or more hot water and steam boilers. The number of heating days in Belgrade per year is 160 to 190, depending on the meteorological conditions.

The heating plant that is the subject of this study consists of three fuel oil and gas fired hot water boilers (VK1, VK2, VK3) and two stem boilers (PK1 and PK2). The steam boilers produce saturated steam, which is used for heating fuel oil, thermal treatment of feed water, combustion air pre-heating for hot water boilers, for preparation of sanitary hot water, local heating etc. Technical characteristics of hot water and steam boilers are presented in Tables 1 and 2, respectively.

The hot water boilers are designed as free-standing water-tube boilers. The furnace and flue gas ducting walls are the membrane type. The membranes are made of pipes that are connected to each other with welded strips. In that way, the impermeability of the boiler is ensured along the flue gas tract all the way to their exit from the boiler. The boiler works with overpressure from the flue gas side, which is provided by the operation of two centrifugal fans of fresh combustion air and regulation of the position of the blades of the air fans. The fans overcome all pressure losses along the flue gas tract in the boiler. The flue gas outlet is from the back side of the boiler, through a flue gas duct and concrete stack.

On the front side of the boiler there are two (in case of VK1 and VK2) or four (in case of VK3) combined rotary burners for firing of natural gas and alternatively fuel oil. The boiler primary uses natural gas as a fuel, due to the built-in economizer, which is a tubular water heater arranged in two packages in flue gas ducting. The

Table 1. Technical characteristics of hot water boilers

Boiler unit	VK1	VK2	VK3
Manufacturer	Minel, Yugoslavia	Minel, Yugoslavia	Djuro Djakovic, Yugoslavia
Type	TE 350	TE 350	5607
Year of production	1986	1986	1979
Heating duty	58 MW	58 MW	116 MW
Max. pressure/temperature	22 bar/180 °C	22 bar/180 °C	26.5 bar/200 °C
Air fan capacity	40,000 m^3/hr	40,000 m^3/hr	83,600 m^3/hr
Air fan static pressure	5.4 to 6.3 kPa	5.4 to 6.3 kPa	7.44 kPa
Air fan motor power rate	110 kW	110 kW	290 kW
Number of burners	2	2	4
Burner manufacturer	Saacke, Germany	Saacke, Germany	Saacke, Germany
Burner type	GMG 300	GMG 300	GMG 300

Table 2. Technical characteristics of steam boilers

Boiler unit	PK1	PK2
Manufacturer	Minel, Yugoslavia	Minel, Yugoslavia
Type	TE-109	TE-109
Year of production	1981	1986
Capacity	10 t/hr	10 t/hr
Max./working pressure	7.8/12.88 bar	7.8/12.88 bar
Fuel consumption	810 m^3/h	810 m^3/h
Boiler efficiency natural gas/fuel oil	87/86%	87/86%
Burner manufacturer	Saacke, Germany	Saacke, Germany
Burner type	SKVG-A 82–22	SKVG-A 82–22

flow of flue gas over the economizer is in cross-current arrangement. The flue gas temperature at the boiler outlet is 180 °C when firing natural gas, or 200 °C when firing fuel oil. The flow of water and flue gas in the economizer is in counter-current arrangement, which ensures efficient heat exchange. The construction of VK3 boiler is similar to VK1 and VK2 construction with the differences of the number of burners (four instead of two) and the number of economizer packages (three instead of two).

The economizer is used to ensure the additional amount of heating duty that will be delivered to the district heating system by lowering the flue gas temperature when firing the boilers with natural gas. The design flue gas temperature at the outlet of the economizer is 70 °C, which reduces heat loss in the exhaust gases and increases the efficiency of the boiler by approx. 5%. The economizer is in operation only when the boiler is fired with natural gas. When the boiler is fired with fuel oil, the damper valves in the flue gas tract are closed so that the entire amount of flue gases from the boiler goes directly into the stack. When firing the boiler with natural gas, the damper valves on the economizer are open, and the damper valve on the bypass line of the economizer is closed so that the entire amount of flue gas from the boiler passes through the economizer before entering the stack. Additional flow resistances on the flue gas side that occur during the operation of the economizer are overcome by a built-in flue gas fan.

The hot waters boilers are supplied with feed water of constant temperature 105 °C when firing natural gas, or 130 °C when firing fuel oil. The water enters the boiler from the bottom of the furnace and is evenly distributed in the lower longitudinal headers of the front, rear and side pipe screens through several vertical pipe headers. The water than flows upward passing through the furnace screens and collects in the upper screen headers which are interconnected. The upper longitudinal (side) headers are partitioned behind the junction with the middle vertical headers, which enables the descent of water through the front and middle vertical headers into the two lower longitudinal headers, which are located below the header of the furnace screen. Water from these and from one transversely placed headers flows upward the membrane walls of the flue gas ducting to the upper headers (two longitudinal and one transverse). From there, water further descends through the rear vertical headers to the lower headers of the economizer's pipe packages. The water from this header enters the pipe packages and the heating coils. Further, water enters the main outlet header through connecting pipes, from where the heated water leaves the boiler. The circulation of water through the boiler is forced, provided by the operation of boiler water pumps.

Steam boilers are three-pass shell and fire-tube horizontal gas boilers. The boiler has three flue gas passages in which the flue gases are under pressure. Fuel is burnt in a fire-tube, which has the function of the boiler furnace. Leaving the fire-tube, the gases go through the rear turning chamber into the flue gas pipes of the second gas passage, which are located under the fire-tube. Through the front turning chamber, the flue gases are channelled into the third gas passage, which is formed by flue gas pipes located on the sides of the boiler, under the water level inside the boiler shell. From the third gas passage, the flue gases go directly to the boiler stack. The circulation of water through the boiler is driven by natural convection due to the density difference. Water located next to the fire-tube is heated primarily through radiation. Colder water is located closer to the cylindrical walls of the boiler, thus receiving a smaller amount of energy. Burner control is done automatically on consumption demand.

A turnkey project for reducing NOx emissions of boilers VK1, VK2, VK3, PK1 and PK2 was carried out. The project requirements defined the scope of retrofitting design, while the scope of project documentation was defined by legal acts. The proposed technical solution is in line with the company strategy for optimal reduction of nitrogen oxide emissions from heating plants operated by PUCBE. The main aim of the project was to retrofit the existing combustion systems and burners on gas fired boilers in order to comply with national regulations, by the introduction of FGR system.

The project was also carried out in line with EU directives for large combustion plants with the following objectives:

1. Define and implement technical solution for reducing NOx emissions,
2. New equipment to be installed on existing burners,
3. Calculations for the air-flue gases tract,
4. Checking the availability of existing fans,
5. Installation of new fans with frequency regulators for flue gas recirculation,
6. Maintain existing oil burners,
7. Maintain the existing burner management system (BMS),
8. Replacement of automation system,
9. Replacement of worn-out double solenoid valves for gas in front of burners on all boilers,
10. Replacement of defective probes for measuring O_2 concentration in flue gas.

Before the reconstruction project was implemented the measured NOx emission values on hot water boilers when operation with natural gas, in the entire range of burner loads, were in range from 120 to 180 mg/m^3. The goal was to reduce these values, below 100 mg/m^3 to comply with the limiting reference value defined by the

Decree on reference values for emissions of pollutants into the air (Official Gazette of the Republic of Serbia, No. 06/2016 [29]).

The reconstructed system had to ensure safe, reliable and economical operation of the boilers. Equipment was installed in such a way to enable reading, testing, inspection, cleaning and maintenance without the risk of bodily injury and without the need to dismantle adjacent equipment. Equipment with a limited service life was designed to be easily accessible for replacement and repair. The design of the entire installation should minimize the harmful effects on the environment. During the construction and operation of the plant, the least harmful impact on the environment had to be ensured. The housing of the fan and flue gas dampers within the recirculation ducts was designed as gas-tight, heat-resistant and to allow for the flue gas to flow through all internal parts with maximum corrosion resistance. The outer lining of the flue gas ducts is insulated so that the temperature of the outer surface of the duct does not exceed 30 °C.

2.4 Applied FGR System

FGR, as mentioned before, involves moving one part of the cooled flue gases from the flue gas tract after leaving the furnace and bringing it back into the furnace in order to achieve a lower combustion temperature. One of the most important factors influencing the formation of nitrogen oxides is the combustion temperature in the furnace. The lower combustion temperature results in a significant reduction in the amount of nitrogen oxides formed during the combustion process. By using this system the increase of excess air and combustion control problems are avoided. The flue gases leaving the furnace have a lower static pressure than the flue gases in the furnace, but a higher static pressure than the combustion air at the fan intake. For this reason, there are two possible solutions for flue gas recirculation.

The first is to transfer the flue gas from the end of the flue gas tract directly to the furnace, whereby the existence of a flue gas recirculation fan is necessary in order to overcome the existing pressure difference.

Another option is to transfer the flue gas to the air fan suction line. In this case, the existence of a recirculation fan is not necessary, since the movement of the flue gas will be enabled by the difference between the static pressure of the flue gas in the boiler gas tract and the suction line of the combustion air fan. This method requires lower investment costs as it does not involve the installation of a recirculation fan. However, the disadvantage of this method is that there must be a certain margin of safety regarding the capacity of combustion air fan (about 10%) to prevent reduction of the boiler heating duty, since the combustion air fan has to transfer not only combustion air but also recirculated flue gas.

In the case of VK1, VK2 and VK3 the capacity of combustion air fans are close to the maximum rated capacity, so any percentage reduction in boiler heating duty would represent an unacceptable loss. Hence, in this case the method with recirculation of flue gas by additional flue gas fan was selected.

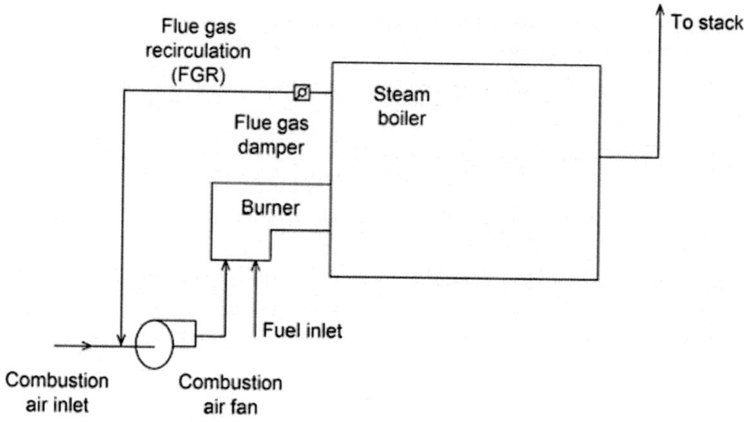

Fig. 4. A schematic view of FGR systems in steam boilers

For steam boilers (PK1 and PK2), the method without flue gas fan was chosen, by taking the flue gas from the front turning chamber, after the second gas passage, to the suction line of the existing combustion air fan (Fig. 4). The flue gas flow is enabled by the pressure difference between the flue gas in the front turning chamber and the combustion air fan intake. To achieve the required NOx emission levels, the flue gas circulation rate of 10% of the total amount of flue gas volume flow at a certain load was applied.

System Design for VK1 and VK2 Boilers (58 MW)

Fans for recirculation of flue gas of boilers VK1 and VK 2 are located in the boiler room, from the back side of the boilers, just above their flue gas ducts. Recirculated flue gas is taken from the underground horizontal flue gas duct, from which is led to the suction line of the flue gas recirculation fan by the vertical Ø500 mm duct and manually operated damper.

From the flue gas recirculation fan, the flue gas is conducted horizontally through a Ø500 mm duct to the front side of the boiler and split in two branch Ø350 mm ducts. Each boiler burner (2 pcs.) is supplied with combustion air through one air duct. Connection of flue gas into the air ducts is done just before the burner and after the air damper of the burner through the 600 × 150 mm duct. On the recirculated flue gas ducts Ø350 mm, just before the connection into the air ducts, flue gas dampers Ø350 mm are installed in order to control the flue gas recirculation rate.

In order to reduce flue gas fan load, the recirculation flue gas duct is connected to the combustion air duct after the air damper. This configuration reduces the pressure drop in ducting. For this reason, air ducts were redesigned and the existing air dampers were replaced with new ones (740 × 470 mm at VK1 and 1110 × 360 mm at VK2).

Maximal gas consumption per boiler is 6,430 m^3/hr (at 0 °C and 101.3 kPa), thus the maximal flue gas flow rate is 71,823 m^3/hr (at 0 °C and 101.3 kPa). With the recirculation rate of 10%, the amount of flue gases that is recirculated is 7,182.3 m^3/h (at 0 °C and 101.3 kPa).

Technical characteristics of the installed flue gas recirculation fans at boilers VK1 and VK2, with adopted margin of safety, are as follows: centrifugal fan with flow rate of 10,000 m^3/hr at 180 °C, and static pressure at outlet of 6.6 kPa. The fan is driven by electric motor with rated power of 37 kW and speed of 2955 rpm.

Design Solution for Boiler VK3 (116 MW)

The flue gas recirculation fan on boiler VK3 is located outside of the boiler room. Separation of the flue gas that is recirculated is done from the flue gas duct that comes out of the ground and goes to the heat utilizer. The part of flue gas flow is taken before the heat utilizer and led through a Ø750 mm duct and manually operated damper towards the suction line of the flue gas fan.

From the discharge line of the flue gas recirculation fan, the flue gas is led to the front side of the boiler VK3 and split in two branch Ø500 mm ducts. Each air duct supplies two burners at the same boiler level. Each of the Ø500 mm flue gas ducts branches into two Ø350 mm ducts just before the connection with air duct. The connection is done just before the burner and after the air damper of the burner through the 950 × 150 mm duct. On the recirculated flue gas ducts Ø350 mm, just before the injection into the air ducts, flue gas dampers Ø350 mm are installed in order to control the flue gas recirculation rate.

In order to reduce flue gas fan load, the recirculation flue gas duct is connected to the combustion air duct after the air damper. For this reason, air ducts were redesigned for the lower row of burners and the existing air damper was replaced with new one (1200 × 510 mm) while for the upper row of burners the duct was raised by 600 mm and the existing dampers were retained.

Maximal gas consumption of the boiler is 13,016 m^3/hr (at 0 °C and 101.3 kPa), thus the maximal flue gas flow rate is 145,388.7 m^3/hr (at 0 °C and 101.3 kPa). With the recirculation rate of 10%, the amount of flue gases that is recirculated is 14,538.8 m^3/h (at 0 °C and 101.3 kPa).

Technical characteristics of the flue gas recirculation fan of boiler VK3, with adopted margin of safety, are as follows: centrifugal fan with flow rate of 23,000 m^3/hr at 180 °C, and static pressure at outlet of 6.3 kPa. The fan is driven by electric motor with rated power of 75 kW and speed of 2975 rpm.

Design Solution for Steam Boilers

Recirculated flue gas flow is taken from the front turning chamber of the steam boilers through the horizontal Ø168.3 × 4.5 mm duct, and led to the suction line of combustion air fans to the provided connection pipe DN 150.

A separate segment for connecting the DN150 recirculation duct is attached to the combustion air fan intake of each boiler. The control damper (DN150) is located on the recirculated flue gas duct just before it is connected with the combustion air fan intake. The purpose of the control damper is to regulate the recirculation flow rate.

With the determined flue gas flow rates and ducting dimensions, resulting values of flue gas speed are in range 21.88 to 27.44 m/s, which are within the recommended values given by burner manufacturer.

The materials used for fabricating ducting were S235JRG2 steel sheets with dimensions 758 × 4 mm, 508 × 4 mm and 358 × 4 mm, while for fabricating circular cross section ducts the P235GHTC1 steel pipes and elbows with dimension Ø168.3 × 4 mm were used.

3 Results and Discussion

After combustion system reconstruction, process parameters setting and trial operation, the measurement of NOx emission from hot water and steam boilers was carried out in 2019 by accredited laboratory. A representative volume of flue gas was sampled from each boiler at constant flow from flue gas duct. The filter was used to remove powdery substances from the sampled gas before entering the analyser. The configuration of the gas sampling and conditioning system also included a cooling system for the removal of water vapour by condensation. The measuring equipment setup at steam boiler PK1 is shown in Fig. 5.

Fig. 5. Measuring equipment setup at steam boiler PK1 [30].

For flue gas analysis, the automatic gas analyser HORIBA PG-350E was used. The analysis is based on cross-flow modulation chemiluminescence detection for NOx components, while the CO detector operates with a cross-flow modulation non-dispersive infrared (NDIR) absorption method and the O_2 unit uses the paramagnetic method. The gas is sampled through a sampling line into the reaction chamber of the analyser, where it is mixed with excess ozone to determine the concentration of nitrogen oxides. The emitted radiation (chemiluminescence) is proportional to the concentration of NO present in the sampled gas. The emitted radiation is filtered by means of an optical filter and converted into an electrical signal by means of a photomultiplier. To determine the concentration of nitrogen dioxide, the sampled gas enters through a converter where nitrogen dioxide is reduced to nitrogen monoxide and analysed as previously described. The technical characteristics of gas analyser are shown in Table 3. The measuring methodology for NOx components complies with relevant standard SRPS EN 14792:2017 [31].

The measurements were carried out when the boilers operated at 55.5%, 77.7% and 100% of nominal load. Operating parameters of the emission reduction system are essential for its efficiency. At the time of measurement, flue gas recirculation was performed at recirculation rate of 10%. The results of measurements are shown in Tables 4, 5, 6, 7 and 8 for hot water boilers VK1, VK2, VK3 and steam boilers PK1 and PK2, respectively.

Table 3. Technical characteristics of flue gas analyser HORIBA PG-350E

Specification	Values
Measuring range O_2	0–5/10/25 vol%
Measuring range CO	0–200/500/1000/2000/5000 ppm
Measuring range NOx	0–25/50/100/250/500/1000/2500 ppm
Repeatability	± 0.5% of full scale (NOx: ≥ 100 ppm range / CO: ≥ 1000 ppm range) ± 1.0% of full scale (Except above)
Linearity	± 2.0% of full scale
Drift	± 1.0% of full scale / day
Response Time (T90)	10–30 s
Sample gas flow rate	Approx. 0.5 l/min

Table 4. Results of measurements for hot water boiler VK1

Component	Unit	Boiler load (%)			Reference values
		55.5%	77.7	100%	
O_2	%	3.69 ± 0.03	2.65 ± 0.02	2.16 ± 0.02	–
CO	mg/m^3	< 2	< 2	< 2	100
NOx as NO_2	mg/m^3	91.2 ± 3.6	96.2 ± 3.7	94.6 ± 3.7	100

Table 5. Results of measurements for hot water boiler VK2

Component	Unit	Boiler load (%)			Reference values
		55.5%	77.7	100%	
O_2	%	2.64 ± 0.02	2.18 ± 0.02	1.36 ± 0.01	–
CO	mg/m^3	< 2	< 2	< 2	100
NOx as NO_2	mg/m^3	91.6 ± 3.6	92.8 ± 3.6	93.0 ± 3.6	100

Table 6. Results of measurements for hot water boiler VK3

Component	Unit	Boiler load (%)			Reference values
		55.5%	77.7	100%	
O_2	%	3.31 ± 0.03	3.23 ± 0.03	2.65 ± 0.02	–
CO	mg/m^3	10.1 ± 0.3	11.8 ± 0.3	13.7 ± 0.4	100
NOx as NO_2	mg/m^3	85.3 ± 3.3	95.4 ± 3.7	93.9 ± 3.7	100

Table 7. Results of measurements for steam boiler PK1

Component	Unit	Boiler load (%)			Reference values
		55.5%	77.7	100%	
O_2	%	3.87 ± 0.22	3.05 ± 0.17	1.56 ± 0.09	–
CO	mg/m^3	< 2	< 2	< 2	100
NOx as NO_2	mg/m^3	108.4 ± 4.2	102.0 ± 4.0	99.0 ± 3.9	150 (110*)

Table 8. Results of measurements for steam boiler PK2

Component	Unit	Boiler load (%)			Reference values
		55.5%	77.7	100%	
O_2	%	4.07 ± 0.23	2.98 ± 0.17	2.47 ± 0.14	–
CO	mg/m^3	< 2	< 2	< 2	100
NOx as NO_2	mg/m^3	101.7 ± 4.0	97.9 ± 3.8	92.8 ± 3.6	150 (110*)

*Note: Reference value for nitrogen oxides expressed as nitrogen dioxide was additionally lowered to 110 mg/m^3 by the operator

The presented mass concentrations of pollutants (CO and NOx) were expressed at normal conditions (at 0 °C and 101.3 kPa), at dry flue gas basis and reference oxygen content of 3%.

The measured NOx emissions were in range from 85.3 to 96.2 mg/m^3 and from 92.8 to 108.4 mg/m^3 for hot water and steam boilers, respectively.

The measured concentration of CO and NOx in flue gas did not exceed the reference values defined by the national legislative. Emissions of NOx were by 20 to

50% lower compared to the measured values before the combustion system reconstruction. Emissions of NOx are higher at steam boiler which is expected because of the higher flue gas temperature at furnace and boiler exhaust. However, since the measured values for hot water boilers are close to the reference values, it can be concluded that there is a potential for further improvements by increasing the flue gas recirculation rate.

4 Conclusion

In this paper the effectiveness of the flue gas recirculation (FGR) system that is applied to reduce NOx emissions of the utility hot water and steam boilers in district heating plant is experimentally investigated. With the combustion systems reconstruction and application of FGR technology the emissions of NOx did not exceed the reference values defined by the national legislative. However, since the measured values for hot water boilers are close to the reference values, it can be concluded that there is a potential for further improvements by increasing the flue gas recirculation rate.

References

1. Directive 2001/77/EC of the European Parliament and of the Council of 27 September 2001 on the promotion of electricity produced from renewable energy sources in the internal electricity market.
2. Directive 2003/87/EC of the European Parliament and of the Council of 13 October 2003 establishing a scheme for greenhouse gas emission allowance trading within the Community.
3. Directive 2003/30/EC of the European Parliament and of the Council of 8 May 2003 on the promotion of the use of biofuels or other renewable fuels for transport.
4. Directive 2009/28/EC of the European Parliament and of the Council of 23 April 2009 on the promotion of the use of energy from renewable sources.
5. Directive 2012/27/EU of the European Parliament and of the Council of 25 October 2012 on energy efficiency.
6. Directive 2006/32/EC of the European Parliament and of the Council of 5 April 2006 on energy end-use efficiency and energy services.
7. Directive 2004/8/EC of the European Parliament and of the Council of 11 February 2004 on the promotion of cogeneration based on a useful heat demand in the internal energy market.
8. Directive 2009/125/EC of the European Parliament and of the Council of 21 October 2009 establishing a framework for the setting of ecodesign requirements for energy-related products.
9. Directive 2010/30/EU of the European Parliament and of the Council of 19 May 2010 on the indication by labelling and standard product information of the consumption of energy and other resources by energy-related products.
10. Directive 2010/31/EU of the European Parliament and of the Council of 19 May 2010 on the energy performance of buildings.
11. Directive (EU) 2018/2002 of the European Parliament and of the Council of 11 December 2018 amending Directive 2012/27/EU on energy efficiency.
12. Directive (EU) 2018/844 of the European Parliament and of the Council of 30 May 2018 amending Directive 2010/31/EU on the energy performance of buildings and Directive 2012/27/EU on energy efficiency.

13. Directive (EU) 2018/2001 of the European Parliament and of the Council of 11 December 2018 on the promotion of the use of energy from renewable sources.
14. Tao, Y., et al.: Emission characteristics of NOx, CO, NH_3 and VOCs from gas-fired industrial boilers based on field measurements in Beijing city. China Atmos. Environ. **184**, 1–8 (2018)
15. Yongyu, Z., Chunhua, W., Xu, C.: Combustion characteristic study with a flue gas internal and external double recirculation burner. Chemical Engineering & Processing: Process Intensification 162, (2021).
16. Baolu, S., Jie, H., Hongwei, P., Satoru, I.: Effects of internal flue gas recirculation rate on the NOx emission in a methane/air premixed flame. Combust. Flame **188**, 199–211 (2018)
17. Matsumoto, R., Iio, T.: Low NOx combustion of DME by means of flue gas recirculation. Journal of Power Energy Systems **2**(3), 1074–1084 (2008)
18. Habib, M., Elshafei, M., Dajani, M.: Influence of combustion parameters on NOx production in an industrial boiler. Comput. Fluids **37**(1), 12–23 (2008)
19. Nhan, H., Kwon, M., Kim, S., Park, J.: CFD investigation of NOx reduction with a flue-gas internal recirculation burner in a mid-sized boiler. J. Mech. Sci. Technol. **33**(6), 2967–2978 (2019)
20. Yaojie, T., et al.: NOx reduction in a 40 t/h biomass fired grate boiler using internal flue gas recirculation technology. Appl. Energy **220**, 962–973 (2018)
21. Jankes, G., Trninić, M., Stamenić, M., Simonović, T., Tanasić, N., Labus, J.: Biomass gasification with CHP production: A review of state of the art technology and near future perspectives. Therm. Sci. **16**(suppl. 1), 115–130 (2012)
22. Jun, L., Xiaoyu, L., Sheng, Y., Quangong, L., Weishu, W.: Influence of flue gas recirculation on the performance of incinerator waste heat boiler and NOx emission in a 500 t/d waste-to-energy plant. Waste Manage. **105**, 450–456 (2020)
23. Milivojević, A., Adžić, M., Gojak, M., Stamenić, M., Adžić, V.: Analysis of the performance of a low-power atmospheric burner for gas appliances for households and their impact on the emission and stability of the burner. Therm. Sci. **25**, 1891–1903 (2021)
24. Adžić, V., Milivojević, A., Stamenić, M., Adžić, M.: Investigation of CO_2 diluted methane and propane swirling premixed flames using CH* chemiluminescence imaging. Therm. Sci. **23**, 1511–1521 (2019)
25. BMEPB (Beijing Municipal Environmental Protection Bureau) (2015). Standard of boiler air pollution in Beijing. http://www.bjepb.gov.cn/eportal/fileDir/oldfile/bjepb/resource/cms/2015/06/2015060311453388513.pdf, Last Access: 3 June 2015.
26. Best Available Techniques (BAT) Reference Document for Large Combustion Plants. Industrial Emissions Directive 2010/75/EU (Integrated Pollution Prevention and Control), https://publications.jrc.ec.europa.eu/repository/handle/JRC107769.
27. Genić, S., Stamenić, M., Živković, B., Čantrak, Đ, Nikolić, A., Brdarević, L.J.: Handbook for training of energy managers for industrial energy. Republic of Serbia, University of Belgrade - Faculty of Mechanical Engineering, Ministry of Mining and Energy (2017).in Serbian
28. Jankes, G., Stevanović, D., Stamenić, M.: A review of development of new heating system in industrial furnaces, Proceedings GAS 2003, Vrnjačka Banja, 22.-25.06.2003, in Serbian
29. Decree on reference values for emissions of pollutants into the air (Official Gazette of the Republic of Serbia, No. 06/2016, in Serbian.
30. Report on the measurement of emissions of pollutants into the air, Laboratory „Occupational safety and environmental protection Ltd." Belgrade, 2019, in Serbian.
31. ***, SRPS EN 14792:2017 - Stationary source emissions - Determination of mass concentration of nitrogen oxides - Standard reference method: chemiluminescence.

Advanced Procedure for Making Vibro Motor Coupling of Basket Crusher by Welding and Plasma Cutting

Aleksandra Mitrović, Đorđe Matić, Zorana Golubović, and Aleksandar Sedmak

Abstract The aim of this paper was to improve the technological procedure, operations and construction of the vibro-motor coupling of the crusher bucket from steel grade C45. The work defines important materials, approaches and methods of welding and plasma cutting. Problems with time-consume and additional finances which are appearing in manufacturing processes in mechanical engineering are also presented in detail in the paper. Also, analysis and solution of the problem are shown. The connection between the two couplings is not made of pieces, but is cut from the board in the appropriate quality. During the fabrication, in addition to machining on a lathe, making a wedge on a nut machine for making wedges, drilling holes, threading, plasma cutting operations and welding operations, the heat treatment process was applied to obtain required material properties, cold galvanizing process, and also process of rubber molding located between two couplings is presented. Nowadays, plasma cutting and welding are highly reliable methods of metal processing that immediately bring profit to customers. The great importance of the proposed and successfully realized idea of a finished piece fabrication has its practical application.

Keywords Plasma cutting · Turning · Heat treatment · Steel · Welding

1 Introduction

Processing of the piece by scraping, on the way that it has been done till now is considerably time-consuming process. One of the most important factors in mechanical engineering is time of production, and as that, it is very important factor that needs to be considered and resolved [1, 2]. In order to make better approach to subject, factor of time of production is observed in literature and practice, with

A. Mitrović (✉) · Đ. Matić
Department of Computer-Machine Engineering, The Academy of Applied Technical Studies Belgrade, 11000 Belgrade, Serbia
e-mail: aleksandra.mitrovic@visokatehnicka.edu.rs

Z. Golubović · A. Sedmak
Faculty of Mechanical Engineering, University of Belgrade, 11120 Belgrade, Serbia

N. Mitrovic et al. (eds.), *Current Problems in Experimental and Computational Engineering*, Lecture Notes in Networks and Systems 323,
https://doi.org/10.1007/978-3-030-86009-7_18

solving the practical problem [3, 4]. There are numerous researches that have dealt with the analysis of different deformation behaviors, integrity assessments and failures [5–14]. Welding is on the most efficient processes of connecting parts in solid state, and because of that it has a broad range of applications in different industries. Main goal is to solve problems which are characteristic for steel welding by melting processes. Distinctive joints can be made by joining two different base materials, and because of these possibilities, applications of welding are various, i.e., automotive industry, ship building, airspace industry, building construction, or process industry [15]. There is a need for constant upgrading of welding technology precisely because of its widespread applications [16]. Welding zone, because of its complex thermo-mechanical state is mostly in plastic state, and due to that characteristic it can be categorized as heat affected zone where experimental determining of temperature is very difficult. Better control of the welding process can be achieved by the analysis of these complex thermo-mechanical conditions which occur during welding [17].

Consistent production of high-quality welds demands experienced staff. Knowledge of proper selection of welding parameters for a required task is needed in order to get good quality of weld, which can be further identified by correct bed geometry, micro-structure and the amount of spatter [18].

The aim of this paper is to reduce the time and costs of fabrication the coupling of the vibro-motor of the crusher basket by welding and plasma cutting. The connection between the two couplings is not made from the piece, but is cut from the board in the appropriate quality, i.e., C45, which is also the material of the coupling body. During the fabrication, in addition to machining on a lathe, making a wedge on a nut machine for making wedges, drilling holes, threading, plasma cutting operations and welding operations, the heat treatment process was applied to improve material properties, cold galvanizing process, and also process of rubber molding located between two couplings is presented.

Material properties are changed minimally in plasma cutting, because there is no significant heat generation during plasma cutting. When it comes to lathe machining, most of the required elevations, except for the internal openings, are "open" elevations, so the dimensions may deviate if necessary. The required quality of the material is necessary to be accomplished and retained, and it must not deviate depending on the type of steel, the material from which the rubber is cast, and the type of welding electrode. Process of surface protection (galvanizing) is an important part because it prevents rapid corrosion of the material, which can further lead to rotting of the rubber placed between the joints [19, 20].

Influence of changes in material properties in welded joint zones is adjoint to the influence of constraint (existing in macroscopically heterogeneous and homogenous structures) [21]. Many studies covered different fracture and failure analysis of welded joints which are appearing under different exploitation conditions [23–26]. It is necessary to have wide knowledge of different aspects because of the multidisciplinary approach in welding and joining materials [27].

The manufacturing process, described in this paper, largely reduces the cost and time of fabrication, and thus the manufacturer becomes more competitive in the market. Experimentally, the place where the upper and lower part of the joint are

a b

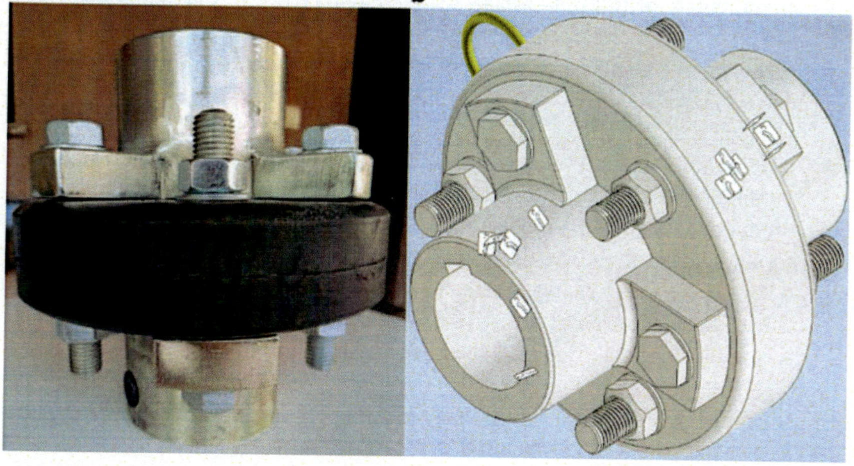

Fig. 1. **a** Assembly formed by welding with a rubber element; **b** Assembly model formed by welding with a rubber element.

connected does not suffer a large load, so welding can be performed without problems and without checking the welded joint, because the rubber part of the joint is of the greatest importance when it comes to shock mitigation (Fig. 1).

2 Experimental Procedure

This paper presents the welding of a joint from two parts—the connection of the lower part (clamp) and the upper part (hub). The joint is made in a ratio of 1: 1 and the type of steel is C45. The workpiece in the form of a clamp was cut with plasma from a plate 25 mm thick, with dimensions 250 mm × 500 mm, with the attached certificates confirming the required type of steel. The drawing of the piece was done by using software AutoCAD [28], and the drawing is saved as a version of Auto Cad 2000 dxf, which as such is "imported" into the program for writing G-code, and further program "Fast Nest" made the requested piece from the code.

The workpiece in the form of a hub was placed on a circular saw (Pehaka, Germany) whose cutting strip length is 5400 mm with a maximum cutting diameter of 450 mm, and was cut to 75 mm. The preparation measures Ø120 × 75 mm. After the saw, the piece goes to the universal lathe "Potisje PA 501", (SFRY production), with a maximum machining of 500 mm in diameter and 1000 mm in length, whose motor is 7.5 kW, and the chuck is Ø250 mm/3. In the first clamping, the first drilling of the hole in the center of the piece along the entire length was done, with a drill Ø50. After that, the opening was widened to Ø68 mm, which was grinding premeasure. Processing is divided into 6 passes of 3 mm. Also, in the first clamping, the alignment of the forehead was done (frontal processing), and longitudinal processing of the outer diameter from Ø120 mm to Ø112 mm—a pre-measure for fine processing. In the second clamping, a "shoulder" with measurements 10 mm × 90 mm was made due to a better connection when welding for the "claw".

After that, a channel was made on the hub on a machine for making an internal channel for a wedge (nut machine) (Fromag, Germany). The machine is used for the production of standard internal channels, with width from 5 to 32 mm, and maximum production length 400 mm. With the installation of a knife holder and a suitable knife for the same tool, the standard width of the channel for the required diameter is 20 × 4.9 mm. When the channel is made, then hole for the fuse is drilled on the opposite side of the channel. The hole Ø10.5 was drilled 20 mm from the end on the flat side where the thread M 12 × 1.5 was made by hand, or by screw tap.

After processing the hub to the pre-final measure, plasma cutting of the clamp from the plate follows, followed by turning. When cutting the clamp, the machine first makes the 3 inner openings with Ø21 mm on the rods, and an opening in the middle Ø60 mm, after which the outer contour was cut. Changes which occur in material when it is cut by plasma have minimal influence on its structure. When machining the clamp, the middle hole is first widened from Ø60 mm to Ø68 mm, with processing divided into 4 passes of 2 mm. This is followed by machining the hole, i.e., countersinking, for a better connection between the hub and the clamp and preparation for welding, where the hole is machined to a size of Ø90.5 mm × 10.5 mm.

After machining to the pre-final measure, the pieces passed heat treatment in the furnace at a temperature of 860 °C for 25 min, where immediately after removal from the oven they are abruptly added to cooling, special hardening, oil (Famkal 22, Serbia). It gives optimal quenching results in the temperature range from 30 °C to 50 °C, and it does not form a precipitate during hardening. Quenching is achieved by adding it to the oil. The pieces need to be degreased, cleaned and the base surface should be sanded for measurement. The Rockwell method gave 28 HRC. The pieces were then placed in a 485 °C kiln for an hour, where 23 HRC was obtained, since it was the goal. When the set hardness was reached, the pieces were taken for welding, which is performed with a base electrode Ø4 mm EVB 50 (Elektrode Jesenice d.o.o.). This is a tested electrode for welding steels and steel castings with strength of about 610 N/mm^2. Welds are tough even at low temperatures and are resistant to cracking, with excellent welding, technological characteristics and a stable arc. Efficiency is about 118%.

After welding, normalization of the welds took place for an hour, at the same temperature 485 °C as was the case with the tempering. The joint is further processed on the lathe (Potisje PA 501), where the final shape is obtained. Also, chemical protection of the piece from rust, cold galvanizing was done, after which the last machining was followed by grinding of the inner opening of the coupling from Ø69.5 mm to Ø70 mm.

The first step in the technical preparation of a piece is to check the dimensions on the drawing and consult with existing errors (if any). The piece was prepared in the software "Auto Cad", it is important to save it as "DXF 2000", which is a version that can later be imported into the program for creating code. Such a version of the saved file is required by the program itself and may be different depending on the program which generates G-code. In this case, the code generator "Fast Nest" requires a saved version of the "DXF 2000" file. Figure 2 shows a piece drawn in "Auto Cad".

Before the piece is "imported", it is necessary to set several parameters related to: internal and external input and output, the initial distance of the piece in relation to the plate, the distance between the pieces, marking the order and entering data on material type and characteristics plates. When adding a piece, there are also options that show some very important parameters that are of great importance in further work, i.e., parameters based on which the time of making a piece can be calculated, and they are: Length of cut per piece, Number of entries within a piece, Weight of a piece and the area occupied by the piece. Also, the same parameters are very important when determining the cutting price, whether the cutting is determined by the cutting length or the production time. The required number of pieces can also be entered in the options table on the board, the joint cut line can be determined (if possible), and the position of the inner or outer entrance can be changed. After all the set parameters are adjusted before adding the pieces, the pieces should be packed on the board for the best possible use, if they are not automatically packed well. After all operations performed during Nesting, i.e., G-code creation (Fig. 3), the obtained data is sent via the network to the machine, or is physically transferred via USB.

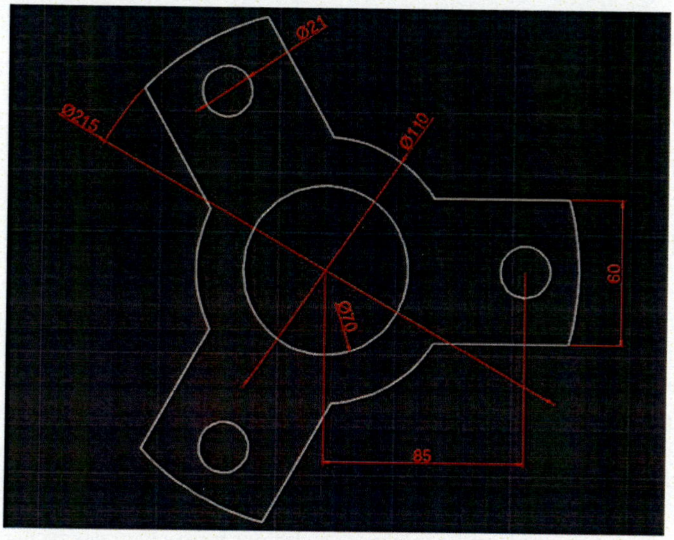

Fig. 2. Piece drawn in the program AutoCAD 2014.

Fig. 3. **a** Appearance of packed pieces. **b** G-code completed.

3 Results and Discussion

3.1 Plasma Clamp Cutting

Before cutting starts, it is necessary to check the several parameters: direction of the board in relation to the holders, guidance of the gun in relation to the board, cleanliness of the board, cleanliness of the nozzle and electrode - cathode, gas pressure, height of the gun when cutting and how the board is cut. These parameters affect the quality of the cut during plasma cutting, and if they are all set, there are no problems with cutting.

The gas pressure in the plasma apparatus must be constant and proportionally adjusted in relation to the thickness (with not large changes), as well as the cutting height which must be constant in order for the cut to be the same along its entire length and without the appearance of furrows. When adjusting the distance of the burner from the material, the height from which the burner "breaks" the material should be taken into account, because if the height is too high cut will be irregular from the start, i.e., it will be chamfered, and if the height is too low, the nozzle and the electrode (cathode) will be damaged very quickly. The direction of the burner must

Fig. 4. The appearance of a cut piece.

also be adjusted, and if it is not adjusted well, the cut will be chamfered, although all other parameters have been adjusted. Condition of the boards (i.e. rust) influence the plasma cutting, because when cutting, slag will be thrown from the bottom surface of the plate and skip parts, which is not good for the nozzle and the electrode, and on that way a bad cut is obtained that often cannot be used later. Rusty pieces can be cleaned or wiped, and when it comes to sheets that are cold rolled, they are coated with oil before cutting. After the piece is cut (Fig. 4), which lasted about 2 min, the measurement and quality control of the cut should be done.

Cutting the upper part of the coupling—hub on the saw. For purpose of this research the saw Pehaka 450 was used, the work piece was Ø120 mm in diameter with needed final dimension to be cut to a length of 75 mm. The characteristics of the saw "Pehaka 450" are: automatic machine of high performance, high accuracy, compact, precise, linearly guided hand, cutting up to a width of 450 mm, cutting tape length 5400 mm, cutting tape thickness 1 mm and water cooling.

Universal lathe "Potisje PA 501". Turning was performed on the universal lathe "Potisje PA 501". The characteristics of the lathe are: Maximum spindle speed 2000 rpm, Maximum machining diameter 500 mm, Maximum machining length 1000 mm, Maximum bar passage - through the entire length of the lathe and Motor power 7.5 kW.

Turning hub processing. After the pieces were cut to the required dimension, turning took place.

Tools and tool marks (turning). Knives for external processing, longitudinal and transverse are knives whose supports have a square cross-section and according to the quality of the surface they process, they belong to knives for rough processing. With a knife for longitudinal machining (Hahn & Kolb, Germany), the tile is interchangeable. Turning knife for internal processing have support with square cross section and interchangeable tile.

Process of scraping. The workpiece was processed on the CNC lathe Potisje PA 501 with all necessary steps: drilling hole with a drill Ø50 mm, widening hole with a knife for internal processing to Ø68 mm, which is a pre-measure for fine processing. Transverse processing was performed with a knife for transverse processing to a given length (pre-measurement). Longitudinal processing followed on the length that was possible in the first operation from a diameter of Ø120 mm to a diameter of Ø112 mm. In the second procedure, longitudinal processing was performed on the entire length of the piece to a diameter of Ø112 mm and a "shoulder" was performed for strength and stability when the piece is joined to the clamp on a diameter of Ø90 × 10 mm. Final processing steps meant knocking down the corners, align the rods in alignment with the hub with the knife for final processing, closing the diameter to the final dimensions and obtaining a smooth surface.

Order of operations (first clamping). Hub processing begun with drilling a central hole Ø50 mm through the entire length of the 75 mm piece, with a drilling speed of 265 rpm. When drilling the hole, a Ø50 mm drill bit was used with the constant presence of coolants and lubricants. After the drilling, the frontal processing was done—the alignment of the piece.

The frontal processing (Fig. 5) was done with a knife for transverse (frontal) rough processing at a speed of 350 m/min, which is followed with the first clamping, and longitudinal processing in the length of 40 mm to a diameter of 112 mm was also done, i.e., making the pre-measure for fine processing (Fig. 6). In order not to repeat the same operations, the processing was done on both pieces in the same order (on both hubs).

Sequence of operations—second clamping. Several operations were performed in the second clamping, and the first among them was the expansion of the inner opening from Ø50 mm to Ø68 mm with a knife for internal roughing along the entire length of the opening (Fig. 7). This pre-measure for fine processing was made at a speed of 420 rpm.

Fig. 5. Frontal processing—alignment of pieces.

Fig. 6. Longitudinal processing.

Fig. 7. Expansion of the inner opening by rough machining.

This was followed by longitudinal roughing which included the processing of the outer diameter at Ø112 mm, and the making of the "shoulder", and the removal of the edge on the inner side of the "shoulder" in order to achieve a stronger and more stable weld and the formation of a stronger weld root (Fig. 8a)). This processing was done at a speed of 500 rpm. After the longitudinal roughing, the inner opening was widened by fine machining from Ø68 mm to Ø69.5 mm, which is, as such, a grinding pre-measure (Fig. 8b)).

Processing of the lower part of the joint—coupling. When processing the lower part of the joint—coupling, all operations were performed in one clamp, with a knife for internal processing. The opening was widened from Ø60 mm to Ø68 mm, expansion by fine processing to pre-measure Ø69.5 mm, the "shoulder" was made to a diameter of Ø90.5 mm and 10.5 mm deep, and the outer edge was lowered as it

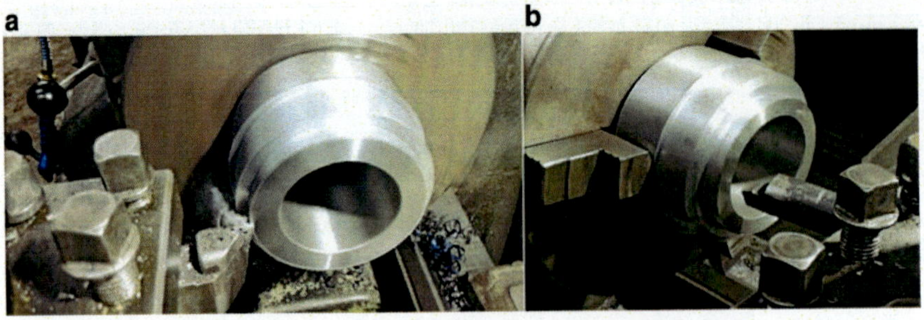

Fig. 8. a Longitudinal roughing. b Expansion of the inner opening.

Fig. 9. Processed lower part of the joint—coupling.

was the case at the hub in order to obtain the strongest possible weld root. Figure 9 shows the lower part of the coupling, on which all the necessary operations were performed before carrying on quenching.

3.2 *Making a Wedge on the Hub*

Machine Fromag (Germany) was used to make internal wedge channels (nut machines). The machine is specific because it is fabricating only standard channels for certain diameters, and channels for particular diameters. The maximum length of the channel is 400 mm, the maximum diameter on which the channel can be word made is Ø130 mm, and the width of the channel can be from 5 to 32 mm. The quality obtained after processing is the same as with grinding.

Mounting the knife with the support on the main pillar. When mounting a knife with support, the choice of the knife is important, because for each diameter that is fabricated, precise width of the channel is determined. On the main support, there are steps that go from the bottom to the top, each of the following is 5 mm larger, and a piece is placed on them. There are also knives of different lengths and different manufacturers, but the mostly used knives are the ones produced by the manufacturer who produced the machine. When making the inner channel, for a diameter of Ø70 mm, the standardized width of the channel is 20 mm × 4.9 mm depth.

Mounting pieces. After mounting the knife with the bracket on the main support of the machine, careful mounting of the piece is taking place (Fig. 10). If the inner openings are of diameter e.g. Ø68 mm, it is necessary to use inserts, which allow the piece to come across a step on a diameter Ø70 mm.

When mounting the piece, another very important thing is that the piece must be clamped evenly with the piece holder. Firstly, it is tightened with screws that are placed horizontally, i.e., on the side, and then with screws that are placed on top of the holder. After mounting the piece, the stroke of the knife was determined, which is automatic - the length of the piece with an additional 2 mm for the output of the knife. When knife took a certain path, the piece was tangled and the cutting depth on the vernier was adjusted.

Channel creation. When all the parameters were set, and the piece was tightened, processing followed. The channel was made with a full width of 20 mm, where the depth of cut was 0.02 mm per pass. As with lathe machining, coolants and lubricants were constantly present during channel construction, as shown in Fig. 11.

At the end of the processing, the piece was removed from the main support, wiped of grease and the dimensions of the obtained channel were checked (Fig. 12).

Fig. 10. Mounting the piece on the main support.

Fig. 11. Construction of the channel with the constant presence of coolants and lubricants.

Fig. 12. View the completed channel on a piece.

3.3 Drilling Fuse Holes and Threading

The holes were drilled on a pole drill (Wesselmann, Germany), which has measuring verniers and a movable table on which there is a clamp in which the piece is placed. The drill has a working pace of 300 mm along the height of the pole and can drill with a drill with maximum of Ø40 mm. The motor has a power of 4 kW, and has 3 drilling speeds. Primarily, the center of the hole is marked, chirped, and drilled (Fig. 13a)). The diameter of the hole was Ø10.5 mm as a preparation for the M12 × 1.5 threads which was hand-made with a notch (Fig. 13b)).

Fig. 13. a Drilling holes; **b** Handmade threading.

3.4 *Heat Treatment of Materials*

The material was heat-treated in the basic heat treatment plant. Furnaces in heat treatment are mostly in constant operation, because when ignited, 16 kW is required.

Tempering pieces. Depending on the type of heat treatment and the type of material, the temperature at which the material is heated is set and the time the material spends in the furnace is determined. When inserting the pieces into the oven, a tool with a long handle was used, and the time required to insert the pieces should be as short as possible, so that the oven loses as little heat as possible. The pieces were heat treated at 860 °C for 25 min. After heating, the pieces are inserted into the oil baths. The pieces taken out of the furnace were immediately inserted into the quenching bath, where certain hardness was obtained by sudden cooling. The time spent in the oil was about 5 min, after which the pieces were degreased and cleaned.

Hardness test. Before measuring on the pieces, it is necessary to grind the base surface (abutting) and the measuring surface. Hardness was tested by the Rockwell method as it is mentioned before.

Dismissal. Value which was obtained when hardness test was done is 28 HRC, and the aim was to obtain 22-23 HRC. In this case it was proved that pieces are the best with hardness of 22–23 HRC, so in order to obtain that value, the method of release was done.

Fig. 14. Piece after the normalization process.

3.5 *Welding*

Welding was performed by manual electric arc process (MAW) where a temperature of about 3200 °C was formed. It is welded with the "forward–backward" technique, and the joint that is formed depending on the mutual position of the piece is the "T joint", this weld was made from several passes. The current used in welding is direct current, and it is welded with a strength of 125 A. After welding, a visual inspection of the weld was performed by an experienced welder. The welded joint showed satisfactory characteristics by visual inspection.

Heat treatment procedure was used to remove all unfavorable changes caused by welding. The kiln used for this process is the same kiln as for tempering. The temperature during the heat treatment was 485 °C, and the time spent in the oven was 60 min. Heat treatment significantly improves the mechanical properties of steel such as yield strength, tensile strength, elongation, cross-sectional narrowing and toughness. In Fig. 14, the piece is shown after the normalization is done.

3.6 *Fine Machining*

In the fine machining operation, the outer diameter of the hub was measured to the extent of 110 mm, and at the same time the clamp rods were aligned with the hub at 90° (Fig. 15).

During the mentioned processing, the radius on the weld $r = 8$ mm was made. Also, in order to improve aesthetics, the weld is grounded with a grinder in places where it was not possible to approach with a turning knife. After the mentioned operation, the rods were straightened from the outside and the edge at the opening was chamfered—15 mm × 35°, which is shown in Fig. 16.

Fig. 15. Fine machining of the outer diameter and straightening of the rods.

Fig. 16. Straightening the rods on the outside and lowering the edges.

3.7 Insertion of Drilled Holes

When the fine processing of the outer contours was completed, the holes were inserted by drilled with a drill with a final diameter of Ø50 mm (Fig. 17a, b).

3.8 Surface Protection by Cold Galvanizing

The cold galvanizing treatment was performed by immersing the pieces in baths for rinsing and degreasing, and then in baths containing zinc salts and zinc anodes which are decomposed by the electrolysis process and applied to the metal. Cold treatment allows very thin sheets to be galvanized without deforming, or sheets with small openings without the risk of closing them when immersed in the solution, in this case a threaded hole drilled for the fuse - M12 × 1.5. After degreasing, the pieces are washed to achieve complete electrolysis. Figure 18 shows a galvanized piece.

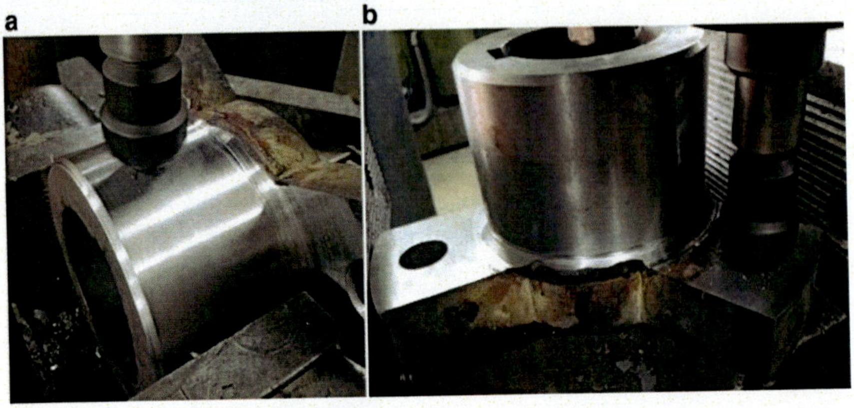

Fig. 17. **a** Countersink bolt hole; **b** Countersink the hole for connecting the piece to the rubber.

Fig. 18. Appearance of a galvanized piece.

Fig. 19. Pressing tool.

3.9 Making a Rubber Part of the Coupling

When it comes to the rubber coupling, it is very difficult to find it in free sale, so a special tool for pressing and baking raw rubber is made for these types of needs. The tool is made of 42CrMo4 steel to dimensions that are 2% larger than the specified dimensions, in order to shrink the tire during cooling. Figure 19 shows the tool into which the tire is inserted.

The tool consists of two parts, the upper—cover, and the lower into which the material is inserted, and which contains spikes for guiding and directing between the two tools, and on which inserts are also inserted where the openings will be required.

The rubber is composed of 95% rubber and 5% sulfur, and can be found in boards of 1000 mm × 2000 mm, or in strips of width from 10 to 1000 mm of certain lengths, sometimes up to 100 m, depending on the width. Due to the large dimensions of the tire, it is necessary to cut it and pack it in a tool. The pressing process itself is performed by placing the tool on the press, the lower tool stands on the hot plate, on which it is heated at the same time, and the upper tool squeezes the piston. In most cases, about 10% more rubber is added than needed, due to the shrinkage of the tire during cooling, and because the liquid rubber heats up, and forms a mass. The baking temperature during production is 150 °C. The lower and upper tools are not fully assembled during baking and this is another reason why it is necessary to make the tool 2% larger (Fig. 20).

Production time per piece is about 45 min. The tire from which the excess should be removed is shown in Fig. 21a). The final dimensions of the tire in technical drawing are shown in Fig. 21b).

Important aspects in modern industry sector are quality and cost control of products. There are numerous advantages of welding and its applications and for that reason, it is widespread technology [29]. However, it is hard and not very reliable to predict behavior of welded joints, due to numerous reasons, i.e., chemical and structural non-homogeneity, changes which occur in wall geometry at welded joint, defects which are forming due to welding, residual stresses. When different steels are

Fig. 20. The pressing process.

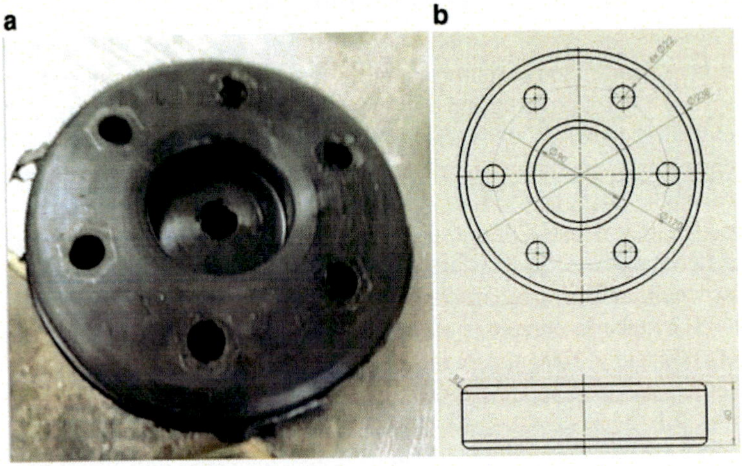

Fig. 21. **a** Made tire; **b** Final dimensions of the tire.

used for making the joints, because of their different physical and mechanical properties, it is more complicated to predict their behavior during exploitation process. Also, because physical and mechanical properties of used steels are different, there is the difference in heterogeneous and homogenous joints [30]. For casting and forging processes, it is necessary to make first tools, either for forging or casting. For that reason, the most important advantages of the improved process is making a coupling compared to the process of obtaining the same by casting and forging.

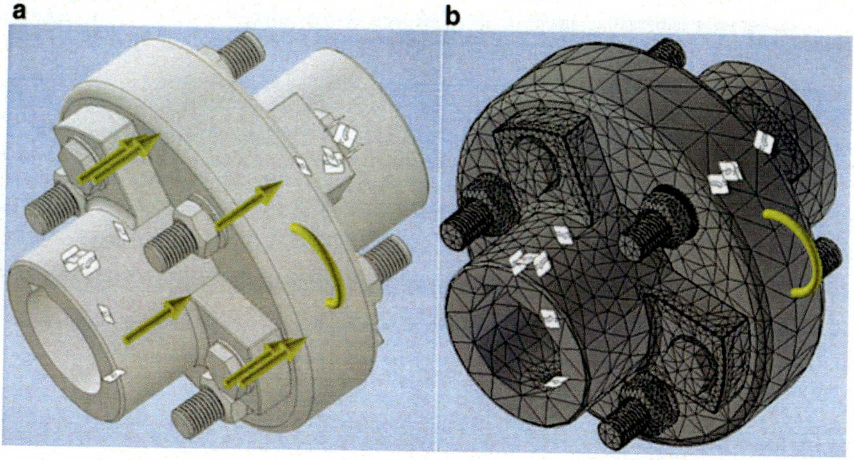

Fig. 22. Numerical model. a Model loading, b Model mash.

4 Numerical Modeling

Finite element method (FEM) is used as a tool for numerical analysis of vibro motor coupling of basket crusher subjected to two load types: force of 1500 N and moment of 50 Nm (as shown in Fig. 22a). FEM software package used in this paper is SolidWorks 2016. FEM model consists of 224,709 solid tetrahedral elements and 128,076 nodes. Mechanical properties of steel are: Yield Strength 207 MPa, Ultimate Tensile Strength 345 MPa, Young's Modulus 220 GPa, Poisson's Ratio 0.275, and for rubber are: Yield Strength 10.34 MPa, Ultimate Tensile Strength 6.5 MPa, Young's Modulus 0.003 GPa and Poisson's Ratio 0.49.

Since the shorter joints were welded to the hub, numerical simulations showed the effect of the working load on these joints.

Figure 23 shows Von Mises stress results for force of 1500 N and moment of 50 Nm. Von Mises scale in MPa is given on ordinate. Highest Von Mises stress value in the model is 45.2195 MPa.

As it can be seen from the numerical model, the rubber element fully absorbs all loads and the coupling elements are minimally loaded.

5 Conclusion

In this paper, significant changes have been applied when it comes to the method of production, time of production, price of materials. The processes that contributed to this were plasma cutting of the lower part of the joint and welding for the upper part of the joint. Also, it is important to mention the better availability of materials, meaning it is easier to get plate material and material with a smaller diameter Ø110 mm,

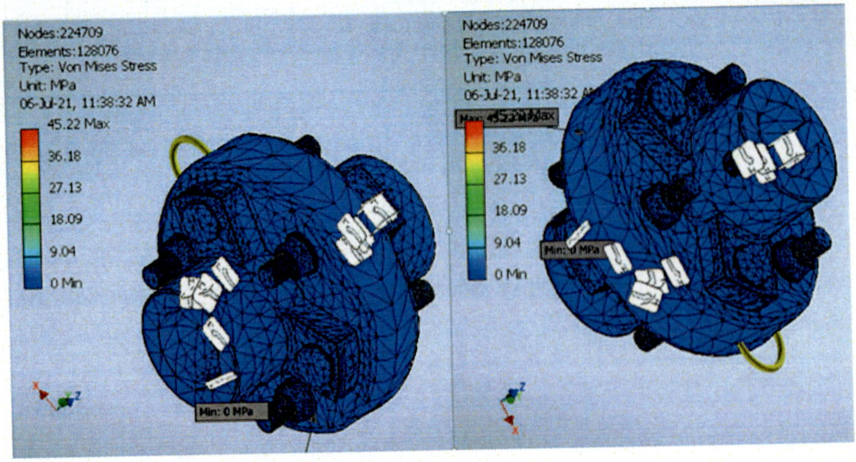

Fig. 23. Von Mises stress results for force of 1500 N and moment of 50 Nm.

compared to the original procurement of material with a larger diameter Ø220 mm. Previous procedure implied that about 50% of the material fell off while turning and milling, and thus pieces were obtained with more hard work.

The welded joint does not suffer a large load, and the additional safety of the welded joint is secured by the construction of the "shoulder" at the place where it was welded, i.e., at the upper and lower joints. The improved coupling fabrication process does not affect the shock-absorbing rubber fabrication process, and the tire does not wear more intensely than the previous coupling fabrication process.

The advantages of the improved process of making a coupling compared to the process of obtaining the same by casting and forging is reflected in the fact that for both of these processes it is necessary to first make tools, either for forging or casting. However, such production is not possible in all cities/towns, and the technical equipment also differs among factories and workshops. Therefore, this way of making the coupling is flexible, it is not necessary to make tools for casting and forging, and therefore it is more cost-effective.

Numerical model showed that the rubber element fully absorbs all loads and the coupling elements were minimally loaded.

Practical application of proposed and successfully realized method of finished parts fabrication is important achievement of this paper. The company that manufactures couplings is seriously considering the solution presented in the paper, i.e., to continue to work in the future with the improved process of making couplings of vibro motor crusher basket by welding and plasma cutting.

Acknowledgements The authors of this paper would like to express gratitude to the Academy of Applied Technical Studies Belgrade for the support.

References

1. Nikolić, M.: Electrode coated welding. Society for the Advancement of Welding in Serbia, Belgrade (1987)
2. Jovanović, M., Lazić, V.: Casting and Welding Technology. University of Kragujevac, Faculty of Engineering Sciences, Kragujevac (2015)
3. Grabulov, V., Blačić, I., Radović, A., Sedmak, S.: Toughness and Ductility of High Strength Steels Welded Joints. Structural Integrity and Life **8**(3), 181–190 (2008)
4. Casavola, C., Giordano, V., Pappalettere, C.: Mechanical Behaviour of as-Welded And Smooth Cord Titanium Welded Joints. Structural Integrity and Life **12**(2), 137–142 (2012)
5. Šarkoćević, Ž, Rakin, M., Arsić, M., Sedmak, A.: Fabrication of high strength seam welded steel tubes and quality indicator testing. Struct Integr Life **8**, 81–98 (2008)
6. Marenić, E., Tonković, Z., Skozrit, I.: On the calculation of stress intensity factors and J-integrals using the submodeling technique. J Pres Ves Technol **132**(041203), 1–12 (2010)
7. Medjo, B., Rakin, M., Arsić, M., Šarkoćević Ž., Zrilić M., Putić S.: Determination of the load carrying capacity of damaged pipes using local approach to fracture, Mater Trans – JIM 53, 185–190 (2012).
8. Rakin, M., Arsić, M., Medjo, B., Šarkoćević, Ž, Ivanović, I., Sedmak, A.: API J55 steel casing pipe with an initial surface crack under internal pressure – determination of fracture parameters. Key Eng Mater **488–489**, 577–580 (2012)
9. Jevremović, I., Debeljković (Mitrović), A., Marc, S., Mohsen, A., Nešić, S., Misković-Stanković, V.: The mixture of dicyclohexilamine and oleylamine as corrosion inhibitor for mild steel in NaCl solution saturated with CO_2 under both continual immersion and top of the line corrosion. J Serb Chem Soc **77**(8), 1047–1061 (2012)
10. Dimić, I., Arsić, M., Medjo, B., Stefanović, A., Grabulov, V., Rakin, M.: Effect of welded joint imperfection on the integrity of pipe elbows subjected to internal pressure. Tech Gaz **20**, 285–291 (2013)
11. Dimić, I., Međo, B., Rakin, M., Arsić, M., Šarkoćević, Ž., Sedmak, A.: Failure prediction of gas and oil drilling rig pipelines with axial cracks, In: Proc Mater Sci 3, 955–960 (2014) (20th European Conference on Fracture ECF20, Trondheim).
12. Zhang, B., Ye, C., Liang, B., Zhang, Z., Zhi, Y.: Ductile failure analysis and crack behavior of X65 buried pipes using extended finite element method. Eng Fail Anal **45**, 26–40 (2014)
13. Lavigne, O., Gamboa, E., Costin, W., Law, M., Luzin, V., Linton, V.: Microstructural and mechanical factors influencing high pH stress corrosion cracking susceptibility of low carbon line pipe steel. Eng Fail Anal **45**, 283–291 (2014)
14. Medjo, B., Rakin, M., Gubeljak, N., Matvienko, Y., Arsić, M., Šarkoćević, Ž: Sedmak, A: Failure resistance of drilling rig casing pipes with an axial crack. Eng. Fail. Anal. **58**, 429–440 (2015)
15. Veljić, D., Rakin, M., Medjo, B., Mrdak, M., Sedmak, A.: Temperature Fields in Linear Stage of Friction Stir Welding - Effect of Different Material Properties. Therm. Sci. **23**(6B), 3985–3992 (2019)
16. Khanna, OP: A text book of welding technology, Dhanpat Rai Publications Ltd., P3 (2006).
17. Veljić, D.M., Sedmak, A.S., Rakin, M.P., Bajić, N.S., Medjo, B.I., Bajić, D.R., D.R., Grabulov, V.K. : Experimental and Numerical Thermo – Mechanical Analysis of Friction Stir Welding of High – Strength Alluminium Alloy. Therm. Sci. **18**(1), S29–S38 (2014)
18. Tewari, S.P., Gupta, A., Prakash, J.: Effect Of Welding Parameters on the Weldability of Material. Int. J. Eng. Sci. Technol. **2**(4), 512–516 (2010)
19. Wang, H.-S.: Effect of Welding Variables on Cooling Rate and Pitting Corrosion Resistance in Super Duplex Stainless Weldment. Mater. Trans. **46**(3), 593–601 (2005)
20. Ilyashchenko, D.P., Chinakhov, D.A., Verkhoturova, E.V., Lavrova, E.V.: Stability of MMA Welding with Protective Coatings. Structural Integrity and Life **20**(1), 33–36 (2020)
21. Younise, B., Rakin, M., Gubeljak, N., Medjo, B., Sedmak, A.: Effect of material heterogeneity and constraint conditions on ductile fracture resistance of welded joint zones – Micromechanical assessment. Eng. Fail. Anal. **82**, 435–445 (2017)

22. Younise, B., Rakin, M., Gubeljak, N., Medjo, B., Sedmak, A.: Numerical simulation of constraint effect on fracture initiation in welded specimens using a local damage model. Struct. Integr. Life **11**, 51–56 (2011)
23. Rakin, M., Medjo, B., Gubeljak, N., Sedmak, A.: Micromechanical assessment of mismatch effects on fracture of high-strength low alloyed steel welded joints. Eng. Fract. Mech. **109**, 221–235 (2013)
24. Zerbst, U., Ainsworth, R.A., HTh., Beier, Pisarski, H., Zhang, Z.L., Nikbin, K., Nitschke-Pagel, T., Münstermann, S., Kucharczyk, P., Klingbeil, D.: Review on fractur and crack propagation in weldments - a fracture mechanics perspective. Eng. Fract. Mech. **132**, 200–276 (2014)
25. Carpinteri, A., Ronchei, C., Scorza, D., Vantadori, S.: Fracture mechanics based approach to fatigue analysis of welded joints. Eng. Fail. Anal. **49**, 67–78 (2015)
26. Tu, H.Y., Schmauder, S., Weber, U.: Simulation of the fracture behavior of a S355 electron beam welded joint by cohesive zone modeling. Eng. Fract. Mech. **163**, 303–312 (2016)
27. Radua, D., Sedmak, A., Băncilă, R.: Determining the crack acceptability in the welded joints of a wind loaded cylindrical steel shell structure. Eng. Fail. Anal. **91**, 341–353 (2018)
28. AutoCAD, Autodesk, San Rafael, California: https://www.autodesk.com/
29. Raweni, A., Majstorović, V., Sedmak, A., Tadić, S., Kirin, S.: Optimization of AA5083 Friction Stir Welding Parameters Using Taguchi Method. Tehnički vjesnik **25**(3), 861–866 (2018)
30. Jovičić, R., Sedmak, S., Samardžić, I., Sedmak, A., Tatić, U., Milošević, M.: Possibilities of Predicting the Behaviour of Ferrite-Austenite Welded Joints in Pressure Equipment During Exploitation. Tehnički vjesnik **24**(5), 1361–1368 (2017)

Mathematical Model of Patient Support System in Medical Linear Accelerators for External Beam Radiation Therapy

Ivan M. Buzurovic, Slavisa Salinic, and Vladimir Misic

Abstract In this study, a mathematical model of the patient support system (PSS) in medical linear accelerators was presented. The PSS consists of a treatment table capable of performing vertical, horizontal and rotational movements to position patients to the required location. The mathematical model includes the set of differential equations of motion. The model was developed for two basic designs of the lifting mechanisms: (a) a PSS mounted above the floor, and (b) a PSS mounted in a deep pit. In addition, this model includes dynamic equations of motions for the linear accelerator gantry and treatment head. The proposed model can be utilized for various tasks, such as adaptive or robust control design, tracking and motion compensation or collision avoidance assessments. The use of the proposed mathematical model can potentially improve real-time tracking of target-volumes when delivering a highly conformal radiation dose at full duty cycle while minimizing irradiation to healthy tissues and sparing critical organs during radiation treatments with linear accelerators. Furthermore, the equation of motion can be used to confirm the position of the mechanical isocenter of the linear accelerator. Such results can be compared against the radiation isocenter; thus, possible deviations can be tested more frequently without additional resources.

Keywords Mathematical modelling · Linear accelerator · Radiation therapy

I. M. Buzurovic (✉)
Harvard Medical School, Harvard University, Boston, MA 02115, USA
e-mail: ibuzurovic@bwh.harvard.edu

S. Salinic
Faculty of Mechanical and Civil Engineering, University of Kragujevac, Kraljevo, Serbia

V. Misic
University of Pittsburgh Medical Center, Pittsburgh, PA 15213, USA

N. Mitrovic et al. (eds.), *Current Problems in Experimental and Computational Engineering*, Lecture Notes in Networks and Systems 323,
https://doi.org/10.1007/978-3-030-86009-7_19

1 Introduction

A medical linear accelerator (LA) is the device commonly used for external beam radiation treatments for patients diagnosed with cancer. The LA is a device that uses high-frequency electromagnetic waves to accelerate charged particles such as electrons to high energies through an accelerator (linear) tube. Patient support system (PSS) is one of the most important systems within the mechanical subsystem of the LA. It consists of treatment couch which can be moved to the required position by vertical, horizontal and rotational movements.

Recently, a rapid growth and technological development were noticed in the field or radiation therapy. External beam radiation therapy (EBRT) is a part of that process, [1, 2]. Development and implementation of advanced treatment techniques such as intensity modulated radiation therapy (IMRT) [3], or volumetric modulated arc therapy (VMAT) [4], are already a standard of treatment for many malignancies. These techniques initially included dynamic motion of the multi-leaf collimator (MLC) leaves during the treatment delivery, as in IMRT. Further development included the real-time motion of the collimator and gantry of the linear accelerator without treatment interruptions, as in the VMAT technology. Therefore, the various motion tracking and compensation techniques require dynamical motion of the all machine components involved in the radiation treatments such as dynamically moving MLC, couch motion compensation, and real-time imaging of the patients during treatments [5–19], etc. Dynamic model approach is becoming more utilized in other radiation therapy modalities [20]. The robotic treatment couches in EBRT are used for patient positioning. During traditional radiation delivery, the couch maintains its fixed position with respect to the machine isocenter to allow the accurate delivery of the radiation dose to the target volumes. However, it reasonable to conclude that this approach will undergo significant changes when considering real-time tracking techniques using the robotic treatment couches [6–8, 18].

Real-time tracking of the target in radiation therapy can be accomplished in different ways, [7]. This study deals with the method related to the motion compensation using a treatment couch. In this approach, the robotic treatment couch moves in the opposite direction of the target in order to make the target steady with respect to the radiation beam. Consequently, the irradiated volume decreases since the target appears to be steady with respect to the absolute coordinate system. Several research groups have reported the results related to the motion compensation utilizing treatment couches, [6–8, 11, 21–24]. The implementation and experimental results of the real-time tracking technique using 4D target tracking was presented in [7]. The authors modified the existing treatment couch resulting in a provision to perform continuous motion during treatment. In [11], the authors analyzed the dynamical behavior of the system to evaluate the design specifications of the robotic couch for effective motion compensation. They concluded that the time delay (latency) should be considered for proper tracking. In [21] a compensation strategy was proposed using an output-feedback model predictive controller to track the target motion. The system simulation was presented with a clinical data set. The detailed mathematical modeling procedure, a decentralized dynamic-based control approach and a prediction module for a 3 degree-of-freedom (DOF) robotic couch were reported in [8]. The authors concluded that the adaptive control could be an adequate approach to regulate such systems due to the variability in the payload of the system. The possibilities of the different commercial couches to be used in tracking were evaluated in [6]. Despite the differences in design, it was shown that the 6 DOF hexapod and 3 DOF robotic couches with decomposed dynamics were capable of tracking with a high level of accuracy. The implementation of the motion compensation to both the commercial robotic couch and the research platform was studied in [22]. In this approach, a couch deflection was measured so that the tracking errors could be minimized. A novel respiratory compensation system was designed, and its performances were evaluated, [23]. The authors concluded that the robotic systems could be used to offset the organ displacement caused by respiration. In [24], the authors reported a prototype robotic system for tracking purposes. This system did not include a prediction analysis.

Therefore, development of a rigorous mathematical model of the PSS was crucial for further investigation of various topics in EBRT including adaptive or robust control design, tracking and motion compensation, collision avoidance [25] and other various quality assurance tasks. In this article we present differential equations of motion for a PSS mounted above the floor, and PSS mounted in a deep pit together with differential equations of motion for the gantry and treatment head of the LA.

2 Mathematical Model

We presented the equation of motion for one of the most used designs of the PSS, so this model can be utilized for further research investigations. Energy-based Lagrange approach was utilized. A set of generalized coordinates that completely characterize the possible motion of the system's components was adopted. This choice eliminates the need for the constraint force to enter the resultant system of equations.

2.1 *Equations of Motion for a PSS Mounted Above the Floor*

A mechanism shown in Fig. 1a contains of the turntable 1 of mass m_1, the beam 2 of mass m_2 and length L connected to the turntable at the point O_1 by a revolute joint, the motor 3 with ball screws, the beam 4 of mass m_4 and the length L connected to the beam 2 by a revolute joint, and the treatment couch with patient 5. The coordinate frame $O\xi_1\eta_1\zeta_1$ is fixed to the turntable in the manner shown in Fig. 1a while the inertial coordinate frame $Oxyz$ is positioned in the manner shown in Fig. 1b. In addition, the local coordinate frames $C_2\xi_2\eta_2\zeta_2$, $C_4\xi_4\eta_4\zeta_4$, and $C_5\xi_5\eta_5\zeta_5$, are fixed to bodies 2, 4, and 5, respectively.

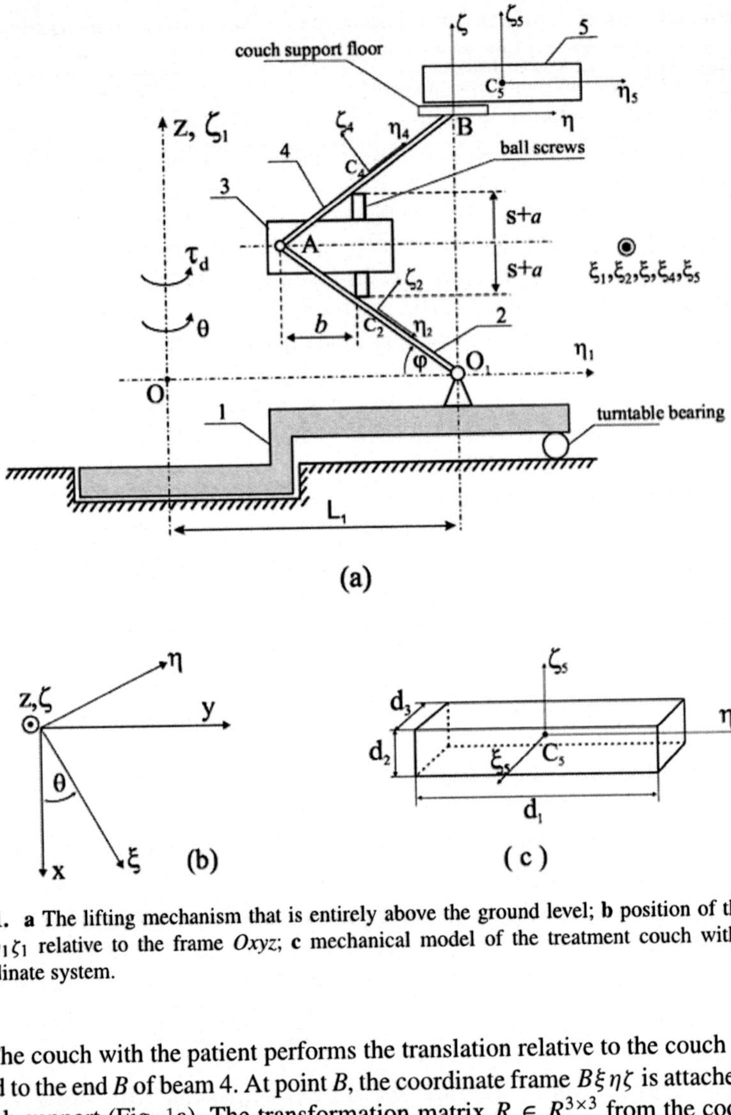

Fig. 1. **a** The lifting mechanism that is entirely above the ground level; **b** position of the frame $O\xi_1\eta_1\zeta_1$ relative to the frame *Oxyz*; **c** mechanical model of the treatment couch with patient coordinate system.

The couch with the patient performs the translation relative to the couch support fixed to the end *B* of beam 4. At point *B*, the coordinate frame $B\xi\eta\zeta$ is attached to the couch support (Fig. 1a). The transformation matrix $R \in R^{3\times3}$ from the coordinate frame $O\xi_1\eta_1\zeta_1$ to the inertial frame *Oxyz* has the form (Fig. 1b):

$$\boldsymbol{R} = \begin{bmatrix} cos\theta & -sin\theta & 0 \\ sin\theta & cos\theta & 0 \\ 0 & 0 & 1 \end{bmatrix}. \tag{1}$$

The motor governs the ball screws that are responsible for the vertical motion of the couch with the patient in a manner that the end of beam 4 moves along the vertical line, passing through point O_1 (length b is a constant quantity).

Displacement s generated by the ball screws (Fig. 1a) with a single thread is determined as:

$$s = nh \tag{2}$$

where n is the number of revolutions for each ball screw, and h is the lead of the threads.

According to this, the rotation angle of the ball screws is given as follows:

$$\psi = 2\pi n = \frac{2\pi s}{h}. \tag{3}$$

Note that geometric parameter a introduced in Fig. 1a represents a constant quantity. ll the vectors and matrices, whose elements are given in $C_i\xi_i\eta_i\zeta_i$ local frame, will be denoted by the corresponding superscript (i). The position of the system with respect to the inertial frame is determined by the generalized coordinates $q = [\theta, \varphi, \xi, \eta]^T$ where θ represents the angle of rotation of the turntable about the z axis, φ is the relative rotation of beam 2 with respect to the turntable, and ξ and $\eta\eta$ are the relative Cartesian coordinates of the center of mass, C_5, of the couch with the patient in the frame $B\xi\eta\zeta$. For $\xi = \eta = 0$ the point C_5 is on the direction O_1B. The kinetic energy of the turntable is:

$$T_1 = \frac{1}{2}J_1\dot{\theta}^2 \tag{4}$$

where J_1 is the mass moment of inertia of the turntable about the z axis. The kinetic energy of the beam 2 is determined as:

$$T_2 = \frac{1}{2}m_2\dot{\boldsymbol{r}}_{C_2}^T\dot{\boldsymbol{r}}_{C_2} + \frac{1}{2}\omega_2^{(2)T}\boldsymbol{J}_{C_2}^{(2)}\omega_2^{(2)} \tag{5}$$

where $\boldsymbol{r}_{C_2}$ is the position vector of the center of mass C_2 with respect to the inertial frame given as:

$$\boldsymbol{r}_{C_2} = [-L_1 sin\theta, L_1 cos\theta, 0]^T + \boldsymbol{R}\left[0, -\frac{L}{2}cos\varphi, \frac{L}{2}sin\varphi\right]^T \tag{6}$$

whose time derivative is denoted by $\dot{\boldsymbol{r}}_{C_2}$; $\omega_2^{(2)} = \left[-\dot{\varphi}, -\dot{\theta}sin\varphi, \dot{\theta}cos\varphi\right]^T$ is the angular velocity of beam 2, and $\boldsymbol{J}_{C_2}^{(2)} = diag\left(\frac{m_2L^2}{12}, 0, \frac{m_2L^2}{12}\right)$ is the inertial tensor referred to the center of mass of beam 2. It is assumed that the mass of the motor 3 is concentrated at the point A. In accordance to this assumption, the kinetic energy of the motor is:

$$T_3 = \frac{1}{2} m_3 \dot{\boldsymbol{r}}_A^T \dot{\boldsymbol{r}}_A \tag{7}$$

where $\mathbf{r}_A$ is the position vector of point A with respect to the inertial frame given by:

$$\boldsymbol{r}_A = [-L_1 sin\theta, L_1 cos\theta, 0]^T + \boldsymbol{R}[0, -Lcos\varphi, Lsin\varphi]^T \tag{8}$$

whose time derivative is denoted by $\dot{\boldsymbol{r}}_A$. The beam 4 has the kinetic energy determined by the following expression:

$$T_4 = \frac{1}{2} m_4 \dot{\boldsymbol{r}}_{C_4}^T \dot{\boldsymbol{r}}_{C_4} + \frac{1}{2} \omega_4^{(4)T} \boldsymbol{J}_{C_4}^{(4)} \omega_4^{(4)} \tag{9}$$

where $\boldsymbol{r}_{C_4}$ is the position vector of the center of mass C_4 with respect to the inertial frame given as:

$$\boldsymbol{r}_{C_4} = [-L_1 sin\theta, L_1 cos\theta, 0]^T + \boldsymbol{R}\left[0, -\frac{L}{2} cos\varphi, \frac{3L}{2} sin\varphi\right]^T \tag{10}$$

whose time derivative is denoted by $\dot{\boldsymbol{r}}_{C_4}$, $\omega_4^{(4)} = \left[\dot{\varphi}, \dot{\theta} sin\varphi, \dot{\theta} cos\varphi\right]^T$ is the angular velocity of beam 4, and $\boldsymbol{J}_{C_4}^{(4)} = diag\left(\frac{m_4 L^2}{12}, 0, \frac{m_4 L^2}{12}\right)$ is the inertial tensor referred to the center of mass of beam 4. Finally, the kinetic energy of the treatment couch with the patient is given by:

$$T_5 = \frac{1}{2} m_5 \dot{\boldsymbol{r}}_{C_5}^T \dot{\boldsymbol{r}}_{C_5} + \frac{1}{2} \omega_5^{(5)T} \boldsymbol{J}_{C_5}^{(5)} \omega_5^{(5)} \tag{11}$$

where $\boldsymbol{r}_{C_5}$ is the position vector of the center of mass C_5 with respect to the inertial frame given as:

$$\boldsymbol{r}_{C_5} = \boldsymbol{R}\left[\xi, \eta + L_1, \frac{d_2}{2} + 2Lsin\varphi\right]^T \tag{12}$$

whose time derivative is denoted by $\dot{\boldsymbol{r}}_{C_5}$, $\omega_5^{(5)} = \left[0, 0, \dot{\theta}\right]^T$ is the angular velocity of the treatment couch with the patient, and $\boldsymbol{J}_{C_5}^{(5)} = diag\left(\frac{m_5(d_1^2+d_2^2)}{12}, \frac{m_5(d_2^2+d_3^2)}{12}, \frac{m_4(d_1^2+d_3^2)}{12}\right)$ is the inertial tensor referred to the center of mass of the couch with patient. Here, the treatment couch with patient are modeled as a uniform rectangular parallelepiped of edges d_1, d_2, and d_3 and mass m_5 (Fig. 1c). For $\xi = \eta = 0$ the point C_5 is on the axis ζ. Consequently, the kinetic energy of the system is:

$$T = \sum_{i=1}^{5} T_i = \frac{1}{2}\dot{q}^T M \dot{q} \tag{13}$$

where $M(q) \in R^{4\times4}$ is the mass matrix whose elements are:

$$M_{11} =$$

$$= \frac{1}{12}\left(\begin{array}{c} 12J_1 + 2m_2L^2 + 12m_2L_1^2 + 6m_3L^2 + \\ 12m_3L_1^2 + 2m_4L^2 + 12m_4L_1^2 + m_5d_1^2 + m_5d_3^2 + 12m_5L_1^2 + \\ 24L_1m_5\eta + 12m_5\eta^2 + 12m_5\xi^2 - 12LL_1(m_2 + 2m_3 + m_4)cos\varphi + \\ 2L^2(m_2 + 3m_3 + m_4)cos2\varphi \end{array}\right) \tag{14}$$

$$M_{22} = \frac{1}{12}\left(\begin{array}{c} 4m_2L^2 + 12m_3L^2 + 16m_4L^2 + \\ 24m_5L^2 + 12(m_4 + 2m_5)L^2cos2\varphi \end{array}\right), \tag{15}$$

$$M_{33} = M_{44} = m_5 \tag{16}$$

$$M_{13} = M_{31} = -m_5\eta - m_5L_1 \tag{17}$$

$$M_{14} = M_{41} = m_5\xi \tag{18}$$

$$M_{12} = M_{21} = 0,\ M_{23} = M_{32} = 0,\ M_{24} = M_{42} = 0,\ M_{34} = M_{43} = 0 \tag{19}$$

The potential energy of the mechanism considered is

$$\Pi = \left(\frac{m_2}{2} + m_3 + \frac{3m_4}{2} + 2m_5\right)gLsin\varphi. \tag{20}$$

The total virtual work of the control forces and torques are given as:

$$\delta A = \tau_d\delta\theta + F_\xi\delta\xi + F_\eta\delta\eta + 2\tau_M\frac{2\pi}{h}\delta s \tag{21}$$

where τ_d is the driving torque for the motion of the turntable, F_ξ and F_η are the internal control forces responsible for translational motion of the couch with the patient along the directions parallel to axes ξ and η, respectively, τ_M is the internal driving torque of the motor for revolutions of ball screws. From Eq. (21), the generalized control forces vector is obtained as follows:

$$\mathbf{Q}^{(c)} = \left[\tau_d,\ \frac{4\pi\tau_M b}{h\cos^2\varphi}\delta\varphi,\ F_\xi,\ F_\eta\right]^T \tag{22}$$

where the relation between the displacement s and the angle φ

$$\tan\varphi = \frac{s+a}{b} \Rightarrow \delta s = \frac{b}{\cos^2\varphi}\delta\varphi \tag{23}$$

is considered. Note that a and b are constant geometric parameters of the mechanism (Fig. 1a). Using Lagrange's equations [26, 27]:

$$\frac{d}{dt}\left(\frac{\partial T}{\partial \dot{\boldsymbol{q}}}\right)^T - \left(\frac{\partial T}{\partial \boldsymbol{q}}\right)^T = \boldsymbol{Q}^{(g)} + \boldsymbol{Q}^{(c)}, \tag{24}$$

the differential equations of motion of the mechanism are obtained, which can be written in the following matrix form:

$$\boldsymbol{M}(\boldsymbol{q})\ddot{\boldsymbol{q}} + \boldsymbol{\Gamma}(\boldsymbol{q},\dot{\boldsymbol{q}}) = \boldsymbol{Q}^{(g)} + \boldsymbol{Q}^{(c)} \tag{25}$$

where

$$\boldsymbol{Q}^{(g)} = -(\partial\Pi/\partial\boldsymbol{q})^T =$$

$$= \left[0, -\left(\frac{m_2}{2} + m_3 + \frac{3m_4}{2} + 2m_5\right)gLcos\varphi, 0, 0\right]^T, \tag{26}$$

and $\boldsymbol{\Gamma}(\boldsymbol{q},\dot{\boldsymbol{q}}) = [\Gamma_1,\ldots,\Gamma_4]^T \in R^{4\times1}$ is a column matrix whose elements are given by:

$$\Gamma_i = \sum\nolimits_{j=1}^{4}\sum\nolimits_{k=1}^{4}[jk,i]\dot{q}_j\dot{q}_k, \quad i = 1,\ldots,4, \tag{27}$$

where:

$$[jk,i] = \frac{1}{2}\left(\frac{\partial M_{ij}}{\partial q_k} + \frac{\partial M_{ik}}{\partial q_j} - \frac{\partial M_{jk}}{\partial q_i}\right), \qquad i,j,k = 1,\ldots,4 \tag{28}$$

are Christoffel symbols of the first kind [26], or in its developed form:

$$\Gamma_1 = 2m_5\eta\dot{\theta}\dot{\eta} + 2m_5L_1\dot{\theta}\dot{\eta} + 2m_5\xi\dot{\theta}\dot{\xi} + +\frac{1}{12}\dot{\theta}\dot{\varphi}\begin{bmatrix} 12LL_1(m_2+2m_3+m_4)sin\varphi - \\ 4L^2(m_2+3m_3+m_4)sin2\varphi \end{bmatrix}, \tag{29}$$

$$\Gamma_2 = -(2m_5 + m_4)L^2\dot{\varphi}^2 sin2\varphi + \frac{1}{24}\dot{\theta}^2\begin{bmatrix} -12LL_1(m_2 + 2m_3 + m_4)sin\varphi + \\ 4L^2(m_2 + 3m_3 + m_4)sin2\varphi \end{bmatrix}, \tag{30}$$

$$\Gamma_3 = -2m_5\dot{\theta}\dot{\eta} - m_5\xi\dot{\theta}^2, \tag{31}$$

$$\Gamma_4 = 2m_5\dot{\theta}\dot{\xi} - m_5\eta\dot{\theta}^2. \tag{32}$$

Therefore, Eq. (25) with elements (26–32) represents dynamic equations of motion for the PSS mounted above the floor.

2.2 Equations of Motion for a PSS Mounted in a Deep Pit

A mechanism shown in Fig. 2 consists of the turntable 1 of mass m_1 carrying the treatment couch lift drive 2 with a lead screw 3, the cylindrical pillar with couch support 4 of mass m_4, and the treatment couch with patient 5. In addition, the treatment couch with patient are modeled as a uniform rectangular parallelepiped of edges d_1, d_2, and d_3, having mass mass m_1 (as in Fig. 1c).

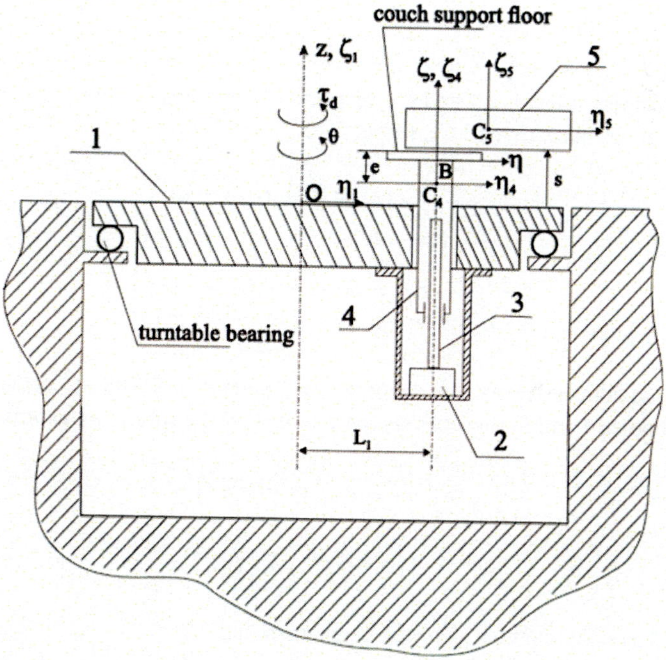

Fig. 2. The lifting mechanism that is mounted in a deep pit.

The coordinate frame $O\xi_1\eta_1\zeta_1$ is fixed to the turntable in the manner shown in Fig. 1 while the inertial coordinate frame $Oxyz$ is positioned with respect to frame $O\xi_1\eta_1\zeta_1$ as it is shown in Fig. 1b. The lead screw has a single thread. By rotating the lead screw the vertical translation (denoted with s of the pillar with respect to the turntable, as in Fig. 1a) is achieved. The relations between the displacement s, the lead of threads h, and the rotation angle of the lead screw ψ, are determined by Eqs. (2) and (3). Additionally, the local coordinate frames $C_4\xi_4\eta_4\zeta_4$ and $C_5\xi_5\eta_5\zeta_5$ are fixed to bodies 4 and 5, respectively, while the coordinate frame $B\xi\eta\zeta$ is attached to the couch support at the point B in a manner shown in Fig. 2. The treatment couch with patient performs translation relative to the couch support. The transformation matrix $\boldsymbol{R} \in R^{3\times3}$ from the coordinate frame $O\xi_1\eta_1\zeta_1$ to the inertial frame $Oxyz$ has the form given in Eq. (1). Note that L_1 and e are constant geometric parameters of the mechanism (as in Fig. 2). Position of the system is determined by the generalized coordinates $\boldsymbol{q} = [\theta, s, \xi, \eta]^T$ where θ represents the angle of rotation of the turntable around the vertical z axis, where ξ and η are the relative Cartesian coordinates of the center of mass, C_5, of the couch with patient in the frame $B\xi\eta\zeta$. For $\xi = \eta = 0$ the point C_5 is on the axis ζ. The kinetic energy of the considered mechanism is:

$$T = \frac{1}{2}J_1\dot{\theta}^2 + \frac{1}{2}m_4\dot{\mathbf{r}}_{C_4}^T\dot{\mathbf{r}}_{C_4} + \frac{1}{2}\omega_4^{(4)T}\mathbf{J}_{C_4}^{(4)}\omega_4^{(4)} + \frac{1}{2}m_5\dot{\mathbf{r}}_{C_5}^T\dot{\mathbf{r}}_{C_5} + \frac{1}{2}\omega_5^{(5)T}\mathbf{J}_{C_5}^{(5)}\omega_5^{(5)} \quad (33)$$

where J_1 is the mass moment of inertia of the turntable together with the couch lift drive and the lead screw about the vertical z axis; $\boldsymbol{J}_{C_4}^{(4)} = diag\left(J_{C_4\xi_4}, J_{C_4\eta_4}, 0\right)$ and $\boldsymbol{J}_{C_5}^{(5)} = diag\left(\frac{m_5(d_1^2+d_2^2)}{12}, \frac{m_5(d_2^2+d_3^2)}{12}, \frac{m_4(d_1^2+d_3^2)}{12}\right)$ are inertial tensors referred to the centers of mass C_4 and C_5, respectively, $\boldsymbol{r}_{C_4}$ and $\boldsymbol{r}_{C_5}$ are position vectors of the centers of mass C_4 and C_5, respectively, given by:

$$\mathbf{r}_{C_4} = \mathbf{R}[0, L_1, s - e]^T, \quad (34)$$

$$\mathbf{r}_{C_5} = \mathbf{R}\left[\xi, \eta + L_1, \frac{d_2}{2} + s\right]^T, \quad (35)$$

whose time derivatives are denoted by $\dot{\boldsymbol{r}}_{C_4}$ and $\dot{\boldsymbol{r}}_{C_5}$, $\omega_4^{(4)}$ and $\omega_5^{(5)}$ are angular velocities of bodies 4 and 5, respectively, determined by the following expressions:

$$\omega_4^{(4)} = [0, 0, \dot{\theta}]^T, \quad (36)$$

$$\Omega_5^{(5)} = [0, 0, \dot{\theta}]^T. \quad (37)$$

The potential energy of the mechanism is:

$$\Pi = m_4 g(s - e) + m_5 g\left(s + \frac{d_2}{2}\right) \tag{38}$$

while the total virtual work of the control forces and torques is given as follows:

$$\delta A = \tau_d \delta\theta + F_\xi \delta\xi + F_\eta \delta\eta + 2\tau_M \frac{2\pi}{h} \delta s \tag{39}$$

where τ_d is the driving torque iducing the motion of the turntable; F_ξ and F_η are the internal control forces under which influence rotational motion of the treatment couch and patient was performed. Direction of that motion is parallel to the axes ξ_5 and η_5, respectively; τ_M is the internal driving torque allowing the rotation of the lead screw. The generalized control forces vector are:

$$\mathbf{Q}^{(c)} = \left[\tau_d, 2\tau_M \frac{2\pi}{h}, F_\xi, F_\eta\right]^T. \tag{40}$$

Consequently, the differential equations of motion of the mechanism can be written in form of Eq. (25), where elements of the matrices $\boldsymbol{M}(\boldsymbol{q})$ and $\boldsymbol{\Gamma}(\boldsymbol{q}, \dot{\boldsymbol{q}})$ are:

$$M_{11} = \frac{1}{12}\begin{bmatrix} 12J_1 + 12L_1^2(m_4 + m_5) + \\ +24L_1 m_5 \eta + m_5\left(d_1^2 + d_3^2 + 12\eta^2 + 12\xi^2\right) \end{bmatrix}, \tag{41}$$

$$M_{22} = m_4 + m_5, \qquad M_{33} = M_{44} = m_5, \tag{42}$$

$$M_{13} = M_{31} = -m_5(L_1 + \eta), M_{14} = M_{41} = m_5 \xi, \tag{43}$$

$$M_{12} = M_{21} = M_{23} = M_{32} = M_{24} = M_{42} = M_{34} = M_{43} = 0, \tag{44}$$

$$\Gamma_1 = 2m_5(L_1 + \eta)\dot{\theta}\dot{\eta} + 2m_5\xi\dot{\theta}\dot{\xi}, \Gamma_2 = 0, \tag{45}$$

$$\Gamma_3 = -2m_5\dot{\theta}\dot{\eta} - m_5\xi\dot{\theta}^2, \Gamma_4 = 2m_5\dot{\theta}\dot{\xi} - m_5(L_1 + \eta)\dot{\theta}^2, \tag{46}$$

while the generalized force vector $\boldsymbol{Q}^{(g)}$ has the form:

$$\mathbf{Q}^{(g)} = -\left(\frac{\partial \Pi}{\partial \mathbf{q}}\right)^T = [0, -(m_4 + m_5)g, 0, 0]^T. \tag{47}$$

System (25) can be transformed to the non-linear state-space form as in [8]:

$$\begin{aligned} \dot{\mathbf{x}}(t) &= A(t)x(\mathrm{t}) + B(t)u(\mathrm{t}) + v(t) \\ y(t) &= C(t)x(t) + w(t) \end{aligned} \tag{48}$$

The state-space representation is out of the scope of this article; however, this transformation can be used for the control design.

In the following part, the dynamic equation of motion of the linear accelerator in conjunction with the treatment table are presented. The combination of the motion trajectories from both the treatment table and the gantry of linear accelerator could be used to calculate possible collisions. Patient and environment specifics, such as exact position of the treatment table and size, could be extracted from the previously obtained computed tomography images collected during the simulation phase.

2.3 Equations of Motion of the Gantry and Treatment Head

A simplified model of the gantry and treatment couch is shown in Fig. 3. The gantry rotates around fixed axis y_1 situated in the coordinate system, and it is parallel to y. In Fig. 3, the notations τ_G and θ_G represent the gantry driving torque and the gantry angle of rotation about axis y_1, respectively. Also, with P_G and P_{IC} are denoted, respectively, the characteristic point on the gantry treatment head and the isocenter, where $\overline{P_G P_{IC}} = const. = h_1$. The point P_{IC} belongs to the axis z. The inertial coordinate frame $Oxyz$ shown in Fig. 2 is defined in the same manner as in Sect. 2.1. Note that the point P_G is moving along the circular trajectory with radius h_1 which belongs to the coordinate plane Oxz.

The coordinates of point P_G with respect to the inertial frame are:

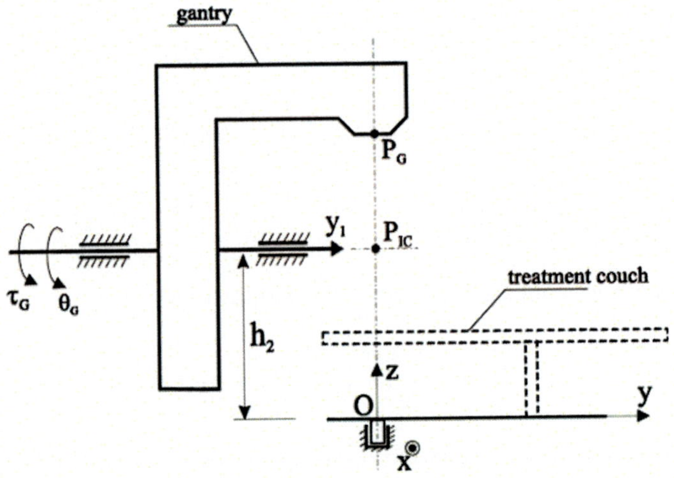

Fig. 3. Simplified model of the linear accelerator with a treatment head (gantry) and treatment couch.

$$x_{P_G} = h_1 sin\theta_G, y_{P_G} = 0, z_{P_G} = h_2 + h_1 cos\theta_G \tag{49}$$

The differential equation of the gantry is:

$$J_G\ddot{\theta}_G = \tau_G \tag{50}$$

where J_G is the mass moment of inertia of the gantry about the axis y_1. The position vector of arbitrary patient point P is given by:

$$\boldsymbol{r}_P = \boldsymbol{r}_{C_5} + \boldsymbol{r}^* \tag{51}$$

where vector $\boldsymbol{r}_{C_5}$ is determined in Eq. (12) and the vector $\boldsymbol{r}^* = \overrightarrow{C_5 P}$ is defined as:

$$\boldsymbol{r}^* = \boldsymbol{R}\left[\xi^* + \xi, \eta^* + \eta, \zeta^* + \frac{d_2}{2}\right]^T \tag{52}$$

where $\xi^* = \text{const.}$, $\eta^* = \text{const.}$, and $\zeta^* = \text{const.}$, are the coordinates of point P with respect to the frame $C_5\xi_5\eta_5\zeta_5$. Based on Eqs. (49), (51), and (52), the coordinates of the point P are:

$$x_P = (2\xi + \xi^*)cos\theta - (2\eta + L_1 + \eta^*)sin\theta, \tag{53}$$

$$y_P = (2\eta + L_1 + \eta^*)cos\theta + (2\xi + \xi^*)sin\theta, \tag{54}$$

$$z_P = 2Lsin\varphi + d_2 + \zeta^* \tag{55}$$

for the mechanism described in Sect. 2.1.

A possible collision between the patient and the treatment head of the linear accelerator can be determined by the analysis of the following condition:

$$\overline{PP_G} = \sqrt{(x_P - x_{P_G})^2 + (y_P - y_{P_G})^2 + (z_P - z_{P_G})^2} \geq \ell \tag{56}$$

where ℓ is an arbitrary prescribed safety constant (such as patient separation or other safety margin). Correlation of the dynamics of the gantry and the dynamics of the PSS mechanism were achieved through the quantity $\overline{PP_G}$.

3 Discussion

In this study we developed a detailed mathematical model of the PSS for medical linear accelerators used in EBRT. The equations of motion included dynamics of the system, and other specifics such as patient positions, safety margin distances and treatment plan parameters. Patient specific parameters can be obtained from the computed tomography images during the simulation. All these parameters can be included in the simulation model for various purposes. They can be summarized in the input xml file, allowing for simple inclusion of all relevant parameters to the dynamic equations of motion. The model can be utilized to guarantee a safety margin between gantry and patient during the radiation treatment. Therefore, the patient safety would not be compromised in any instant of the radiation treatment using linear accelerators. The immediate benefit of the model is that the time-consuming experiments prior to the treatments can be eliminated. Consequently, the clinical resources can be utilized in more appropriate manner.

The proposed methodology can be applied to other clinical procedures such as tracking and motion compensation, [4–9]. In such scenario, the position of the patient can be calculated based on the given equations, and the treatment can be interrupted if the predefined motion limits were reached. The collision avoidance method us suitable for the improvement of the general quality assurance program in radiation treatments [28], including the quality assurance of the isocenter of the linear accelerators and other moving components such as imaging system [29]. In this approach, equation of motion can be utilized to confirm position of the mechanical isocenter of the linear accelerator more rigorously. Such results can be compared against the radiation isocenter, so the possible deviation can be tested more frequently without additional measurements. In addition, the proposed method can be used for the future development of 4π treatment delivery in which the couch motion can be performed in real-time without treatment interruptions. This method would decrease treatment time and potentially improve the patients' dosimetry. Implementation of the proposed technique can potentially improve real-time tracking of the tumor-volume to deliver highly conformal precise radiation dose at almost 100% duty cycle while minimizing irradiation to health tissues and sparing critical organs. This, in turn, will potentially improve the quality of patient treatment by lowering the toxicity level and increasing survival.

The future work will include rigorous simulation of the proposed system using various controllers such as proportional–integral–derivative, adaptive or robust controller. The comparison between these control methodologies and the model-predictive controller (MPC) can potentially lead to minimization of latencies and improved tracking with the treatment table. Prediction module in the MPC approach will be developed to predict couch motion and to compensate errors due to delay in the system response.

4 Conclusion

In this study, a mathematical model of the PSS in medical linear accelerators was presented. The model includes differential equations of motion for the PSS mounted above the floor, the PSS mounted in a deep pit, and the linear accelerator gantry and treatment head. The mathematical model was developed using Lagrange energy-based formulation. This model can be used for investigation of new and non-standard control methodologies, motion tracking and compensation tasks, collision checks, and for quality assurance purposes. Additionally, the model can be used during the design phase of novel linear accelerator for rigorous simulation in pre-fabrication phase. Many variables, such as power of the actuators, patient load and the like can be adjusted for different designs based on the presented dynamic equations.

Appendix

Table 1 summarizes notations used in this article.

References

1. Dawson, L.A., Sharpe, M.B.: Image-guided radiotherapy: rationale, benefits, and limitations. Lancet Oncol. **7**(10), 848–858 (2006)
2. Lo, S.S., Fakiris, A.J., Chang, E.L., Mayr, N.A., Wang, J.Z., Papiez, L. Et al.: Stereotactic body radiation therapy: a novel treatment modality. Nat. Rev. Clinical oncol. **7**(1), 44 (2010)
3. Webb, S.: Intensity-Modulated Radiation Therapy. CRC Press (2015)
4. Palma, D., Vollans, E., James, K., Nakano, S., Moiseenko, V., Shaffer, R. et al.: Volumetric modulated arc therapy for delivery of prostate radiotherapy: comparison with intensity-modulated radiotherapy and three-dimensional conformal radiotherapy. Int. J. Radiation Oncol. Biol. Phys. **72**(4), 996–1001 (2008)
5. Barnes, E.A., Murray, B.R., Robinson, D.M., Underwood, L.J., Hanson, J. et al.: Dosimetric evaluation of lung tumor immobilization using breath hold at deep inspiration. Int. J. Radiation Oncol. Biol. Phys. **50**(4), 1091–1098 (2001)
6. Buzurovic, I., Huang, K., Yu, Y., Podder, T.K.: A robotic approach to 4D real-time tumor tracking for radiotherapy. Phys. Med. Biol. **56**(5), 1299 (2011)
7. Buzurovic, I., Yu, Y., Werner-Wasik, M., Biswas, T., Anne, P.R., Dicker, A.P., Podder, T.K.: Implementation and experimental results of 4D tumor tracking using robotic couch. Med. Phys. **39**(11), 6957-6967 (2012)
8. Buzurovic, I., Podder, T. K., Huang, K., Yu, Y.: Tumor motion prediction and tracking in adaptive radiotherapy. In: 2010 IEEE International Conference on BioInformatics and BioEngineering, pp. 273–278 (2010)
9. Chen, Q.S., Weinhous, M.S., Deibel, F.C., Ciezki, J.P., Macklis, R.M.: Fluoroscopic study of tumor motion due to breathing: facilitating precise radiation therapy for lung cancer patients. Med. Phys. **28**(9), 1850–1856 (2001)
10. Cho, B., Poulsen, P.R., Keall, P.J.: Real-time tumor tracking using sequential kV imaging combined with respiratory monitoring: a general framework applicable to commonly used IGRT systems. Phys. Med. Biol. **55**(12), 3299 (2010)

Table 1. Notations.

Symbol	Quantity
m	Mass
I	Axial moment of inertia
L, b	Length
φ, ψ	Angles
n	Number
h	Lead of the thread
θ	Angle of turntable rotation
C	Center of mass
$Oxyz$	Inertial coordinate frame
Oxy	Vertical plane
$O\xi\eta\zeta$	Moving coordinate frame
T	Kinetic energy
J	Mass moment of inertia, inertia tensor
$\boldsymbol{R}$	Transformation matrix
$\boldsymbol{r}$	Position vector
ω, Ω	Angular velocity
$\boldsymbol{M}(\boldsymbol{q})$	Mass matrix
F	Inertial control forces
$\boldsymbol{q}$	Generalized coordinates
τ	Torque
$\boldsymbol{Q}$	Generalized control force vector
Γ	Christoffel symbols
s	Displacement
Π	Potential energy
A	Virtual work, state matrix
B	Control matrix
C	Output matrix

11. D'Souza, W.D., McAvoy, T.J.: An analysis of the treatment couch and control system dynamics for respiration-induced motion compensation. Med. Phys. **33**(12), 4701–4709 (2006)
12. Erridge, S.C., Seppenwoolde, Y., Muller, S.H., van Herk, M., De Jaeger, K., Belderbos, J.S. et al.: Portal imaging to assess set-up errors, tumor motion and tumor shrinkage during conformal radiotherapy of non-small cell lung cancer. Radiotherapy Oncol. **66**(1), 75–85 (2003)
13. Shirato, H., Suzuki, K., Sharp, G.C., Fujita, K., Onimaru, R., Fujino, M. et al.: Speed and amplitude of lung tumor motion precisely detected in four-dimensional setup and in real-time tumor-tracking radiotherapy. Int. J. Radiation Oncol. Biol. Phys. **64**(4), 1229–1236 (2006)
14. Huang, K., Buzurovic, I., Yu, Y., Podder, T.K.: A comparative study of a novel AE-nLMS filter and two traditional filters in predicting respiration induced motion of the tumor. In: 2010 IEEE International Conference on BioInformatics and BioEngineering, pp. 281–282 (2010)
15. Buzurovic, I., Huang, K., Podder, T.K., Yu, Y.: Comparison between acceleration-enhanced adaptive filters and neural network filters for respiratory motion prediction. In: 11th Symposium on Neural Network Applications in Electrical Engineering, pp. 181–184 (2012)

16. Kamino, Y., Takayama, K., Kokubo, M., Narita, Y., Hirai, E., Kawawda, N. et al.: Development of a four-dimensional image-guided radiotherapy system with a gimbaled X-ray head. Int. J. Radiation Oncol. Biol. Phys. **66**(1), 271–278 (2006)
17. Ozhasoglu, C., Murphy, M.J.: Issues in respiratory motion compensation during external-beam radiotherapy. Int. J. Radiation Oncol. Biol. Phys. **52**(5), 1389–1399 (2002)
18. Podder, T.K., Buzurovic, I., Galvin, J.M., & Yu, Y.: Dynamics-based decentralized control of robotic couch and multi-leaf collimators for tracking tumor motion. In: 2008 IEEE International Conference on Robotics and Automation, pp. 2496–2502 (2008)
19. Podder, T.K., Buzurovic, I., Hu, Y., Galvin, J.M., Yu, Y.: Partial transmission high-speed continuous tracking multi-leaf collimator for 4D adaptive radiation therapy. In: 2007 IEEE 7th International Symposium on BioInformatics and BioEngineering, pp. 1108–1112 (2007)
20. Buzurovic, I.M., Salinic, S., Orio, P.F., Nguyen, P.L., Cormack, R.A.: A novel approach to an automated needle insertion in brachytherapy procedures. Med. Biol. Eng. Comput. **56**(2), 273–287 (2017)
21. Putra, D., Skworcow, P., Haas, O.C.L., Burnham, K. J., Mills, J.A.: Output-feedback tracking for tumour motion compensation in adaptive radiotherapy. In 2007 American Control Conference, pp. 3414–3419 (2007)
22. Haas, O.C.L., Skworcow, P., Paluszczyszyn, D., Sahih, A., Ruta, M., Mills, J.A.: Couch-based motion compensation: modelling, simulation and real-time experiments. Phys. Med. Biol. **57**(18), 5787–5807 (2012)
23. Chuang, H.C., Huang, D.Y., Tien, D.C., Wu, R.H., Hsu C.H.: A respiratory compensating system: design and performance evaluation. J. Appl. Clin. Med. Phys. **15**(3), 307–322 (2014)
24. Jöhl, A., Lang, S., Ehrbar, S., Guckenberger, M., Klöck, S., Meboldt, M., Daners, M.S.: Modeling and performance evaluation of a robotic treatment couch for tumor tracking. Biomed. Eng./Biomed. Technik, **61**(5), 557–566 (2016)
25. Buzurovic, I.M., Salinic, S., Debeljkovic, D. L., & Cormack, R. A.: A Method for Collision Avoidance in 4π External Beam Radiation Therapy. Lecture Notes in Networks and Systems, pp. 359–374 (2019)
26. Greenwood, D.T.: Classical Dynamics. Courier Corporation (1997)
27. Salinic, S., Nikolić, A.: A new pseudo-rigid-body model approach for modeling the quasi-static response of planar flexure-hinge mechanisms. Mech. Mach. Theory **124**, 150–161 (2018)
28. Klein, E.E., Hanley, J., Bayouth, J., Yin, F.F., Simon, W., Dresser, S. et al.: Task Group 142 report: Quality assurance of medical accelerators A. Med. Phys. **36**(9), 4197–4212 (2009)
29. Sun, B., Goddu, S.M., Yaddanapudi, S., Noel, C., Li, H., Cai, B. et al.: Daily QA of linear accelerators using only EPID and OBI. Med. Phys. **42**(10), 5584–5594 (2015)